Automobile Mechanics

[AS PER AICTE MODEL CURRICULUM 2018]

A K BABU
Associate Professor
Department of Automobile Engineering
SRM Easwari Engineering College
Chennai-600089, Tamilnadu, INDIA

S C SHARMA
Former General Manager,
THDC India Limited

T R BANGA
Former Principal,
Government Polytechnic,
Alwar (Rajesthan)

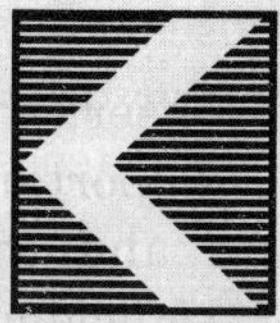

KHANNA BOOK PUBLISHING CO. (P) LTD.
Publisher of Engineering and Computer Books
4C/4344, Ansari Road, Darya Ganj, New Delhi-110002
Phone : 011-23244447-48 **Mobile:** +91-9910909320
E-mail : contact@khannabooks.com
Website: www.khannabooks.com

Dear Readers,

To prevent the piracy, this book is secured with HIGH SECURITY HOLOGRAM on the front title cover. In case you don't find the hologram on the front cover title, please write us to at contact@khannabooks.com or whatsapp us at +91-99109 09320 and avail special gift voucher for yourself.

Specimen of Hologram on front Cover title:

Moreover, there is a SPECIAL DISCOUNT COUPON for you with EVERY HOLOGRAM.

How to avail this SPECIAL DISCOUNT:

Step 1: Scratch the hologram
Step 2: Under the scratch area, your "coupon code" is available
Step 3: Logon to www.khannabooks.com
Step 4: Use your "coupon code" in the shopping cart and get your copy at a special discount
Step 5: Enjoy your reading!

Price : ₹ 350.00

ISBN: 978-93-86173-01-0

Automobile Mechanics
by Dr. A.K. Babu

First Edition: 2018
Reprint: 2021, **2024**

Published by:
Khanna Book Publishing Co. (P) Ltd.
Visit us at: **www.khannabooks.com**
Write us at: **contact@khannabooks.com**
CIN: U22110DL1998PTC095547

To view complete list of books, Please scan the QR Code:

Printed by Print Point

Prof. SANDEEP SANCHETI Ph.D (UK), FIETE, FIE(I), MIEEE
Vice Chancellor

FOREWORD

Research innovations and technological developments in the field of automobile engineering are numerous, from hybrid technologies and energy saving vehicles to brake dynamics; the sole aim being to address key challenges for future mobility and transport.

I am happy that Prof. A.K. Babu, Associate Professor of Automobile Engineering at SRM Easwari Engineering College, Chennai, Tamil Nadu has authored a book on the subject titled "Automobile Mechanics" published by M/s. Khanna Publishing House, Delhi along with Shri S.C. Sharma and Shri T.R. Banga, renowned authors and experts in the field, who have contributed their knowledge and expertise.

The comprehensive, thorough and up-to-date presentation of the principles of automobile engineering in the book, background to theories and essential design of automobile components will assist students both in and out of the classroom.

Having the right study material not only helps the student to be better prepared but also boosts confidence and helping them put in the best possible preparation to face any professional challenge in this domain. I strongly believe that the learning material in this book will help students successfully obtain a strong foundation and master vital concepts; teachers, faculties and engineers will also find the book resourceful and useful to them.

I wish the authors, contributors and the publisher great success; and all my academic fraternity "Happy Reading".

Regards,

SRM Nagar, Kattankulathur – 603 203, Kancheepuram District, Tamil Nadu, India. Phone: +91-44-27455765, Fax: 044-27455764, Mobile: +91-9940036001
E-mail: vc@srmuniv.ac.in, URL: www.srmuniv.ac.in

Research in [illegible] Automobile Engineering [illegible] energy [illegible] the key challenges [illegible]

The primary objective of this book is to present a [illegible] thorough and [illegible] presentation of the principles of automobile engineering and [illegible] to the [illegible] design [illegible] exploration [illegible] especially [illegible] matters [illegible] manner [illegible] him to [illegible] practical [illegible]

The [illegible] of every [illegible] relevant [illegible] solving [illegible] the study [illegible] this [illegible] face [illegible]

As [illegible] engineering [illegible] necessary [illegible] explain [illegible] without [illegible] are provided [illegible] works through the examples [illegible] greater [illegible] To help the student develop [illegible] there are [illegible] knowledge [illegible]

PREFACE

Research innovations and technological developments in the field of Automobile Engineering are numerous, from hybrid technologies and energy saving vehicles to brake dynamics, the sole aim being to address the key challenges for future mobility and transport.

The primary objective of this book is to present a comprehensive, thorough and up-to-date presentation of the principles of automobile engineering and give a background to the theories and essential design of automobile components to graduate students. The easy exploratory, descriptive and explanatory style of the book has been especially formulated to be reader-friendly and to facilitate the quick understanding and easy learning of all students.

A second objective is to present essential information in a lucid manner, to assist the student both in and out of the classroom, enabling him to obtain the requisite knowledge to solve a variety of challenging practical problems.

The third and final objective of the book is to illustrate, by way of several fully worked examples and sample exercises, a number of relevant applications of the theoretical basics. Plenty of practice in solving problems is an excellent way to gain a good understanding of the subject and reinforce/improve learning. Having the right study material not only helps the student to be better prepared but also boosts his confidence and helps him put in the best possible preparation to face an examination.

As mathematics is applied to solve fundamental real-world engineering problems, a good knowledge of mathematics is deemed necessary for every aspiring engineer. Due care has been taken to explain core concepts of the curriculum in a simple and concise manner without sacrificing understanding. Numerous designs and problems are provided at the end of the chapter. It is expected that as the student works through the examples and problems, he or she will develop a greater understanding of the mathematics required for engineering. To help the student develop a sound grasp of the principles covered there are many diagrams, notes and applications as an aid to develop knowledge and facilitate understanding.

The learning material in this book will help students successfully obtain a good/strong foundation and master vital concepts. The structure of the text book is as follows: Chapter 1 is an introduction to the book, covering the basic information on automobiles. Chapter 2 deals with engines and their auxiliary units. Chapters 3-10 cover several aspects of design of automobile components - SI system, background mathematics and advice on problem solving, particularly exam questions. Chapters 11-15 cover essential theory part of support systems for vehicles.

ABOUT THE AUTHORS

A K Babu is currently an Associate professor of Automobile Engineering at the SRM Easwari Engineering College, Tamilnadu, Chennai. A K Babu received his B.E. degree in Mechanical Engineering from University of Madras, in 1994, and M.E. degree in Automobile Engineering from Anna University, in 1997. He has authored and coauthored many academic research papers and published over refereed technical papers in reputed journals and conferences. His current research interests are primarily in the areas of energy conservation and drying of food materials. A K Babu has to his credit over 20 years of professional experience in the field of not only university education but also research and industry. He has previously worked overseas in Malaysia, Ethiopia and Eritrea for more than 5 years. A passionate and dedicator educator, he is also the author of several textbooks based on the syllabus prescribed by different universities.

S C Sharma after graduation in 1966 joined as lecturer in Mechanical Engineering. He had been associated for more than 4 decades in various fields including learning and management of projects in India and abroad in different capacities. While working in Hydro power projects for more than 15 years he has actively associated with various environmental and rehabilitation & resettlement issues and successfully resolved various complicated issues. He has also worked as consultant for matters related to safety, environment and R & R. He has written about a dozen books on subjects related to engineering and management including management of projects, which got successfully accepted in the curriculum of many Technical Universities in India as well as abroad.

T R Banga is Ex-Principal Govt. Polytechnic, Alwar (Rajasthan). He has co-authored many famous text-books and reference books with Sh. S C Sharma which got National and International acceptance and recommendations.

ACKNOWLEDGEMENT

Special thanks [illegible]
because [illegible]
influenced [illegible]
we wish [illegible]
concerned [illegible]

We would like to [illegible]
Dr. Anton [illegible]
[illegible]
Anna [illegible]
[illegible]
this book [illegible]

Pvt. Ltd. [illegible] after graduation in 1966 joined as lecturer in Mechanical [illegible] excellence [illegible] and [illegible] field of [illegible] editing and management of [illegible] book to [illegible] different [illegible] for more than 15 years he was actively associated with [illegible]

ACKNOWLEDGEMENT

Specific acknowledgements for this book are rather difficult to make because a number of standard/popular books and manuals have influenced us, directly or indirectly, in its preparation. Therefore, we wish to express our indebtedness and grateful thanks to all the concerned authors, publishers and manufacturers.

We would like to convey our sincere gratitude and appreciation to Dr.V.Antony Aroul Raj, Easwari Engineering College, Chennai for his unstinting help and constructive comments. We would also like to make particular mention of Dr.G.Kumaresan, Institute for Energy Studies, Anna University for his helpful advice on various technical issues and encouraging support. The painstaking efforts of both experts and their valuable suggestions for improving content helped a great deal to make this book complete.

The authors would also like to thank Khanna Book Publishing Pvt. Ltd, New Delhi for investing their time and effort in ensuring the excellent appearance of the book with clear text, handy size, neat finish and reasonable price. Full credit must go to them for designing the book to make it more readable, inviting and visually appealing to the target audience.

CONTENTS

CONTENTS

10. SUSPENSION SYSTEM 313

11. BRAKE SYSTEM 363

Automobile

The automobile is defined as a self-propelling vehicle. Automobiles are used to transport people and goods from one place to another place. Automobiles generally use gasoline engine or diesel engine as their power source, but technological advances have led to the design of cars that run on electricity and even water.

The modern automobile consists of two distinct sub-assemblies, the body and the chassis.

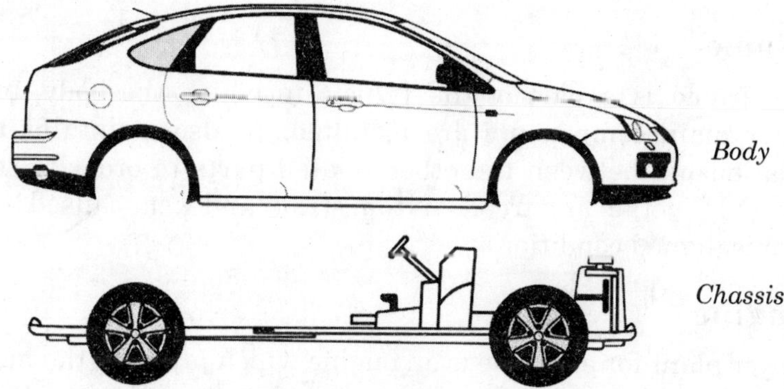

Fig.1.1: *An automobile body and chassis*

BODY

The body is a part of the vehicle that carries the driver and the passengers or cargo. It is made of steel sheets 0.6 mm to 3 mm thick, welded into a box shape. The main function of the body is to provide comfort and protection to the passengers besides giving a good look.

CHASSIS

The chassis is an assembly of a vehicle without body. The chassis includes frame, engine, clutch, gearbox, propeller shaft assembly, final drive, differential, axles, wheels and tyres, steering system, suspension system, braking system, and electrical system.

BASIC STRUCTURE OF AN AUTOMOBILE

Although there are many varieties of automobiles today depending on their usage, their basic constructions are the same. The basic structure of an automobile comprises the following components:

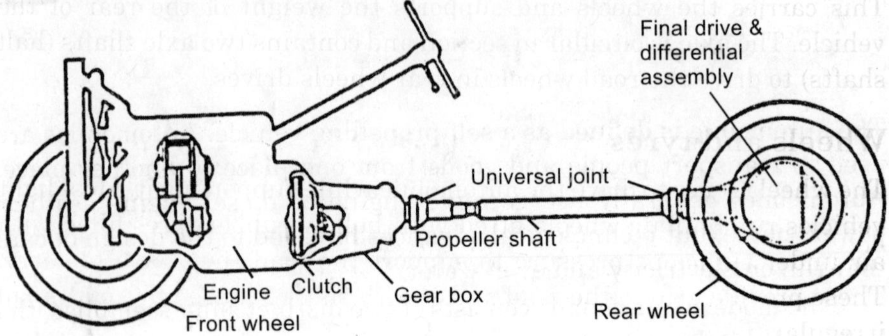

Fig.1.2: *Basic structure of an automobile*

Frame

The frame is a part of the vehicle to which the body, engine, and other main components are mounted. It also maintains the correct relationship between the other related parts in order to secure their normal function and freedom from strain and wear caused by operation in misaligned condition.

Engine

Power plant for a vehicle is an engine which provides the motive power to all the units and the parts those needs power. Engines that are used to power automobiles must be compact and lightweight as they are mounted in a limited space. They must be able to generate great speed and high power output. They must be easy to operate and their noise must be kept minimal. For these reasons, petrol engines and diesel engines are used more often than other types of engines for automobiles

Power train

The mechanism that transmits the power from the engine to the wheels is called the power train. The power train of the most generally used front-engine, rear-drive system comprises essentially of the clutch that serves to engage and disengage the power from the engine to the gearbox, the gearbox that changes the engine speed and torque to that required by wheels, the propeller shaft that transmits the power from the gearbox to the differential, and the differential that divides and transmits the power to the left and right wheels.

Front axle

This arrangement supports the front of the vehicle and is also used for steering.

Rear axle

This carries the wheels and supports the weight of the rear of the vehicle. The axle is circular in section and contains two axle shafts (half shafts) to drive the road wheels in rear wheels drives.

Wheels and tyres

The wheels serve to drive the automobile while supporting it. Most light vehicles run on four wheels fitted with hollow rubber tyres filled with air under sufficient pressure to support the load they have to carry. These provide grip to the road and absorb shocks caused by small road irregularities.

Steering system

The function of this system is to change the direction of vehicle progress.

Suspension system

The purpose of this system is to provide smoother riding by cushioning the road shocks and vibrations and preventing them from being transferred into the body and frame. The suspension system consists mainly of springs to absorb the road shocks and shock absorbers (dampers) to absorb the excessive oscillation of the springs.

Brake system

The brake system is necessary to slow down or stop the running automobile. Parking brake is used for parking the vehicle.

Electrical system

The electrical system is a part of both chassis and body. The system includes the starting, charging, ignition, lighting and horn circuit. Some electrical circuits are for engine operation, some for power transmission and others for lighting and operation of protective devices and accessories.

Accessories

Accessories are used to make driving more safe and comfort. They include car heater, air-conditioner, radio, windscreen wiper, indicators etc.

TYPES OF AUTOMOBILES

The different types of automobiles found on roads are explained in this topic.

TWO WHEELERS

A two-wheeler refers to vehicles that run on two wheels. They include a motorcycle, scooter or moped which uses an internal combustion engine or electric motor as its primary source of propulsion. The two-wheeler segment in India may be divided into three sub segments: mopeds, scooters and motorcycles.

Mopeds

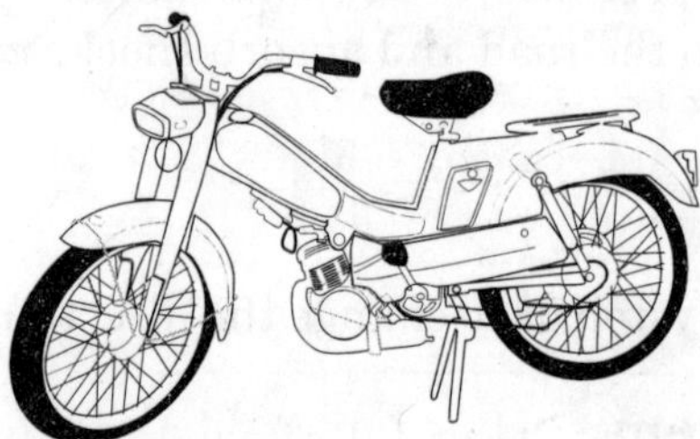

Fig.1.3: *Moped*

The smallest designs are termed mopeds (from motor pedal), and have very small (50 cubic centimeters) engines offering 50 km-per-litre fuel economy.

Motor scooter

Scooters are the most popular types of two-wheelers in the Indian market, especially for ladies and aged people.

Motor scooters have no mechanical elements between the rider's knees, smaller-diameter wheels in most cases and no pedals. The engine is enclosed within the bodywork and usually back close to or alongside the rear wheel. Their engines are usually fan-cooled. The rider sits behind a front panel.

Fig.1.4: *Scooter*

Motor cycle

The normal definition of a motorcycle is a single-track, two-wheeled motor cycle. A motorcycle can be divided into two parts, the 'motor' meaning the engine and transmission, and the 'cycle' composed of the frame, suspension, wheels, and brakes.

Fig.1.5: *Motorcycle*

Motorcycles are produced with both two-stroke and four-stroke-cycle engines and have up to four cylinders. Most of them are air-cooled, though a few are water-cooled. Their engines are generally limited to 1,200CC displacements.

THREE WHEELERS

Three-wheelers in India act as intermediate public transport, a feeder system to public transport in large cities. They cater to the mobility needs of people not owning a private transport mode and inadequately served by the public transport system. They have smaller engine capacities and higher mileage rates than the regular car taxis.

Three wheelers such as auto rickshaws, pick-ups and delivery vans are largely used as commercial vehicles, used to carry passengers and goods. Bajaj, Piaggio, and TVS are some of the brands who make vehicles in this category.

All the three-wheelers usually have maximum speed in the region of 50 to 55 km/h.

Auto rickshaw

A small, three-wheeled vehicle, the auto rickshaw frequently speeds along the streets of Indian cities. It moves small loads of cargo and carries passengers from one place to another. One of the most famous 3 wheelers manufacturer in India is Bajaj.

An auto-rickshaw is generally characterized by a sheet-metal body or open frame that rests on three wheels, a canvas roof with drop-down sides, a small cabin in the front of the vehicle for the driver, and seating space for three in the rear. They are generally fitted with an air-cooled scooter version of a two-stroke engine, with handlebar controls instead of a steering wheel. The triangular form of the auto also makes maneuvering easy, with the front single wheel negotiating the available gap, and the rear two wheels forcing a larger space.

Fig.1.6: *Auto rickshaw*

Auto-rickshaws are light vehicles and have a top-speed of around 50 km/h (about 31 mph) and a cruising speed of around 35 km/h (22 mph).

Pickup

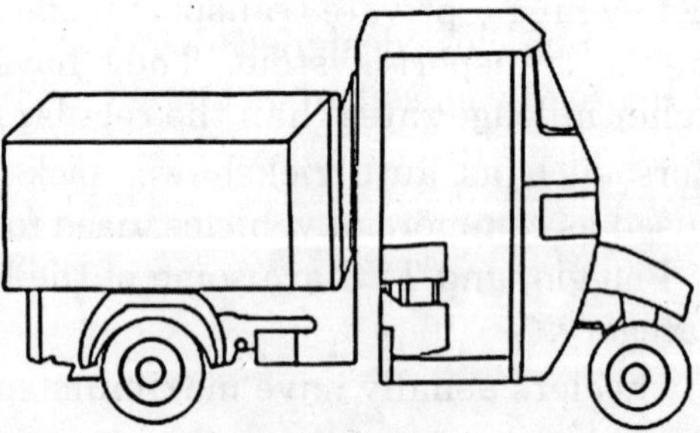

Fig.1.7: *Pickup*

Pickups are designed to carry daily goods. It has a higher load capacity and has an enclosed cabin for driver and load man. They have better powered and high torque engine.

Besides offering more space, it ensures easy loading and the utmost reliability. It is good for short distance and long distance carrying and it has low maintenance and its performance on road is unbeatable.

Delivery van (Cargo van)

Delivery van is one of the most popular three wheelers used for transportation of goods like vegetables, bread, kitchen gas cylinders, etc., from one place to another. Its storage area is completely closed, ensuring complete protection of the goods. Most of these vehicles have a payload capacity of half to one tonne. This 3 wheeler is attributes with fuel efficiency and excellent mileage. The range comprises of Piaggio Truk, Piaggio Half Ton, Mahendra Genio and Tata ACE.

Fig.1.8: *Delivery van*

LIGHT VEHICLES

Vehicles classified in this category have a laden mass of less than three tones. A wide range of different body shapes and sizes come into this light vehicle category. These range from two-seater cars to personnel carriers (mini-buses) and small trucks.

The light motor vehicles designed to carry passengers and sometimes goods are broadly classified as follows:

Sedan / Saloon

Fig.1.9: *Saloon*

'Sedan' and 'Saloon' generally both describe a four-door car with a separate boot. A sedan has four doors and an enclosed compartment to accommodate a row of front and row of rear seats without any partition between the driver an rear-passenger seats. A separate luggage space is made either at the front or the rear, based on the location of the engine. The enclosed trunk offers security, while the rear doors allow easy entry for rear-seat passengers. Most luxury vehicles are four-door sedans because they're more comfortable than most other body styles. The smallest available are sub-compact sedans like the Hyundai Accent and Chevrolet Metro. Slightly larger are compact models like the Honda Civic, Hyundai Verna and Maruti Suzuki Ciaz. Mid-size sedans include the Honda Accord, Toyota Camry, and Hyundai Elantra, while the Ford Crown Victoria and Buick LeSabre are considered large sedans.

Hatchback

Fig.1.10: *Hatchback*

A hatchback / liftback is a passenger car with two or four side doors and a hinged door (which includes the rear window) covering most of the back of the vehicle. This door lifts upwards to give easy access to the rear luggage storage area. The Maruti Alto, Hyundai Eon and the Volkswagen Polo are some examples.

Coupe

Fig.1.11: *Coupe*

Coupe is a French word meaning cut. A coupe is a car that has been cut off diagonally at the back. A sloping rear end helps the car slip through the air more gracefully. Although coupés generally have rear seats, they often (but not always) have only two doors. The rear seats are difficult to access, as the front doors must be used.

Audi R8, Lamborghini Aventador, and Ferrari 488 are few examples of Coupe.

Wagon

A station wagon / estate is a passenger vehicle with a large luggage compartment behind the seats and an extra door at the rear. Wagons have all the same passenger room and driving characteristics as the sedans they're based on, but offer more cargo room. A few of the larger wagons even offer a third-row seat. The roof line of a wagon continues past the rear doors. A few wagon examples are the BMW 3 Series Wagon, Ford Flex and Volkswagen Jetta Sportwagen.

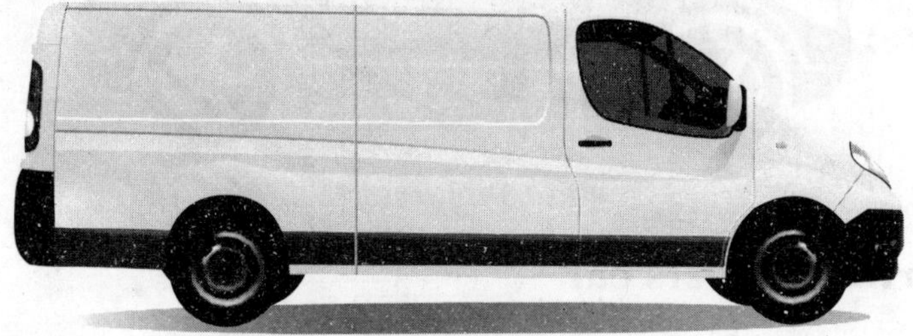

Fig.1.12: *Wagon*

Station wagons offer more stability, and better fuel mileage.

Sport Utility Vehicles (SUVs)

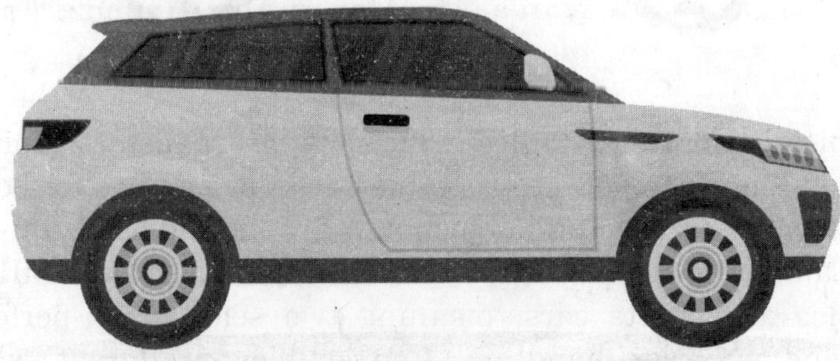

Fig.1.13: *SUV*

Sport-utility vehicles offer four-wheel or all-wheel drive and raised ground clearance in combination with a two- or four-door body. SUVs come in three sizes:

- Small: Honda CRV, Suzuki Grand Vitara, Hyundai Creta.
- Medium: BMW X5, Mercedes-Benz M-Class, Lexus RX300, Isuzu Trooper, Toyota 4Runner, Mahindra Scorpio.
- Large: Toyota Land Cruiser, Land Rover Range Rover.

Crossover

Crossovers are tall wagons and SUVs that are based on a passenger-car platform's architecture (as opposed to a truck's) for improved ride, comfort and fuel economy. They come in all sizes, and many offer a third-row seat. A light-duty four-wheel-drive or all-wheel-drive system is usually optional. A few crossover examples are the Maruti Suzuki S-cross, Renault Koleos, and BMW X1.

Fig.1.14: *Crossover*

Convertible/Sports car

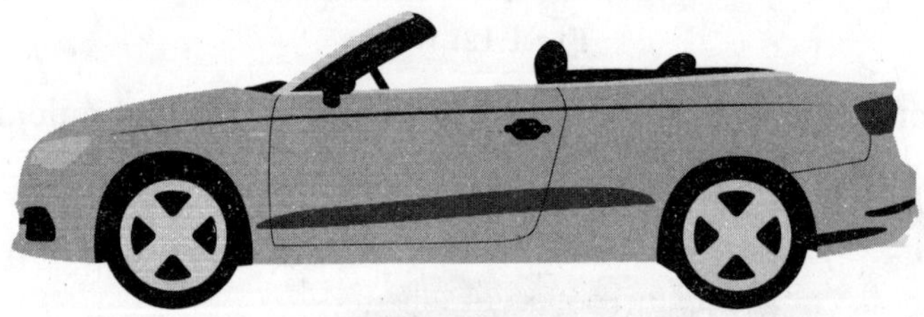

Fig.1.15: *Roadster*

Convertibles, sometimes called cabriolets or roadsters, come with either a soft folding top or a retractable hardtop designed for both daily travel and week-end racing hobbyists. A few examples of convertibles are the Lamborghini Huracan, Jaguar F-type, and Porche 911. Most convertibles are sports cars, meaning two seats, high-performance engines and superior handling. Convertibles are great when the weather is pleasant.

Luxury

Fig.1.16: *Luxury car*

Luxury cars are those with features that goes beyond the standard of a normal car. They have genuine leathers, ultrafine wooden dashboard, handcrafted parts and superior quality.

Luxury cars are more expensive because of their limited production volumes. Only few special people are eligible to buy the car.

Audi, BMW, Cadillac, Lexus, Mercedes-Benz and the like brands vehicles typically achieve luxury status because of their price.

Multi Purpose Vehicles (MPV)

Multi purpose vehicles (MPV) is a vehicle designed primarily for carrying people and their luggage in reasonable comfort. People also refer to such a vehicle as Minivan or Multi-utility vehicle (MUV).

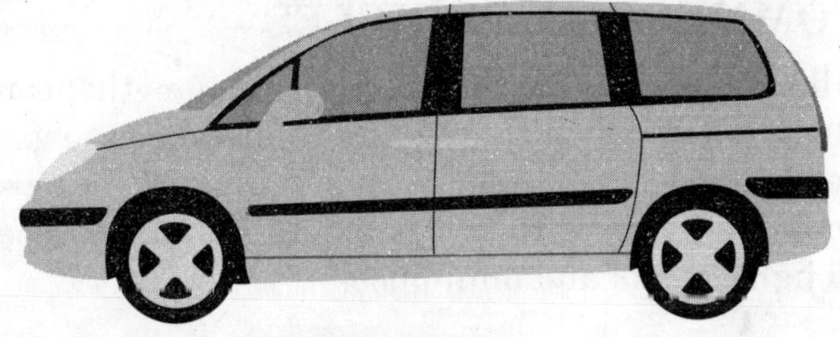

Fig.1.17: *MPV*

Typically, a multi-purpose vehicle has two or three rows of seats which serve for 6 to 8 passengers. In addition, they have a flexible interior spach which means that it can be adjusted as per the requirement. If more cargo needs to be carried, then some of the seats can be folded.

Generally, these vehicles use unitized body construction and thus are lightweight. Improved construction and reduced weight result in better handling and fuel economy.

Due to the upright body style, multi-purpose vehicles have better ground clearance. Also, the upright design style allows the interior space to feel more spacious and roomy.

People-carriers also usually have a significant rear luggage compartment and often have sliding side doors to make it easier to get people in and out of the rear seats, especially in tight spaces such as carparks.

Toyota innova crysta, maruti suzuki ertiga and omni, chevrolet enjoy, honda mobilo, nahindra xylo are a few examples if MPVs available in India.

Hybrid/Electric

A hybrid vehicle has both a gasoline-powered engine and an electric motor that operate in unison and/or independently to propel the vehicle. Some hybrids are based on existing models like the Ford Fusion Hybrid, while others like the Toyota Prius have no gas-power-only version. These vehicles are known for their exceptional fuel economy. Electric vehicles like the Nissan Leaf or on the high end, the Tesla Model S, have no gas engine and must be recharged after driving.

COMMERCIAL VEHICLES

Commercial vehicles are used to transport goods and people safely and efficiently. Commercial vehicles can be divided up not only according to the position of their engine but also according to their use, their body or their trailer type.

LIGHT COMMERCIAL VEHICLES

These small commercial vehicles are used for the conveyance of relatively light goods usually over short distances, e.g. the type of vehicles used by tradesmen to carry their equipment or by shopkeepers to make local deliveries to their customers. Common types are vans, pick-up and light trucks and mini-buses

Vans

Vans are generally used for short-range distribution of goods and passengers. Most vans have a load-carrying capacity of about 0.5 tonnes.

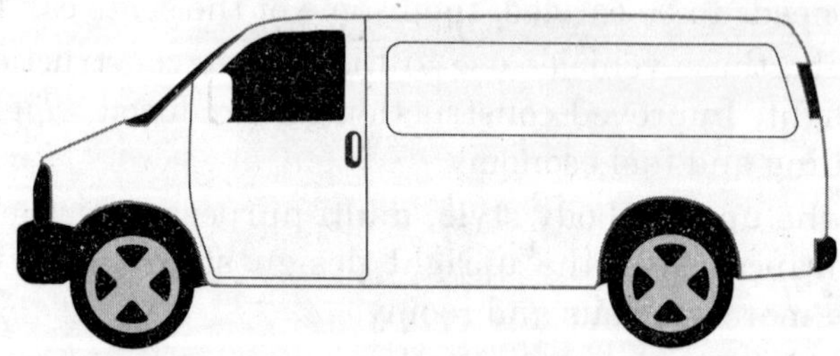

Fig.1.18: *Van*

They have seats in the front for the driver and for only one or two passengers. The engine is usually located over or just in front of the front axle. Hinged or sliding type doors are located on each side opposite the seats. There are double doors at the rear of the van, which open outwards for easy loading. Small vans combine the cab and the body with integral construction. Large vans sometimes have separate cab and body, mounted on an independent chassis frame. The rear axle may have twin road-wheels to have higher load carrying capacity.

Pick-up

When a vehicle is required for the transportation of bulky equipment, e.g. materials carried by builders and decorators, an open body is often preferred. A light vehicle having this partly enclosed body is called a pick-up.

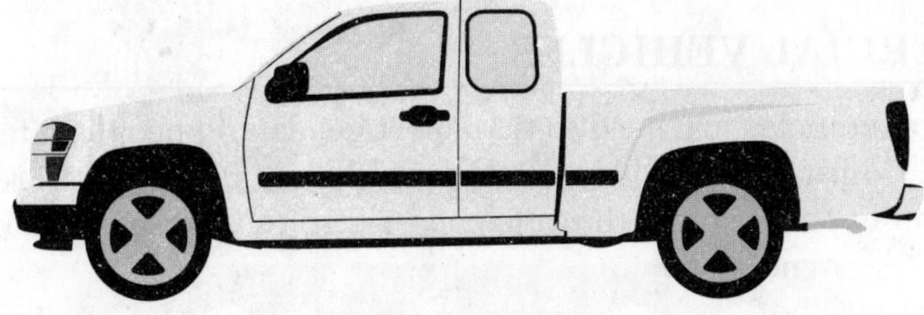

Fig.1.19: *Pickup*

Light trucks

Larger vehicles such as light trucks and minibuses need to carry heavier loads, so they require a stronger construction and have to be specially designed to suit the application. Since low-cost operation is an essential factor, most of these vehicles are fitted with a diesel engine.

Fig.1.20: *Light trucks*

HEAVY COMMERCIAL VEHICLES

Although this book concentrates mainly on light vehicles, a review of some heavy vehicles is included for comparison purposes. Common types are:

- Trucks
- Semi-trailer trucks
- Road trains (truck and trailer outfits)
- Buses
- Special-purpose vehicles

Trucks

A heavy vehicle has a load-carrying capacity in excess of about 3 tonnes. For long-distance transportation of goods a truck is used and these come in various shapes and sizes. Depending on their size and weight, trucks have different numbers of axles and these can be steered or driven.

Fig.1.21: *Trucks*

A simple truck has two axles and four wheels. More number of axles and wheels are added to increase load-carrying capacity. The number wheel hubs and the number of drive axle hubs classify the rigid trucks as follows:

- A four-wheeler (4 × 2) truck with two driving wheels
- A six-wheeler (6 × 4) truck with four driving wheels
- A six-wheeler (6 × 2) truck with two driving wheels
- An eight-wheeler (8 × 4) truck with four driving wheels

Semi-trailer trucks

The term semi-trailer truck refers to a combination of a tractor vehicle and a trailer. This type of truck does not have its own body but instead has a semi-trailer which rests on the semi-trailer coupling (fifth wheel) of the tractor unit.

Fig.1.22: *Semi-trailer*

The semi-trailer has a long rigid chassis with a single-axle, tandem-axle, tri-axle, layout at the rear end. All the trailer axles are dead axles. The front end of the trailer chassis is supported on the rear of the tractor chassis. At this point it is free to swivel about a pivot known as the fifth wheel coupling.

Road train

A road train has a relatively normal tractor unit, but instead of towing one trailer or semi-trailer, it pulls two or more of them.

Fig.1.23: *Road train*

Buses and coaches

These are used for short-and-long distance passenger transport and are referred to as scheduled-service buses and coaches respectively.

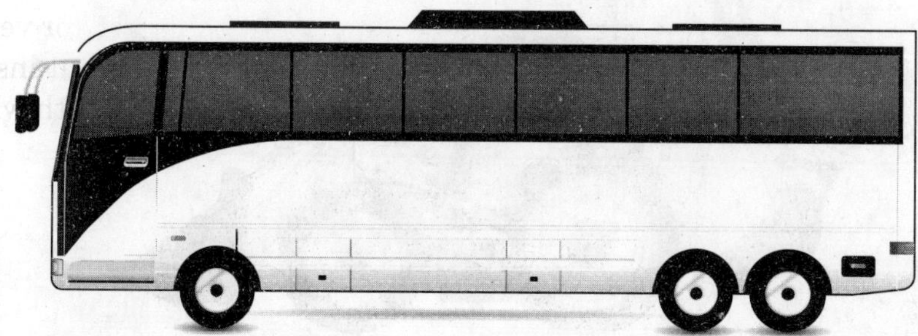

Fig.1.24: *Bus*

Coaches carry passengers travelling on long distance, and hence the interior is designed to provide the best possible comfort and to minimise fatigue. Seats are located facing the front to provide passengers the benefit of looking ahead. For better visibility of passengers large panelled windows are provided on either side extending the full length of the vehicle and across the back seats. There is a door adjacent to the driver. The passenger's doors are located opposite side of the driver's seat one towards the front and the other towards the rear. An emergency door is usually provided towards the centre on the opposite side of passenger's doors.

Articulated bus

Articulated bus has the specific meaning of a bus which is contiguous (can walk from end to end), but bends in the middle (to get around corners). These buses are also referred to as bendy buses.

Fig.1.25: *Articulated bus*

Special-purpose vehicles

These range from custom-built to purpose-built vehicles. These are large vehicles for the transportation of very heavy loads. They also include fire engines and trucks with crane or concrete-pump bodies. Because of their high permissible gross weight and their unusual dimensions, special-purpose vehicles often require a special operating license.

Fig.1.26: *Front-end loader*

Depending upon their utility in various fields they are classified such as:

1. Military applications
2. Aircraft service
3. Industries and power stations

4. Underground boring
5. Car carriers
6. Caravans
7. Construction of dams bridges, roads etc.
8. Mine works
9. Material handling
10. Earth moving equipment
11. Petroleum and L.P.G. carriers

CHASSIS LAYOUT

Chassis layout refers to location of an engine and driveline components and also how power is transmitted from the engine to the drive wheels.

CHASSIS LAYOUT WITH REFERENCE TO DRIVE

REAR WHEEL DRIVE

This layout has the engine in the front, mounted longitudinally, and the drive axle in the rear. The gearbox is usually right behind the engine, and a drive shaft transfers power back to the rear axle. In this arrangement the rear wheels act as the driving wheels and the front wheels swivel to allow the vehicle to be steered.

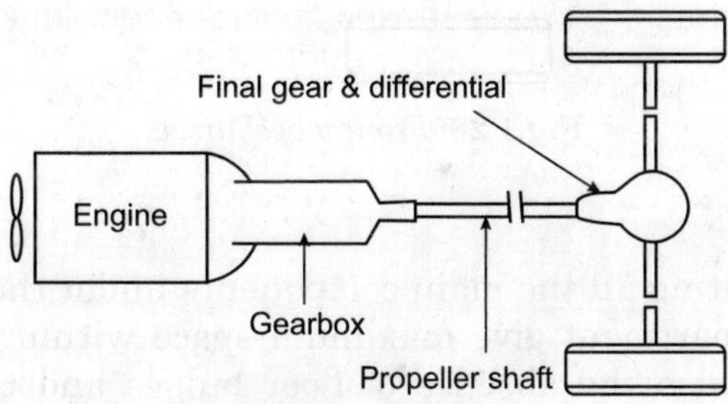

Fig.1.27: *Rear wheel drive*

Advantages

1. Allows a relatively even load to be applied to each road wheel.
2. Spacing out the main components allows the main components to be easily removed and thereby provides for economic maintenance.
3. Load transfer takes place from the front to rear of the vehicle when the car is climbing a hill or accelerating and therefore good traction is obtained.

Disadvantages

1. Increased ground clearance or a raised long bulge (tunnel) to accommodate gearbox and propeller shaft.

Applications

Buses and trucks are usually rear wheel drive vehicles.

FRONT WHEEL DRIVE

In this layout the engine, gearbox, final drive and differential assembly are combined into a single unit. The front wheel drive has both the engine and transaxle in the front. The compactness of the layout has made it very popular for use on cars.

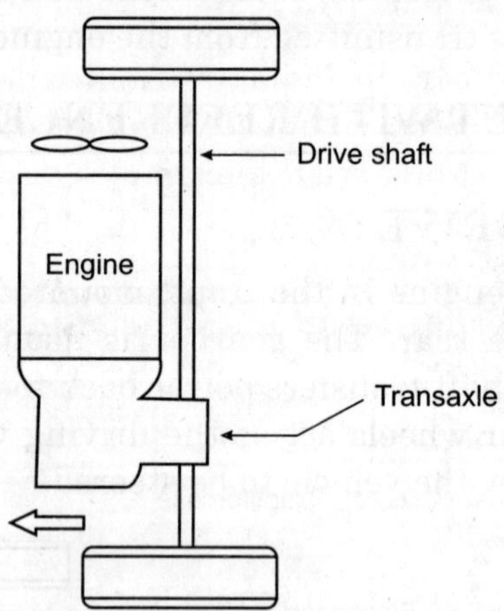

Fig.1.28: *Front wheel drive*

Advantages

1. Accommodating all the main components under the bonnet (hood) in one compartment give maximum space within the car for the occupants, also the absence of floor bulges and tunnel provides more room for the rear passengers.
2. A low flat floor as there is no propeller shaft or normal rear axle.
3. Transverse mounting of the engine simplifies the transmission, because the output shafts from the engine and gearbox move in a similar direction to the wheels. This avoids the need for a bevel-type final drive. Instead a simple reduction gear, incorporating a differential, transmits the power by short drive shafts to the road wheels.
4. By placing the combined engine and transaxle over the driving wheels excellent traction on the driving wheels can be obtained.

5. Using the front wheels for steering allows the driving force to act in the same direction as the wheel is pointing. This feature improves vehicle handling especially in slippery conditions.

Disadvantages

- As the front wheels are used for both driving and steering, the arrangement becomes more complicated.
- Mounting the main units in one assembly sometimes makes it difficult to gain access to some parts.
- Driving wheels have less grip on the road when the vehicle is accelerating and hill-climbing.

Applications

Most of the passenger cars in India are front wheel drive type. Mauti Suzuki Alto, Chevrolet Beat, Hyundai Verna, Hyundai i10, Honda City, Honda Civic, Toyota Corolla are few examples of this type.

FOUR WHEEL DRIVE

This arrangement is safer because it distributes the drive to all four wheels.

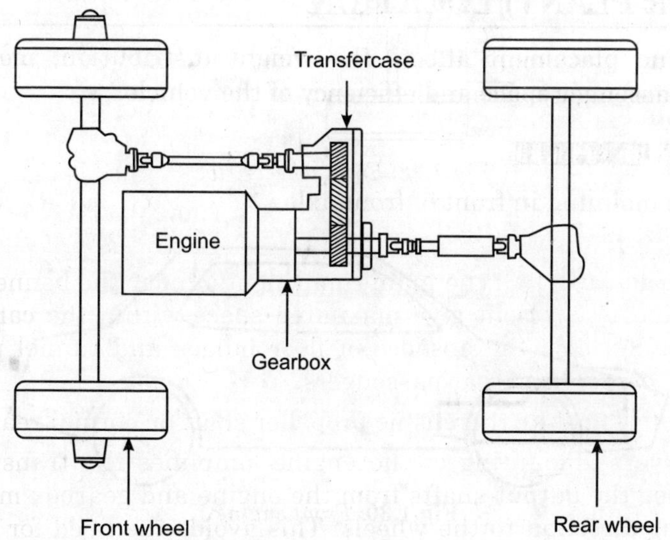

Fig.1.29: *Four wheel drive*

Advantages

1. The sharing of the load between the four wheels during acceleration reduces the risks of wheel spin.

2. Also the positive drive to each wheel during braking minimises the possibility of wheel lock-up.
3. A further advantage of this layout is when the vehicle is driven on slippery surfaces such as snow and mud. When on an icy road or driven cross-country (off-highway), a two-wheel drive vehicle soon becomes undrivable because the loss of grip of one of the driving wheels causes the wheel to spin.

Disadvantages

1. Increased weight of the vehicle
2. Reduced mileage

Applications

Maruti Gypsy, Renault Duster, Toyota fortuner (AWD), Toyota land crusier, Renault Koleos, Mahindra XUV500 (AWD), Skoda Yeti, Mitsubishi Pajero, Honda CRV (AWD), Mahindra Scorpio, Mahindra Bolero, Ford Endeavour, Tata safari Strome, Hyundai Santa Fe (AWD) are four wheel drive vehicles.

CHASSIS LAYOUT WITH REFERENCE TO ENGINE (POWER PLANT) LOCATION

The engine placement affects the weight distribution, moment of inertia, passenger space and efficiency of the vehicle.

FRONT ENGINE

Engine is mounted in front of front axle.

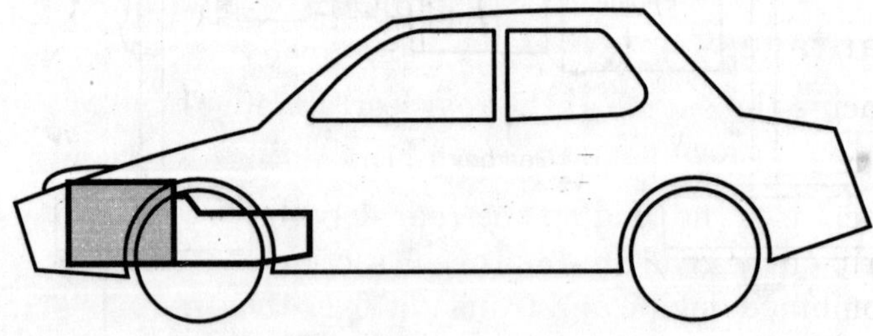

Fig.1.30: *Front engine*

Advantages

1. The large mass of an engine at the front of the car gives the driver protection in the event of a head-on collision.
2. Engine cooling is simpler to arrange.

3. The cornering ability of a vehicle is normally better if the weight is concentrated at the front.
4. Engine can be controlled easily by the driver.

Disadvantages

1. Load transfer takes place from the front driving wheels to rear of the vehicle when the car is climbing a hill or accelerating and therefore traction is reduced.
2. More engine noise
3. Front tyre wear is more

Applications

Maruti Alto, Hyundai Eon, and Nissan micra are front engine cars.

REAR ENGINE

This arrangement is no longer very popular with vehicle designers. The engine is mounted behind a rear axle.

Fig.1.31: *Rear engine*

Advantages

By placing the engine at the rear, certain advantages can be gained.

1. Lower overall height owing to the absence of the propeller shaft
2. Increases the load on the rear driving wheels, giving them better grip of the road i.e. Improved traction of the driving wheels due to combined engine and transaxle over the driving wheels
3. Reduced engine noise
4. The front seats are nearer the front wheels than a front-engined car and the floor is quite flat.

Disadvantages

1. Most of the space vacated by the engine at the front end can be

used for luggage, but this space is usually less than that available at the rear.

2. Suitable for small cars. The heavy engine at the rear has handling issues by making it 'tail heavy'.
3. Prone to oversteer.
4. More difficult to arrange the gearbox and engine controls.
5. Efficient engine cooling is also a problem.
6. Rear tyre wear is more.

Applications

One of the most famous rear-engined cars is the Volkswagen Beetle. In India Tata Nano and Bajai Auto rickshaw are of this type.

CENTRAL (OR) MID ENGINE

The term mid-engine is used because the engine is mounted in front of the rear axle line. Sometimes mid-engine layout describes the location of an engine between the rear and front axles. This situation generally applies to sports cars because the engine sitting gives a load distribution that achieves both good handling and maximum traction from the driving wheels.

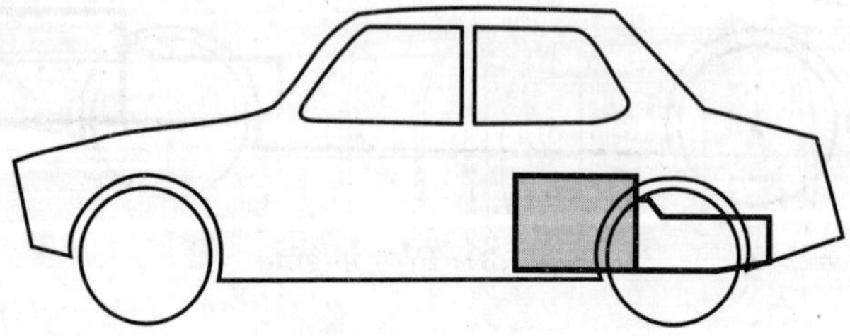

Fig.1.32: *Mid engine*

Advantages

An engine in the center of the vehicle allows better weight distribution and a lower moment of inertia. These cars will have a good center of gravity. This is the reason most supercars have mid mounted engines.

Disadvantages

Reduced passenger space

Accessing to engine for maintenance is difficult.

Applications

Ferrari 488 GTB and Maruti Omni are mid engine vehicles.

VEHICLE WEIGHTS AND DIMENSIONS

Major weights and dimensions of a vehicle are explained below:

WEIGHTS

Payload

The total weight of passengers and cargo that a vehicle carries or can carry.

Gross Vehicle Weight (GVW)

GVW is the weight of the vehicle, plus the weight of all passengers the vehicle is designed to carry, plus the maximum allowable payload (passengers and cargo) or luggage load.

Curb Weight (CW) or Kerb Weight (KW)

CW is the weight of a vehicle wet, meaning with a full tank of fuel and all fluids filled, but without passengers or cargo (luggage).

Model Weight (MW)

MW is the weight of a vehicle wet and with passengers.

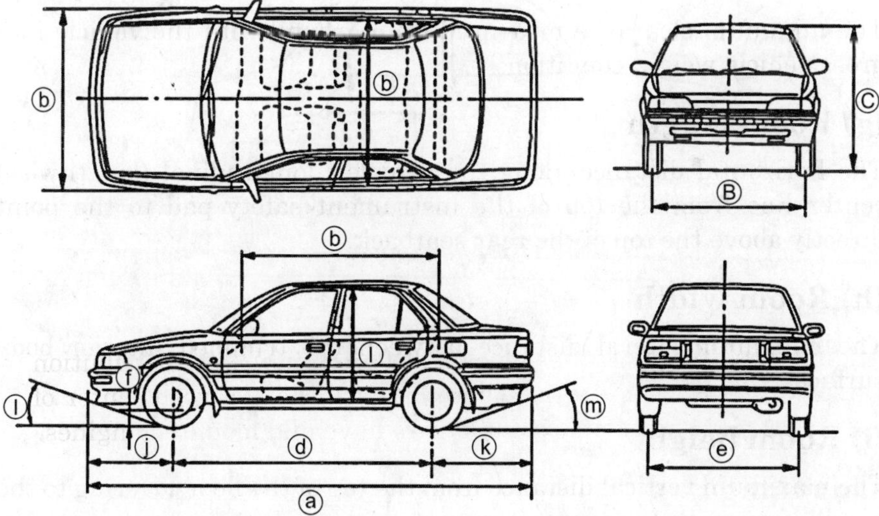

Fig.1.33: *Vehicle weights and dimensions*

DIMENSIONS

(a) Overall length

The distance between the front and rear of the vehicle measured along the longitudinal centre line, including bumper guards, if the vehicle has them.

(b) Overall width

The maximum lateral distance measured between the sides of the vehicle. It includes bumpers, moldings, sheet metal protrusions, etc., measured to the outside of the metal.

(c) Overall height

The maximum vertical distance from the top of the roof to the road surface, measured with the vehicle in a curb weight condition.

(d) Wheelbase

The distance between the center lines of the front and rear axles.

(e) Tread

The distance between the center lines of tyres.

(f) Minimum running ground clearance

The minimum clearance measured from ground with the vehicle in a gross vehicle weight condition.

(g) Room length

The horizontal distance along the vehicle's longitudinal (lengthwise) center line from the top of the instrument safety pad to the point directly above the top of the rear seatback.

(h) Room width

The maximum lateral distance between the trimmed interior body surfaces.

(i) Room height

The maximum vertical distance from the top of the floor covering to the headlining.

(j) Overhang-front

The distance from the center line of the front wheels to the front of the vehicle, including bumper guards if they are standard equipment.

(k) Overhang-rear

The distance from the center line of the rear wheels to the rear of the vehicle, including bumper guards if they are standard equipment.

(l) Angle of approach

The angle formed by the ground and the line tangent to the front tyre static loaded radius arc and the first point of interference (i.e. bumper, guard, gravel deflector, fender or other components excluding license plate).

(m) Angle of departure

The angle formed by the ground and the line tangent to the rear tyre static loaded radius arc and the first point of interference (i.e. bumper, guard, gravel deflector, tail pipe, fender or other components excluding license plate).

MEMBERS OF A BODY

The body of a typical car, depicting various members such as:

Roof

It is the top cover of the car.

Roof headers

Two members extended along the length of the car on which roof is being supported.

Roof rails

Two width-wise members, joining the roof headers and supporting the roof.

Pillars

Vertical members used to attach doors and windows on both the sides.

Rails and Sills

These members bear the weights of roof, pillars and other members of the body.

Bumpers

Crosswise front and rear members which provide additional safety against any impact or collision.

Radiator Yoke

It is a cross member installed just above the front bumper, to whom radiator is attached.

Shear Panels

One top and two sides cover members of the case, in which engine remains.

The body of a car may be fabricated as a single unit or as multiunit, and can be named as one box construction, two box construction or three box construction.

A single unit or one box construction body is stronger, a three box body is easy to assemble and disassemble where as two box body has intermediatory ability.

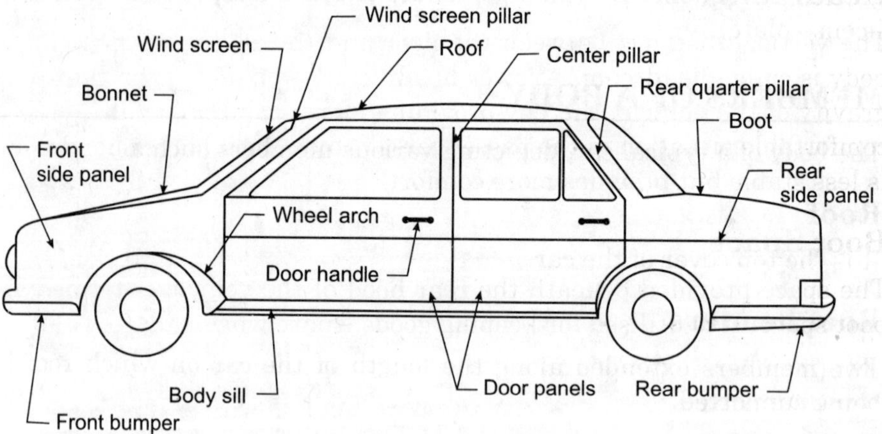

Fig.1.34: *Parts of a car body*

BODY TERMINOLOGY

Cab

It means "cabin". It is the region in which driver sits for driving. It may be an enclosed region as in trucks, deluxe buses etc. or may be an open region as in cars, jeeps etc.

Facia

The frontage of a vehicle, visible to the driver from inside the cab, is called facia. It includes the instrument board, globe box, tool box, tape recorder housing etc.

Dashboard

It is an instrument box panel, and houses fuel level indicator, speedometer, engine temperature indicator, ammeter, voltmeter, odometer etc.

Legroom

It is the space provided for movement of legs of the driver and the passengers. Sufficient space is essential for a comfortable driving and riding.

Shoulder room

It is the clear distance available inside the car along the width. If this distance is sufficient, the passengers will be sitting comfortably.

Headroom

The vertical distance between the floor and the ceiling of the vehicle body is called headroom. A lower headroom lowers down the centre of gravity of the vehicle, thereby providing more stability but it is a less comfortable situation to the passengers. Vehicle with higher headroom is less stable but provides more comfort.

Boot Space

The space provided beneath the rear hood of the vehicle is termed as boot space. It is utilised for keeping goods, spare wheel etc.

Engine

The automobile runs by overcoming the road surface resistance and other resisting factors and for this purpose it is necessary that power be supplied to its wheels. The source of power is called an engine. An *engine* is a device that converts the energy of heat, electricity, wind, atomic power, etc into mechanical energy. An engine that changes the heat energy (thermal energy) into mechanical energy is called a *thermal engine or heat engine.*

CLASSIFICATION OF ENGINES

There are several types of heat engines. Petrol engines (gasoline engines), diesel engines, etc which produce thermal energy internally are called *Internal Combustion (IC) Engines.* In internal combustion engines, the combustion of air-fuel mixture takes place inside the engine cylinder.

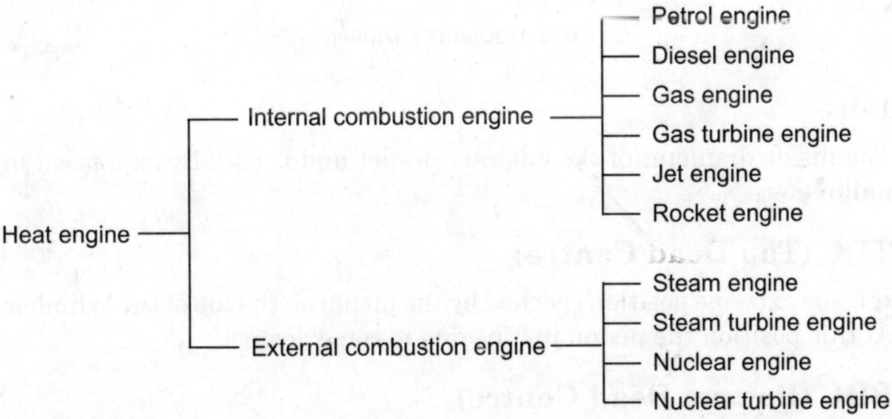

Steam engines, steam turbine engines, etc which produce thermal energy externally are called *External Combustion Engines.* When the fuel is burnt outside the engine cylinder, the engine is an external combustion engine. In a steam engine, steam is raised in a boiler (outside the cylinder) and passed into the cylinder. Steam expands in the cylinder and moves the piston and hence power is transferred to the wheels.

ENGINES FOR AUTOMOBILES

Features specially called for in automotive engines are: engine must be compact and light as it has to be mounted in automobile and at same time be capable of producing high power, must be simple to operate and have minimum troubles, and must run quietly. It was because of these requirements that gasoline and diesel engines are widely used for automotive purpose from among the other type of internal combustion engines.

IC ENGINE TERMINOLOGIES

The following are some of the important terms used in specifying an internal combustion engine.

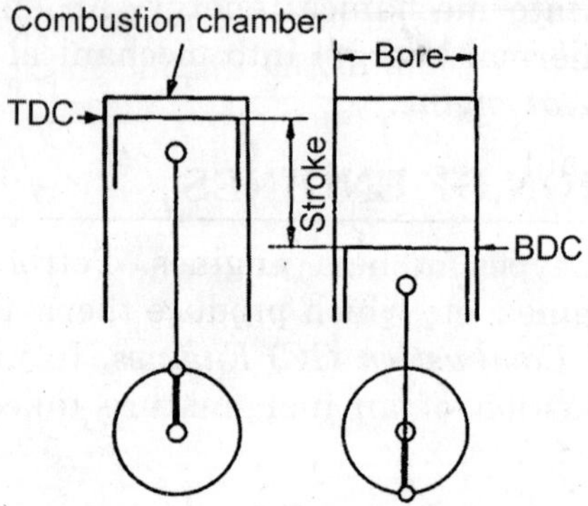

Fig.2.1: *Engine parameters*

Bore

The inside diameter of the engine cylinder and is usually expressed in millimeter.

TDC (Top Dead Centre)

It is the extreme position reached by the piston on the top of the cylinder. At this position the piston just begins to come down.

BDC (Bottom Dead Centre)

It is the bottom most position reached by the piston inside the cylinder. At this position the cylinder volume is at a maximum.

Stroke

The distance between TDC and BDC of the piston travel is called the stroke.

Cycle

The operation in the cylinder from induction of fuel mixture to its compression, ignition, expansion and exhaust of the burnt fuel mixture is called an engine cycle.

Engine capacity, CC

Engine capacity is also called stroke volume, swept volume or piston displacement. The volume covered by the piston as it moves from BDC to TDC or vice versa is engine capacity and indicated in CC (cm^3) or liters.

It does not include the clearance volume or the volume of the combustion chamber. The cylinder volume for a single cylinder engine is higher than the 'CC' of the engine. A 3.3 litres (3300CC) engine basically means that the sum of the displacement volumes of all the cylinders is 3.3 litres. If it is a four-stroke four-cylinder diesel engine, it would turn the crankshaft through two complete revolutions and the four pistons would inhale a total of 3.3 litres of air.

Also the 'CC' specification for any engine is rounded off to nearest value. It is never exact. For example, Maruti 800 has 796CC engine but it is always regarded as an 800CC engine car.

Clearance volume

The volume of the cylinder when the piston is at TDC is refer to as clearance volume. In fact, clearance volume is the combustion chamber volume.

Compression ratio

The compression ratio may be defined as the ratio of the maximum cylinder volume when the piston is at its BDC to the minimum cylinder volume (clearance volume) with the piston at its TDC. That is, the sum of the swept and clearance volumes divided by the clearance volume. The compression ratio measures how much the air/fuel mixture is compressed when the piston reaches the TDC. A ratio of 8:1 means that the volume of the air/fuel mixture is compressed to 1/8 of its original volume when the piston reaches the TDC.

Petrol engines have compression-ratios of the order of 8:1 to 11:1; but to produce self-ignition of the charge, diesel engines usually double these figures and may have values of between 14:1 and 24:1 for naturally aspirated types, depending on the design. To a certain extent, a higher the compression ratio creates a higher combustion gas pressure, resulting in higher output.

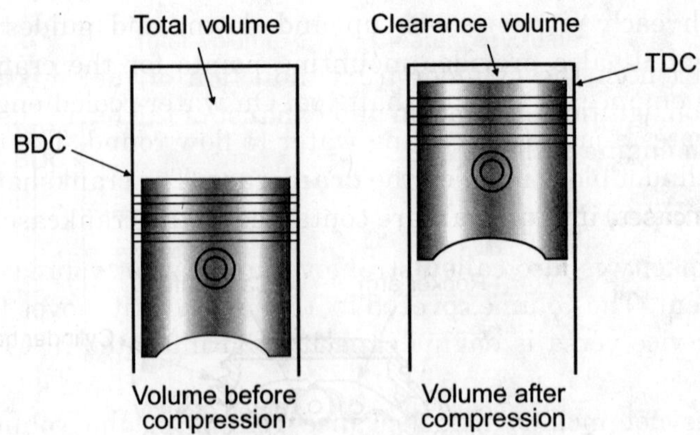

Fig.2.2: *Compression ratio*

$$\text{Compression ratio} = \frac{\text{Volume at BDC}}{\text{Volume at TDC}} = \frac{\text{Total volume}}{\text{Clearance volume}}$$

$$= \frac{\text{Swept volume} + \text{Clearance volume}}{\text{Clearance volume}}$$

$$= \frac{V_S + V_C}{V_C}$$

Note: When the piston is at TDC, a space is left between its top and the bottom of the cylinder. The volume contained in this space is called the clearance volume. The volume enclosed above the piston at BDC consists of the clearance volume plus the swept volume and is called the total volume.

As a general rule, small engines require a higher compression ratio than larger engines. The higher the compression ratio, the more the mixture is compressed. When this compression occurs, the pressure of the mixture increases dramatically. This large increase in pressure helps the mixture burn more completely and produce more power when it is ignited.

BASIC CONSTRUCTION

An engine consists of the engine proper and various auxiliary devices. The engine proper consists of the cylinder block, cylinder head, piston, crankshaft and valve mechanism. Auxiliary devices of the engine are designed to help the operation of the engine. They include fuel lubrication, cooling, intake and exhaust and electrical systems.

This section provides a brief constructional details of an engine construction. The **cylinder block** contains one or more cylinders in which each piston moves up and down, and guides the piston movement. It also provides mounting points for the crankshaft and on some engine, for the camshaft too. On water-cooled engines it also incorporates a jacket for cooling water to flow round. The bottom part of the cylinder block is called the **crankcase**. The crankshaft, camshaft (in some cases) and **oil pan** are contained in the crankcase.

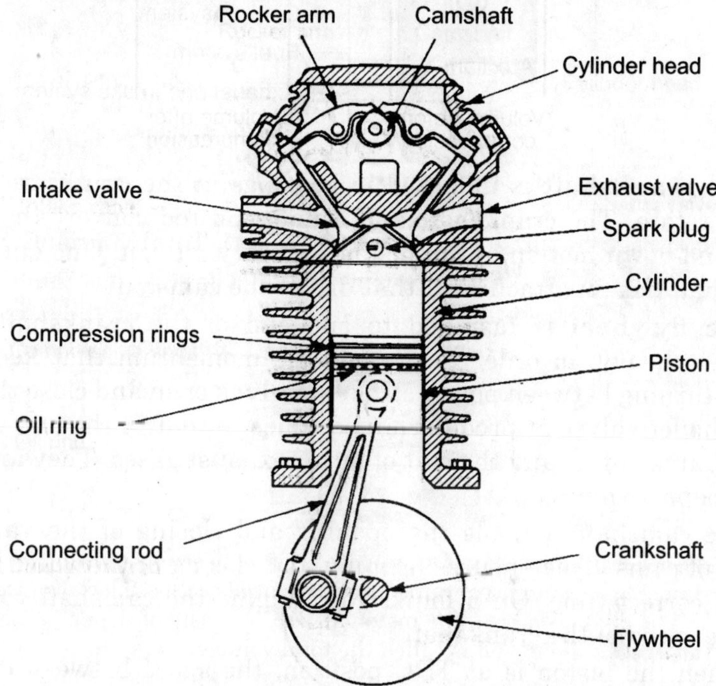

Fig.2.3: *Constructional details of an engine*

Cylinder liners are sleeves that are either cast into the block or installed later (replaceable type). The **cylinderhead** is located over the cylinder block. It contains valves, manifolds and spark plug or fuel injector. A **cylinderhead gasket** is fitted between the cylinder block and the cylinder head to prevent burning gas, coolant and oil from leaking out. A piston moves up and down (reciprocates) inside its cylinder to perform suction, compression, combustion and exhaust. The most important function of the piston is to receive combustion pressure and transmit the force to the crankshaft via the connecting rod. The **connecting rod** connects the piston to the crankshaft and has two bearings. It is attached to the piston by the **gudgeon pin**, which runs in a bearing called the small end. The other, larger end of the connecting rod is connected to the crankshaft by means of the big end bearing.

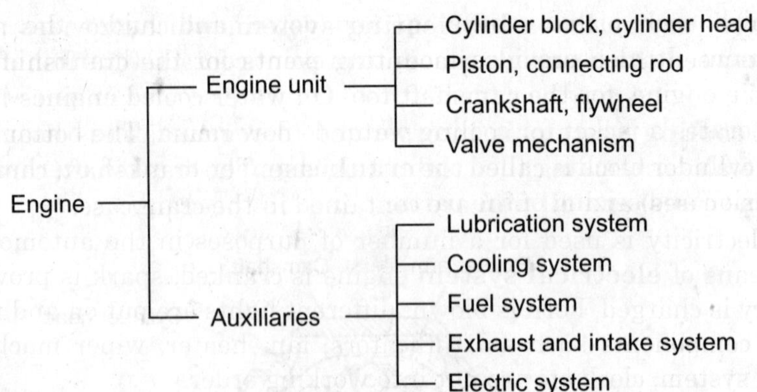

The **crankshaft** is mounted in bearings in the crankcase and is free to rotate. The crankshaft and connecting rod convert the linear or up-and-down motion of the piston to rotary motion. One end of the crankshaft has an attachment that drives the camshaft.

The **flywheel** is fastened to one end of the crankshaft. It is relatively heavy, in order to provide the momentum that keeps the engine turning between power strokes. **Valves** open and close the inlet and exhaust valves at predetermined times to control the entry of the fresh charge of fuel and the exit of burnt exhaust gases. They are often called poppet valves.

The camshaft controls the opening and closing of the valves by means of cams. This ensures opening and closing of the valves at the exactly correct time. On a four stroke engine the camshaft rotates a half the speed of the crankshaft.

When the piston is at TDC position, the space between cylinder head and piston crown (piston top) is known as the combustion chamber.

In a spark-ignition (SI) engine, a mixture of fuel and air from the **carburetor** enters the cylinder through the **intake manifold** and **intake port.** A throttle in the carburetor controls the mass of mixture entering the combustion chamber. The intake valve is located at the junction of the intake port and the cylinder. A **spark plug,** which is located near the top of the cylinder, initiates combustion.

In a compression-ignition (CI) engine, the fuel is injected directly into the combustion chamber through a fuel **injection nozzle.** The quantity of fuel entering is controlled by a fuel control lever which is actuated by an accelerator pedal. Air enters the cylinder from a manifold through a port.

The passage through which the products of combustion leave the combustion chamber consists of the **exhaust port** and the **exhaust manifold.** The exhaust valve is located at the junction of the exhaust port and the cylinder.

Every engine has a **lubricating system,** which provides an oil film between the bearing surfaces to prevent contact, thus minimising wear.

A **cooling system** is incorporated to remove excess heat from the metal surrounding the combustion chamber to avoid overheating, expansion and seizing of the parts.

Electricity is used for a number of purposes in the automobiles. By means of **electrical system** engine is cranked, spark is provided, battery is charged, horn is blown, different lights are put on and many other components such as trafficators, fan, heater, wiper machines, music system, clock etc are put into working order.

IC ENGINE CLASSIFICATION

The internal combustion engines can be classified on the basis of engine design, method of ignition, number of strokes, working cycle, etc.

BASIC ENGINE DESIGN

Reciprocating engine

In a reciprocating engine, the piston moves backwards and forwards in the cylinder, and this motion is converted into a rotary motion by the connecting rod, crankpin and crankshaft.

Rotary engine

In a rotary engine, the piston is a rotor with seals which press against the cylinder wall, and is capable of converting the expansion energy from the combustion process directly into rotary movement. Since there are no reciprocating masses need to be accelerated and decelerated during every working cycle, the engine can run at a higher speed and therefore produce more power for the same weight.

METHOD OF IGNITION

Spark Ignition (SI) engines

The petrol engine is called a spark ignition (SI) engine. A spark is created by the ignition system. A distributor, geared to the camshaft, times and distributes the spark to the spark plug at exactly the correct instant.

Compression Ignition (CI) engines

A diesel engine is a compression ignition (CI) engine. It does not use a spark to ignite the fuel. Basically, when air is compressed and fuel is injected into it, the fuel ignites. The diesel engine was invented by Rudolph Diesel in 1892 in Germany. Diesel engines are used extensively in heavy equipments and were not used in automobiles until the 1930s.

NUMBER OF STROKES

Four– stroke engines

Basically every engine has to repeat the four operations, namely suction, compression, power and exhaust. If the four operations are completed in four strokes of the piston or two revolutions of the crankshaft, the engine is said to be four stroke engines.

Each stroke of the piston consists of 180° of crank rotation and hence a four-stroke cycle is completed through 720° of crank rotation.

Two– Stroke engines

In two-stroke engines the four operations, namely suction, compression, power and exhaust are completed in two strokes of the piston or one revolution of the crankshaft.

Each stroke of the piston consists of 180° of crank rotation and hence a two-stroke cycle is completed through 360° of crank rotation.

WORKING CYCLE

Otto cycle

Most petrol engines operate on the Otto cycle, named after Nikolaus Otto, who invented it. The Otto cycle consists of two constant volume processes and two reversible adiabatic processes.

Diesel cycle

The Diesel cycle is Dr. Rudolph Diesel's original theoretical cycle for the diesel engine. This cycle was introduced in 1893 and it is successfully used in all diesel engines. The Diesel cycle consists of one constant volume process, one constant pressure process and two reversible adiabatic processes.

FUEL

The classification based on the type of fuel used, as below.

- Petrol (gasoline) engines
- Diesel engines
- Compressed Natural Gas (CNG) engines
- Liquefied Petroleum Gas (LPG) engines
- Alcohol (methanol, ethanol) engines

- Fuel oil (kerosene) engines
- Dual fuel engines
- Multi fuel engines (Military vehicle engines)

METHOD OF COOLING

Engines are normally cooled by an air-cooling system or a water-cooling system. Most automotive engines are cooled by the water-cooling system.

Air cooling

It is normally used on *motorcycles* which have an engine exposed to the air stream on its forward motion and no surrounding bodywork.

Water cooling

In a water-cooled system the heat from the cylinders is transferred to the coolant contained in jackets surrounding the cylinders. This coolant then passes through a radiator which transfers heat to the cool air flowing around it. This method of cooling is more efficient in removing excess heat produced by the combustion process.

NUMBER OF CYLINDERS

Multiple cylinders (usually 2, 3, 4, 6, 8 and 12) create better torque, efficiency, and are better balanced compared to single cylinder.

CYLINDER ARRANGEMENT

Generally, the following cylinder arrangements are used:

In-line type

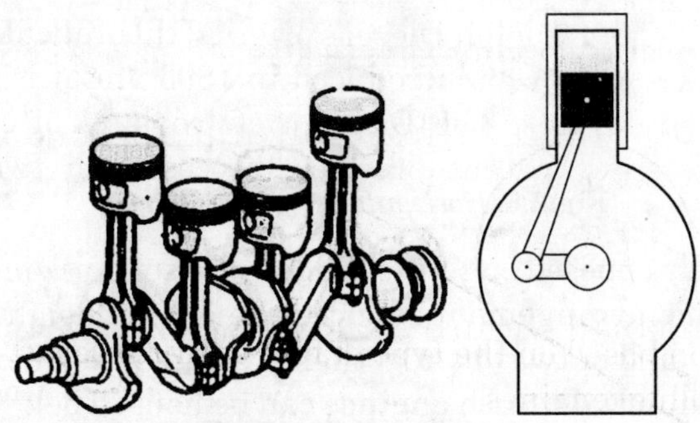

Fig.2.4: *Inline engine*

Cylinders are arranged in a single row. This type is used most frequently because it provides the simplest construction. In-line engines may have

2 to 6, or even 8 cylinders mounted in a straight line, one after the other to obtain smoother performance. Light and heavy motor vehicles usually have in-line engines.

V-type

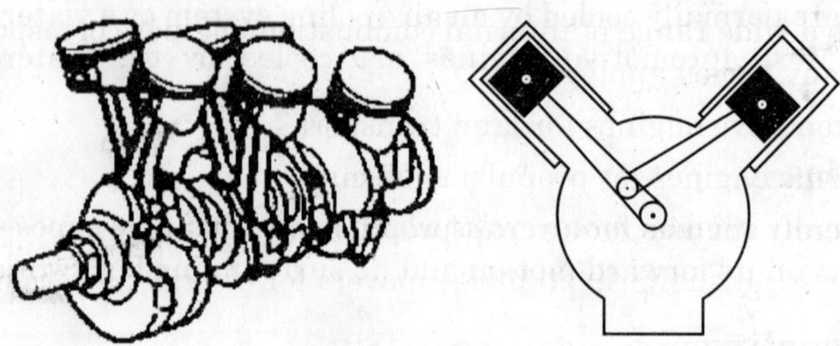

Fig.2.5: *V-engine*

Space considerations often demand that multiple cylinders are mounted into two banks arranged as a 'vee', but operating a common crankshaft. This type allows the overall engine height and length to be reduced. Sport car and bike engines are examples of V-type engines.

Horizontally opposed type

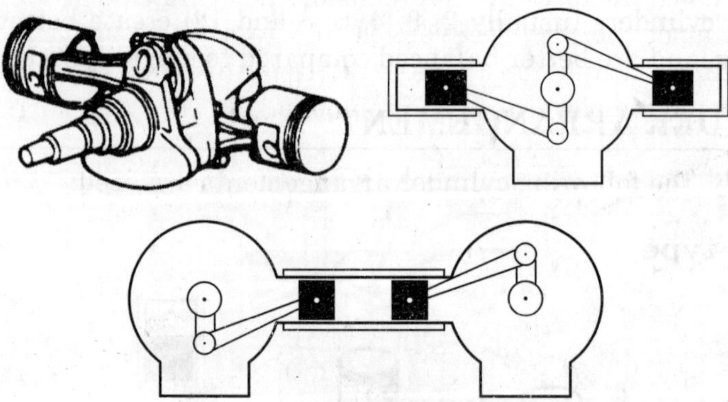

Fig.2.6: *Horizontally opposed engine*

In a horizontally opposed engine, cylinders are set horizontally opposing each other. This arrangement reduces the engine height. The pistons move inwardly and outwardly together providing excellent balance and regular firing intervals. Such engines can be housed below the floor line to increase the space available for people or luggage.

The most notable design was 4-cylinder, opposed piston, air-cooled Volkswagen engine.

ENGINE CAPACITY

The engines can also be classified according to different swept volumes such 100cc engine, 2.5 litres engine, etc.

APPLICATION

There is a wide range of internal combustion engines corresponding to their many varied applications.

- Automotive engines for land transport
- Marine engines for propulsion of ships
- Aircraft engines for aircraft propulsion
- Industrial engines and prime movers for electric power generation

SPARK IGNITION (SI) ENGINE

In spark-ignition engines, a mixture of air and petrol is drawn into the cylinder and is compressed by the upward motion of the piston. At the end of the compression stroke, the air-fuel mixture is ignited with the help of spark plug. When the air-fuel mixture is exploded by sparks, highly pressurised burning gas is produced in the cylinder, This highly pressurised gas pushes down on the piston, which can move up and down (reciprocate) freely inside the cylinder. The reciprocating movement of the piston is converted into a rotary movement by the connecting rod and the crankshaft. It is this rotary movement that supplies motive power to the automobile.

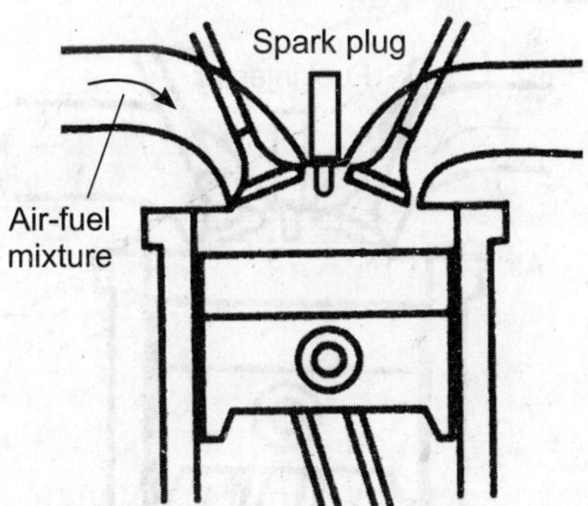

Fig.2.7: *Spark-ignition engine*

Merits

The merits of spark-ignition engines (petrol engines) include:

1. High speed and power output

2. Easy to operate
3. Quite combustion

Demerits

Demerits of spark-ignition engines (petrol engines) are:

1. Petrol engines need to be tuned properly or they will release a lot of carbon monoxide and other poisons.
2. Petrol is an extremely volatile liquid and changes state from liquid to vapour very quickly under normal atmospheric conditions. Petrol vapour is dangerous. It can ignite very easily and will explode under most atmospheric conditions.
3. Sulphur compounds in petrol contribute to acid rain.

COMPRESSION IGNITION (CI) ENGINE

A diesel engine is a compression ignition (CI) engine. It does not use a spark to ignite the fuel. Diesel engines use fuel injectors. Air alone is drawn into the cylinder during intake stroke. As the piston moves up, compression of the air occur and this causes the temperature to rise to the point where ignition of the diesel fuel occurs (self-ignition temperature) as soon as it is injected.

In petrol engines the amount of air entering the engine is changed to control speed and power. In a diesel engine, the amount of air remains the same while the fuel quantity is changed to control speed and power. The mixture can be as rich as 20:1 under load and as lean as about 80:1 at idle.

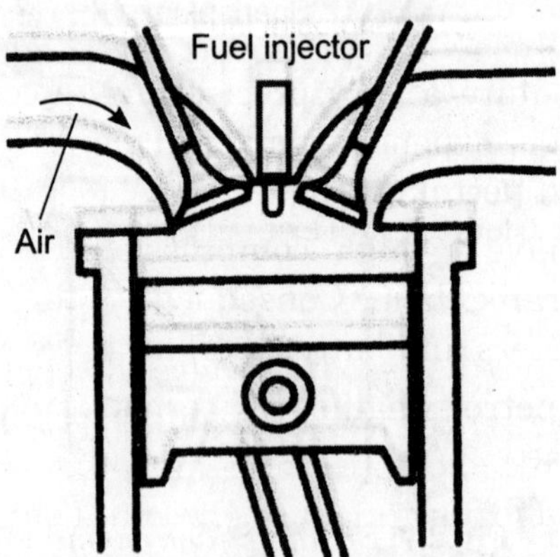

Fig.2.8: *Compression-ignition engine*

It is used in large trucks and other heavy machinery because diesel engines provide more torque on a crankshaft and this is necessary in heavy duty applications.

Merits

The merits of compression-ignition engines (diesel engines) include:

- Diesel engines have greater thermal efficiency. This means they consume less fuel and are more economical than petrol engines.
- The torque of a diesel engine remains unchanged over a wide speed range. This makes diesel engines suited for large vehicles.
- Diesel is more energy dense (heavier weight) than petrol, meaning a tank holding diesel fuel contains more energy than does the same-sized tank filled with petrol.
- Diesel fuel is cheaper than petrol in Indian markets.
- Deliver more power, given the weight of the engine, than petrol.
- More efficient than petrol.
- Diesel combustion produces less CO and HC exhaust than petrol combustion.

Demerits

The demerits of compression-ignition engines (diesel engines) are:

- The maximum pressure of combustion of a diesel engine is nearly twice that of a petrol engine. This means a diesel engine generates greater noise and vibration.
- Since the maximum combustion pressure is so high, diesel engines must be made of highly pressure-resistant materials and must have a very strong structure. This means diesel engines have a greater weight per horsepower than petrol engines. They also cost more to produce.
- Diesel engines require a very precise fuel injection system. This means they are more expensive and need more careful maintenance and servicing than petrol engines.
- Diesel engines have a high compression ratio and require a greater force to crank them. Consequently, they require cranking equipment as starters and batteries of larger capacity.
- Not as good as petrol engines in terms of speed, weight, and vibration and noise.
- More expensive
- Diesel contains more sulphur than petrol does, which contributes acid rain.
- Problems with older diesel engines were their high particulate emissions (soot) and the high temperature of combustion, which produces high levels of NOx emissions.

- Diesels also have starting problems in cold weather and require more frequent oil changes and other maintenance.

ENGINE OPERATING CYCLES

The two main types of reciprocating engine are two-stroke and four stroke engines.

FOUR-STROKE ENGINES

Basically every engine has to repeat the four operations, namely suction, compression, power and exhaust. If the four operations are completed in four strokes of the piston or two revolutions of the crankshaft, the engine is said to be four stroke engines.

Each stroke of the piston consists of 180° of crank rotation and hence a four-stroke cycle is completed through 720° of crank rotation.

Working principle of 4-stroke engines

The cycle of operation for a four stroke engine consists of the following four strokes.

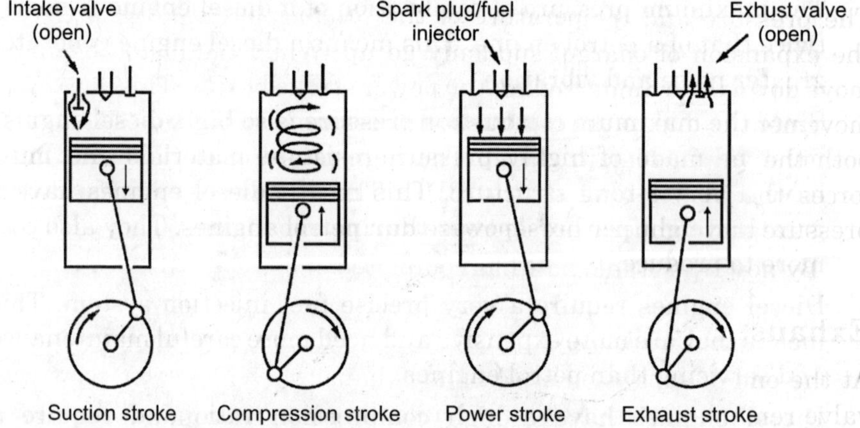

Fig.2.9: *Operation of 4-stroke engine*

Suction stroke (Intake stroke)

The 4-stroke cycle begins with the piston at TDC. To start the intake stroke, the piston moves down the cylinder. The inlet valve is open at this time and the exhaust valve is closed. As the piston moves down, it produces a vacuum in the cylinder. The pressure reduction or depression actually generated will depend on the speed and load experienced by the engine, but below atmospheric pressure. Fresh charge will rush into fill the depression in the cylinder. At the end of the intake stroke, the piston is in the bottom of the cylinder at BDC. The cylinder is filled with a charge (air-fuel mixture if petrol engine or only atmospheric air if diesel engine). Now the inlet valve closes.

Compression stroke

During this stroke, piston moves up from BDC to TDC. Both the inlet and exhaust valves are closed, so the charge has no place to go. It is pushed or compressed into the clearance volume (charge is progressively compressed to one-eighth to one-tenth of the cylinder's original volume in petrol engines). This compression squeezes the air and atomised fuel molecules closer together and increases the charge pressure and temperature. Just before the end of the compression stroke, the air-fuel mixture is ignited with the help of spark plug in petrol engines. Similarly, diesel is injected into the hot compressed air in the cylinder in the form of fine spray. The fuel gets vapourised and self-ignited due to the heat of compressed air.

On completion of this stroke, the crankshaft completes one full revolution.

Power stroke (Combustion and expansion)

The pressure and temperature of the product of combustion (due to the expansion of charge) suddenly go up. When the piston begins to move down the cylinder again, the power stroke begins. This downward movement is due to the combustion pressure. During this stroke also, both the valves remain closed. The high pressure of the burnt gases forces the piston towards the BDC. Thus, power is obtained. Both pressure and temperature decrease during expansion.

By now, the crankshaft has completed one and a half revolutions.

Exhaust stroke

At the end of expansion stroke the exhaust valve opens and the inlet valve remains closed. At this instant, the pressure of the gas is higher than atmospheric pressure. This high pressure forces about 60% of the burnt gases into the atmosphere through exhaust valve. The remainders of the burnt gases are forced out when the piston moves from BDC to TDC.

The crankshaft by now has completed two revolutions.

Each cylinder of a four-stroke engine completes the above four operations in two crank revolutions, one revolution of the crankshaft occurs during the suction and compression strokes and the second revolution during the power and exhaust strokes. Thus, for one complete cycle there is one power stroke in a four stroke engine.

Advantages and disadvantages of four-stroke engines	
Advantages	**Disadvantages**
• Lower HC emissions (less air pollution) • Better fuel economy as the loss due to blowby is small • Lower engine temperatures • Generally more reliable • Less vibration (smoother running) • Wider power band • Smoother operation at low speed as there is no misfiring due to proper exhausting.	• More internal moving parts • Heavier • Generate less power • Lower mechanical efficiency • The valve mechanism is complicated so that there are more parts and greater mechanical noise. • Since there is less number of explosions, the number of cylinders must be increased if smoother operation is desired.

TWO-STROKE ENGINES

In two-stroke engines the four operations, namely suction, compression, power and exhaust are completed in two strokes of the piston or one revolution of the crankshaft.

Each stroke of the piston consists of 180° of crank rotation and hence a two-stroke cycle is completed through 360° of crank rotation.

Working principle of 2-stroke engines

The cycle of operation for a two stroke petrol engine consists of the following two strokes of a piston.

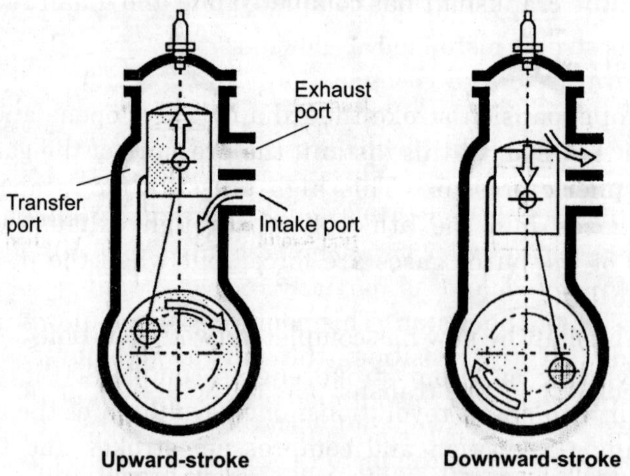

Fig.2.10: *Operation of 2-stroke engine*

Upward stroke (suction and compression)

This stroke begins with the piston at BDC. The upward motion of the piston during this stroke lowers the pressure in the crankcase below atmosphere pressure. At the same time, the intake port is uncovered first by the moving piston and, therefore, a fresh charge is drawn into the crankcase through the inlet port. The crankcase is filled with a charge (air-fuel mixture if SI engines or only atmospheric air if CI engines). As the piston moves upward further, it covers the transfer port stopping the flow of fresh charge into the cylinder. A little later, the piston covers the exhaust port and actual compression of the charge begins. The compression of the charge is continued until the piston reaches to its TDC position. Both the inlet and exhaust ports are covered, so the charge has no place to go. The charge is now compressed into the clearance volume. Just before the end of the compression stroke, the air-fuel mixture is ignited with the help of spark plug in SI engines. Similarly, diesel is injected into the hot compressed air in the cylinder in the form of fine spray. The fuel gets vapourised and self-ignited due to the heat of compressed air.

In this stroke of the piston, the following processes are completed.

- Fresh charge is sucked in the crankcase.
- Compression of the charge is completed.
- Partly scavenging takes place as the piston moves from BDC until it covers exhaust port.

On completion of this stroke, the crankshaft completes a half revolution (180°).

Downward stroke (power and exhaust)

During this stroke, piston moves down from TDC to BDC. The pressure and temperature of the gases increase and the gases push the piston downward producing the power stroke. When the piston uncovers the exhaust port during the downward stroke, the expanded burnt gases leave the cylinder through the exhaust port. A little later, the piston uncovers the transfer port. In this position, the crankcase is directly connected to the cylinder through the transfer port. When the piston moves down, the charge is partially compressed in the crankcase by the underside of the piston. This compression is known as *crankcase compression*. At this position, the partially compressed charge is transferred through the transfer port to the upper part of the cylinder. The exhaust gases are swept out with the help of fresh charge entering the cylinder. The top of the piston is made of a deflected shape, called *deflector*, so the fresh charge entering the cylinder is deflected upward in the cylinder. This serves the double purpose of scavenging the upper part of the cylinder of combustion products and minimizing the escape of fresh charge (short circuit) along with the exhaust gases. The same objective can be achieved without piston deflector by proper shaping of the transfer port. The two-stroke diesel engine does not suffer from these defects. There is no loss of fuel with exhaust gases as the intake charge in diesel engine is air only.

The crankshaft by now has completed one revolution (360°).

Each cylinder of a two-stroke engine completes the above four operations in one crank revolution, half revolution of the crankshaft occurs during the suction and compression strokes and the second half revolution during the power and exhaust strokes. Thus, for each crankshaft revolution there is one power stroke in a two stroke engine.

Note:

- Except that fuel is admitted near the TDC, the working of two-stroke cycle diesel engine is similar to that of the two-stroke petrol engine.
- The process of removing burnt gases from the cylinder with the help of fresh charge is known as *scavenging*.

Advantages and disadvantages of the two-stroke engine	
Advantages	**Disadvantages**
Fewer internal parts due to the absence of valve mechanism Lighter Generates more power Higher mechanical efficiency Smoother operation is possible with lesser number of cylinders as there are twice as many explosions as four-stroke cycle engine. With same displacement and speed, two-stroke cycle engine can develop more power.	High HC emissions (air pollution) Poor fuel economy Higher engine temperatures Generally less reliable Frequent servicing required Narrow power band Time allowed for exhaust is only one-half that of 4-stroke cycle engine so that scavenging is relatively poor that affects engine performance.

COMPARISON OF SI AND CI ENGINES

S.No	Spark Ignition Engines	Compression Ignition Engines
1.	Operates on the principle of Otto cycle	Operates on the principle of Diesel cycle
2.	Petrol (Gasoline) as fuel. High self-ignition temperature desirable.	Diesel as fuel. Lower self-ignition temperature desirable.

3.	Fuel and air introduced as a gaseous mixture in the suction stroke.	Fuel is injected directly into the cylinder at the end of compression stroke.
4.	Fuel is ignited by spark plug	Fuel is ignited by hot compressed air
5.	Low compression ratio (8 to 11). Upper limit of compression ratio fixed by antiknock quality of fuel.	High compression ratio (14 to 24). Upper limit of compression ratio is limited by the rapidly increasing weight of the engine structure as the compression ratio is further increased.
6.	Operates at high speed due to lighter weight.	Operates at low speed due to more weight.
7.	Maximum efficiency is lower due to low compression ratio.	Maximum efficiency is higher due to high compression ratio.
8.	Lighter in weight	Heavier in weight. Due to high compression ratio, operating pressure is more. To withstand such pressure, it has to be stronger.
9.	Less vibration.	More vibration.
10.	Petrol is highly volatile. So, handling is more risky.	Diesel is less volatile. So, handling is not risky.
11.	Less maintenance.	More maintenance.
12.	Initial cost is less due to lighter in construction and less costly fuel system. Operating cost (running cost) is more because of higher petrol price.	Initial cost is more. Operating cost is less due to less price of diesel.

COMPARISON OF 4-STROKE AND 2-STROKE ENGINES

S.No	4–Stroke Engine	2–Stroke Engine
1.	The cycle is completed in 4 strokes of the piston or in two revolutions of the crankshaft. Thus one power stroke is obtained in every two revolutions of crankshaft.	The cycle is completed in 2 strokes of the piston or in one revolution of the crankshaft. Thus one power stroke is obtained in every revolution of crankshaft.
2.	Because of only one power stroke in every two revolutions of the crankshaft, turning movement is not so uniform. So heavier flywheel is needed.	Because of one power stroke for every revolution of the crankshaft, turning movement is more uniform. So lighter flywheel is needed.

3.	Because of only one power stroke in every two revolutions of the crankshaft, for the same power the engine is heavy and bulky.	Because of one power stroke for every revolution of the crankshaft, for the same power the engine is light and compact.
4.	Because of one power stroke in two crank revolutions lesser cooling and lubrication requirements. Lesser rate of wear and tear.	Because of one power stroke in every crank revolution greater cooling and lubrication requirements. Greater rate of wear and tear.
5.	Valves are provided	Ports are provided
6.	Because of the heavy weight and complicated valve mechanism, higher in initial cost.	Because of light weight and simplicity due to the absence of valve mechanism, cheaper in initial cost.
7.	Volumetric efficiency is more due to greater time for induction of fresh charge.	Less volumetric efficiency due to lesser time for induction of fresh charge.
8.	Thermal efficiency is more.	Less thermal efficiency. Due to short circuit, some part of fuel is exhausted which results in higher fuel consumption and lower thermal efficiency.
9.	Mostly liquid cooling system is used	Mostly air cooling system used
10.	Used where efficiency is important such as cars, buses, locomotives, etc	Used where low cost, compact-ness and light weight are important such as motorbikes, scooters, mopeds, etc.

FIRING ORDER

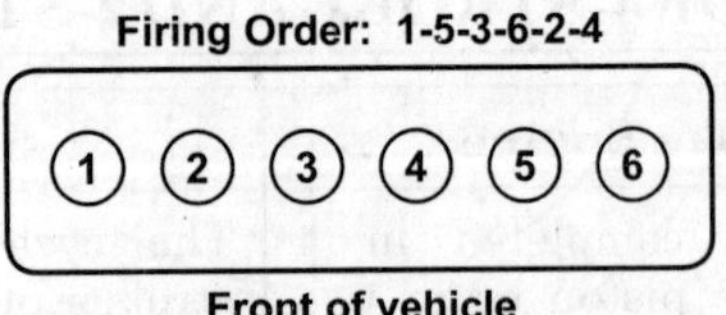

Fig.2.11: *Firing order*

Engines are designed to deliver the power strokes in a certain pattern. Two adjacent cylinders are usually not allowed to fire one after the other. When an end cylinder fires, the next cylinder to fire should be near the center or toward the other end of the crankshaft. The purpose of carefully spreading the power strokes along the crankshaft is to avoid stress on any one part of the crankshaft. If two or three adjacent cylinders are fired one after the other, this places a strain on the crankshaft. It may break. A proper firing order prevents this.

The firing orders for single and multi-cylinder arrangement are as follows.

Single cylinder arrangement	:	1
Twin cylinder arrangement	:	1, 2 (in-line and horizontally opposed)
Three cylinder arrangement	:	1, 2, 3 (in-line)
Four cylinder arrangement	:	1, 3, 4, 2 (in-line)
	:	1, 4, 2, 3 (horizontally opposed)
Five cylinder arrangement	:	1, 2, 4, 5, 2 (in-line)
Six cylinder arrangement	:	1, 5, 3, 6, 2, 4 (in-line)
	:	1, 4, 5, 2, 3, 6 (horizontally-opposed)
Eight cylinder arrangement	:	1, 6, 2, 5, 8, 3, 7, 4 (in-line)

VALVE TIMING AND PORT TIMING DIAGRAMS

VALVE TIMING DIAGRAM

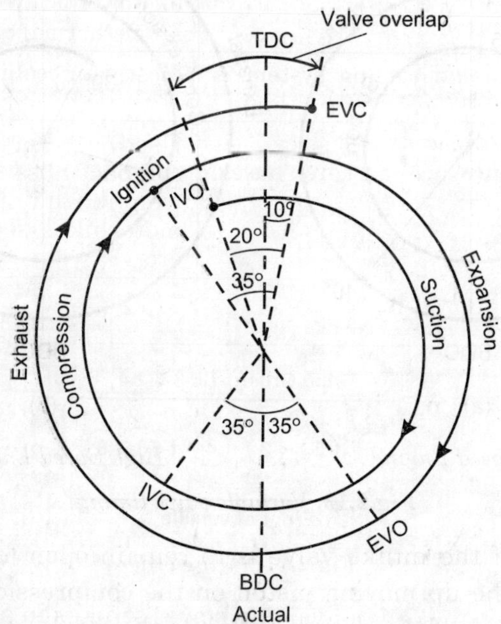

Fig.2.12: *Fixed valve timing*

In the ideal cycle inlet and exhaust valves open and close at dead centres, but in actual cycles they open or close before or after dead centres.

Intake valve timing

The intake valve opens few degrees before the arrival of piston at TDC on the exhaust stroke. This is to ensure that the valve will be fully open and the fresh charge starting to flow into the cylinder as soon as possible after TDC.

As the piston comes down in the suction stroke, the fresh charge is drawn in through the intake valve. When the piston reaches the BDC and starts to move in the compression stroke, the inertia of the entering fresh charge tends to cause it to continue to move into the cylinder. To take advantage of this, the intake valve is closed after TDC so that maximum air or air-fuel mixture is taken in. This is called *ram effect*. The kinetic energy of the charge produces this ram effect which packs more charge into the cylinder during this additional valve opening. Therefore, the intake valve closing is delayed.

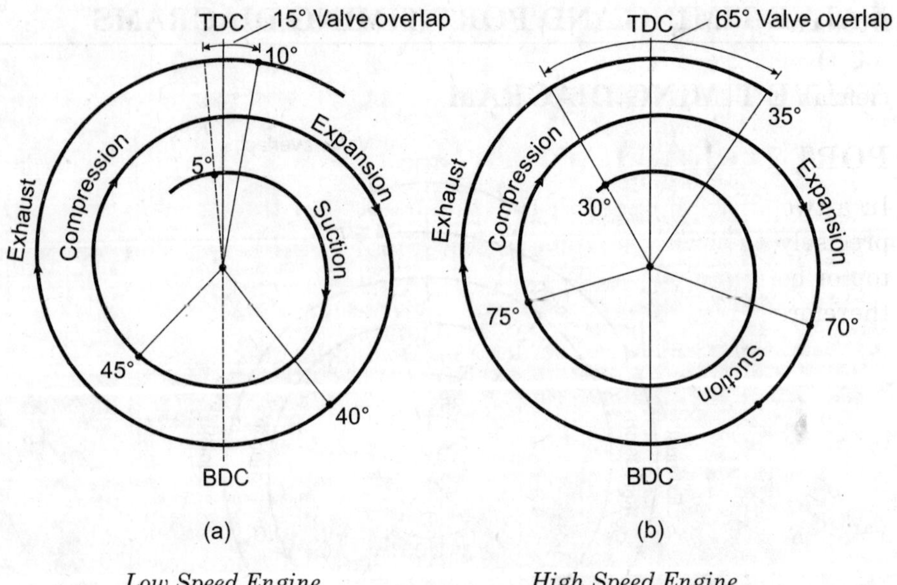

Low Speed Engine *High Speed Engine*

Fig.2.13: *Variable valve timing*

However, if the intake valve is to remain open for too long time beyond BDC, the up moving piston on the compression stroke would tend to force some of the charge back into the intake manifold.

The time the intake valve should remain open after TDC is decided by the speed of the engine. At low engine speed the charge speed is low and so the air inertia is low and hence the intake valve should close relatively early after BDC for a slow speed engine. In high speed engines the charge speed is high and consequently the inertia is high and hence to induct maximum quantity of charge due to ram effect the intake valve should close relatively late after BDC.

Exhaust valve timing

The exhaust valve is set to open before BDC and close few degrees after TDC. Complete clearing of the exhaust from the cylinder is necessary to take in more charge. Earlier opening of the exhaust valve facilitates the removal of burnt gases by virtue of the pressure at this point. The kinetic energy of the fresh charge may also assist the removal of the burnt gases.

It is obvious from the valve timing diagram that both the intake and exhaust valves are open at the same time for some period. This is called *valve overlap*. This overlap should not be excessive otherwise it will allow the burned gases to be sucked into the intake manifold or the fresh charge to escape through the exhaust valve.

Overlapping of the valves makes possible the scavenging of the clearance space, resulting in an increased output.

PORT TIMING DIAGRAM

In a two-stroke engine, the inlet, exhaust and transfer ports open at precisely the same number of degrees of crankshaft rotation before top or bottom dead centre as they close. The resulting timing chart is therefore symmetrical.

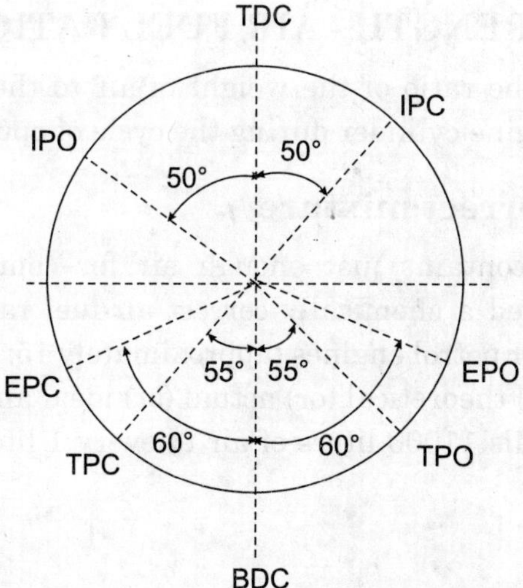

Fig.2.14: *Port timing diagram*

As the piston moves up toward TDC, the transfer port closes first, followed by the exhaust port. This can allow fresh mixture to escape through the exhaust.

As the piston moves down towards BDC, the exhaust port opens first, followed by the transfer port. The pressure drop when the exhaust port opens is so severe that the remaining burnt gas cannot flow back so strongly into the crankcase and mix there with the pre-compressed fresh mixture.

The charging and scavenging period of two stroke diesel engine is greater than the scavenging period of two-stroke gasoline engine. This is because, there is no danger of loss of fuel during scavenging (due to short circuit) of diesel engine as only air is charged.

FUEL SYSTEM

In fuel supply system, *air cleaners or air filters* remove dust, dirt and other foreign particles present in the air and send clean air into engine cylinder. *Fuel filters* filter foreign particles that may be present in the fuel. A *fuel pump* lifts fuel from the *fuel tank* and supplies the same to the carburetor or fuel injector through tubes and hoses (*fuel lines*). Correct quantity of air fuel mixture is supplied to the cylinder by *carburetor or electronic fuel injection system.*

FUEL SYSTEM IN GASOLINE ENGINES

MIXTURE STRENGTH - AIR FUEL RATIO

Air fuel ratio is the ratio of the weight of air to the weight of petrol supplied to an engine cylinder during the cycle of operations.

Chemically correct mixture

A mixture that contains just enough air for complete combustion of the fuel is called a chemically correct air-fuel ratio. This value is normally 14.7:1 for petrol engines (approximately 15: 1). It is also called stoichiometric (or) theoretical (or) actual (or) ideal air-fuel mixture. By volume, that equals 11000 liters of air to every 1 liter of fuel - a ratio of 11000:1.

Rich mixture

A mixture is said to be rich (< 15:1) if it contains more fuel than the fuel quantity in the chemically correct mixture.

Lean mixture

A mixture is said to be lean (> 15:1) if it contains more air than the air quantity in the chemically correct mixture.

AIR-FUEL RATIO REQUIREMENTS FOR SI ENGINES

The approximate proportions of air to fuel (petrol by weight) for different engine operating conditions are:

Driving conditions and fuel ratio		
Driving condition	**Air-fuel ratio**	**Remarks**
Normal operation after warm-up	Theoretical	The theoretical air-fuel ratio is used because the fuel and air mix well at this temperature.
Cold engine	Richer (depending on engine temperature)	When the engine is cold, fuel does not vaporize well and often clings to the walls of the intake manifold and combustion chambers without mixing properly with air. Thus, in order for the actual combustion mixture to approach as close as possible to the theoretical ratio, a larger amount of fuel must be provided.
During starting	Richer	Since the engine cranks at a very low rpm during starting, the air flows at an extremely low speed. For this reason, the fuel and air do not mix well, thus requiring more fuel.
During idling	Slightly richer	Since the engine runs at a low speed at idle, the air also flows at a low speed, slowing the mixing of fuel and air. Slightly more fuel must therefore be supplied in order to decrease the disparity of air-fuel mixtures supplied to the cylinders.
During acceleration	Additional fuel injected	Because fuel is heavier than air, the flow of fuel cannot keep up with the airflow at the beginning of acceleration; this results in a leaner mixture. An additional amount of fuel is therefore injected from the acceleration pump in order to prevent the mixture from becoming too lean.

Under high loads	Slightly richer	Slightly more fuel is provided in order to obtain a higher power output and to protect the engine parts.
Cruising (light load)	Slightly leaner or theoretical	In ail engine that aims for fuel economy, a slightly leaner air- fuel mixture is provided. In an engine that aims for more power, a theoretical air-fuel ratio is provided.

Richer: about 9:1, Slightly richer: 12:1 to 13:1, Theoretical: 15:1, Slightly leaner: about 16:1.

CARBURETION AND CARBURETOR

The process of preparing combustible air fuel mixture in the petrol engine is called *carburetion*. A device which does this process is called *carburetor*. The carburetor atomizes the fuel and mixes it Aith air in correct proportions according to the engine operating conditions.

A SIMPLE (OR) ELEMENTARY CARBURETOR

To understand a modern carburetor, which is a complicated one, it is helpful to first study a simple or elementary carburetor. It consists of a float chamber, venturi, main nozzle, choke valve and throttle valve.

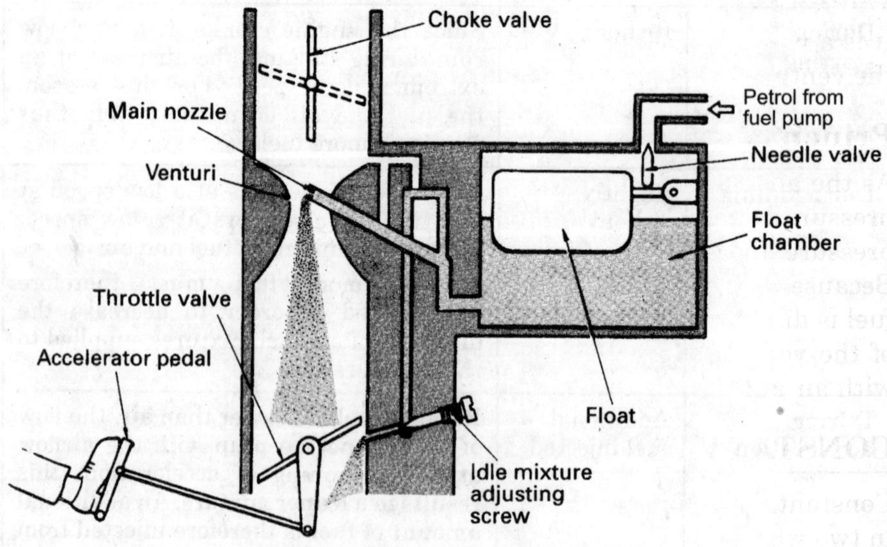

Fig.2.15: *Simple carburetor*

Float chamber (or) Float bowl

The float chamber is a fuel reservoir which maintains constant height of fuel. It consists of a float and needle supply valve. The float is made of either brass, rubber or plastic. The float chamber is vented to the atmosphere to allow the atmospheric pressure to act on the fuel in it.

Venturi (or) Choke tube

Venturi is a tube of decreasing cross section which reaches a minimum at the throat. During suction stroke air is drawn through the venturi. The air passing through the venturi increasing in velocity and the pressure in the venturi throat decreases.

Main nozzle

From the float chamber, the fuel is fed to a main nozzle. The tip of the main nozzle is located in the throat of the venturi.

Throttle valve

The throttle valve is a round disc located between the venturi and the intake manifold. The throttle is connected to the accelerator pedal by a flexible cable. Depressing the accelerator pedal opens the throttle valve and permits an increased amount of air-fuel mixture to reach the engine cylinders. This will increase the power output of the engine and hence the vehicles speed.

Choke valve

It is a valve which is used for controlling the amount of air passing to the venturi and is used when starting a cold engine.

Principle

As the air flows through the venturi, its velocity increases and hence pressure decreases. Now the pressure in the float chamber is atmospheric pressure and at the venturi throat is below the atmospheric pressure. Because of the pressure difference (called *carburetor depression*) the fuel is discharged (atomized) through the main nozzle into the air flow of the venturi. Now the fuel particles are vapourised and get mixed with air and the mixture is supplied to the engine cylinder.

CONSTANT VACUUM CARBURETOR

Constant vacuum or constant depression carburetors are widely used in two wheelers engines. In this type of carburetor, the venturi area is variable and is controlled by a movable piston (sliding piston).

CABLE OPERATED TYPE

The accelerator pedal is connected to the sliding piston through flexible cable to raise and lower the sliding piston. By its change in position, the sliding piston controls the venturi opening size. When the piston is raised, the size of the venturi is enlarged and the amount of air allowed to enter the engine is increased. This causes the engine speed to increase. When the piston is lowered, the venturi size is reduced. That is, the air passage through the venturi shrinks and engine speed is decreased.

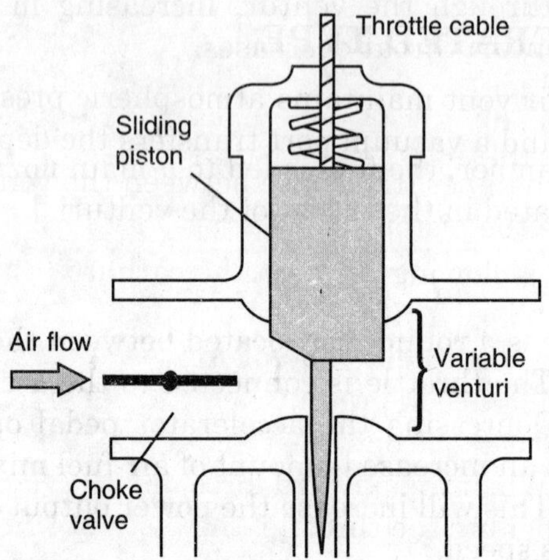

Fig.2.16: *Cable operated type sliding piston*

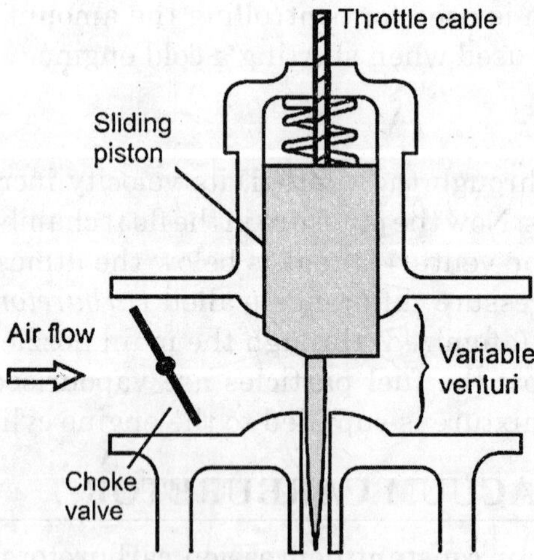

Fig.2.17: *Cold starting*

Cold starting

The choke system is an air restriction system that controls the amount of air available during a cold engine start. This system uses a rider-controlled plate, called a choke valve, to block air to the carburetor venturi at all throttle openings. This plate has either a small hole cut into it or a spring-loaded relief valve, to allow some air into the carburetor venturi. This gives the engine enough air to run but creates a very rich mixture in comparison to the mixture created if the plate were in the open position. The choke valve is located on the air filter side of the carburetor.

VACUUM OPERATED TYPE

In this type, an air vent maintains atmospheric pressure in the space below the piston and a vacuum port transfers the depression (vacuum) from the mixing chamber (the space between the venturi and throttle) to the space above the piston.

When the air is flowing through the carburetor there is a vacuum in the mixing chamber. This vacuum is transferred to the space above the piston and the space below the piston is vented to the atmospheric pressure. This difference in pressure acts on the piston and will give an upward force to oppose the downward force caused by the piston weight and light spring. When the upward force due to air pressure difference in increased by opening the throttle, the piston will rise and the choke (venturi) area will enlarge. Similarly, when the throttle closes, the mixing chamber depression reduces, the piston falls and the choke area decreases.

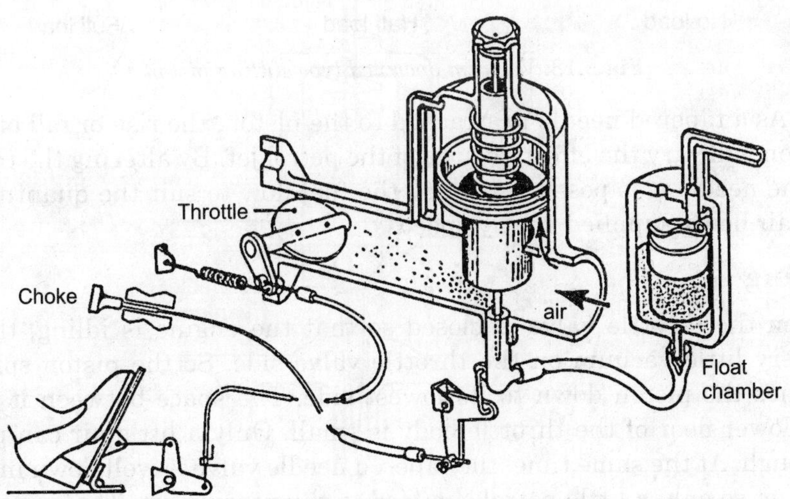

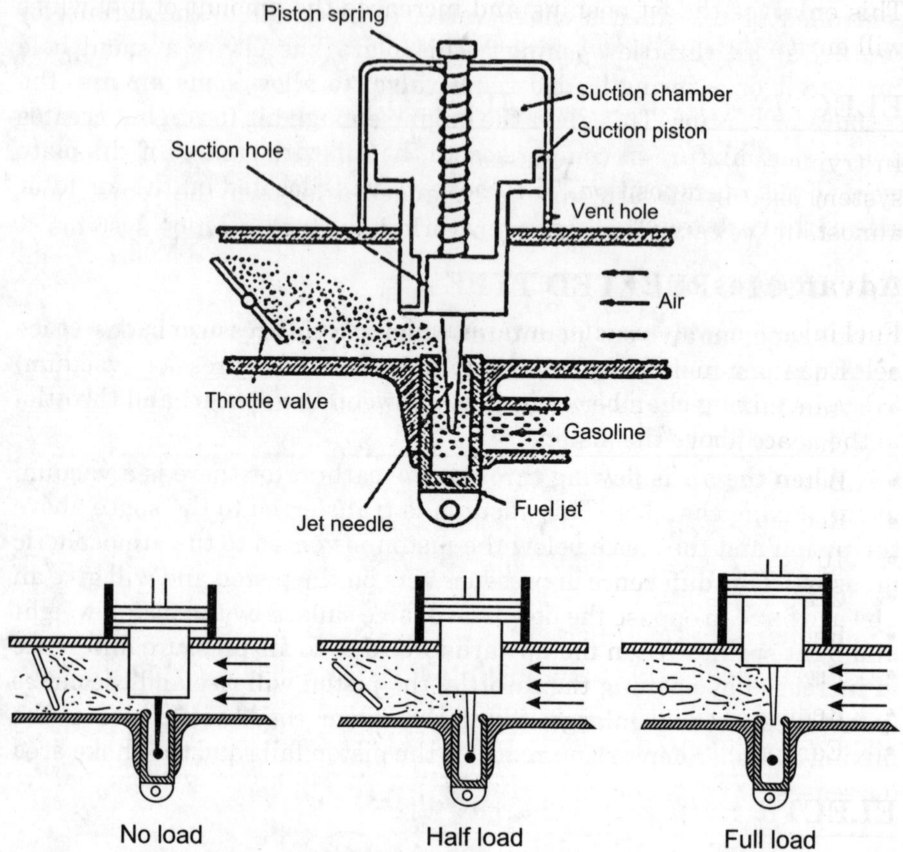

Fig.2.18: *Vacuum operated type sliding piston*

As a tapered needle is attached to the piston, the rise or fall of the piston will vary the effective area of the petrol jet. By altering the taper of the needle it is possible to vary the fuel flow to suit the quantity of the air being supplied at any speed.

Idling

When the throttle valve is closed so that the engine is idling, there is very little vacuum on the throttle valve side. So the piston spring pushes the piston down to its lowest side. The space between it and the lower floor of the throttle body is small. Only a little air can pass through. At the same time, the tapered needle valve is well down in the fuel jet so only a little petrol can feed to the passing air. The resulting mixture is for engine idling.

Cold starting

A lever operated by a cable lowers the fuel jet in relation to the needle.

This enlarges the jet opening and increases the amount of fuel which will enrich the mixture for cold starting.

ELECTRONIC FUEL INJECTION

In trying to keep up with emissions and fuel efficiency laws, the fuel system used in modern cars has changed a lot over the years. Now, almost all the cars sold in all over world have fuel injection systems.

Advantages of EFI

Fuel injection systems offer several major advantages over carburetors.

- Equal air-fuel ratio to all the cylinders
- Accurate air-fuel ratio can be obtained throughout all engine rpm changes
- Better cold starting
- Reduces fuel consumption with no loss of engine performance
- 10% advance of fuel economy over carbureted systems
- Quick response
- Reduce pollution levels
- Precise control of engine operation
- Reduce design consideration in the manufacture of inlet manifold
- Earlier default detection

ELECTRONIC FUEL INJECTION SYSTEMS

There are three basic types of gasoline fuel-injection systems, throttle body injection (TBI), multi-point injection (MPFI) and gasoline direct injection (GDI).

SINGLE-POINT INJECTION

It is also known as throttle body injection (TBI), central fuel injection (CFI) or digital fuel injection (DFI). The system has one or two small injection valves positioned just above the throttle valves in the throttle body. The throttle body is similar to the carburetor throttle body. There they spray fuel into the air stream passing through. The fuel pressure for TBI is usually on the order of 2-3 bar.

The difference is in how many places fuel (under pressure) is injected into the engine. In single-point system, fuel enters the intake air at only one place. This is usually the throttle body on the intake manifold.

The drawback of this system is maldistribution is not eliminated totally.

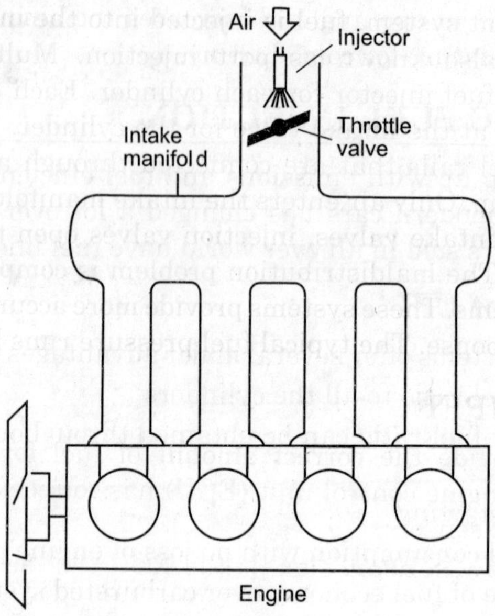

Fig.2.19: *Single-point injection*

Note: Intake manifold supplies different mixture strength to the engine cylinders for the same operating condition is called 'maldistribution'.

MULTI-POINT FUEL INJECTION (MPFI)

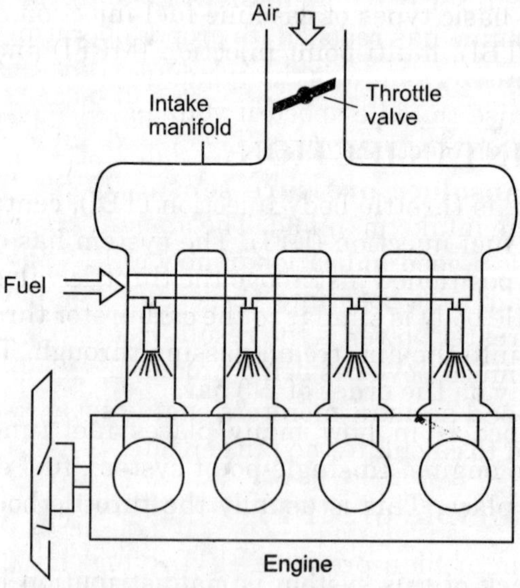

Fig.2.20: *Multi-point fuel injection (MPFI)*

In the multi-point system, fuel is injected into the intake air for each cylinder. It is also known as port injection. Multi-point injection systems have a fuel injector for each cylinder. Each injector is placed so that it points at the intake valve for the cylinder. The injectors are connected to fuel rails that are connected through a fuel filter to an electric fuel pump. Only air enters the intake manifold. Then as the air approaches the intake valves, injection valves open to spray fuel into the air stream. The maldistribution problem is completely eliminated with MPFI systems. These systems provide more accurate fuel metering and quicker response. The typical fuel pressure runs between 3-7 bar.

Sensors in MPFI

In order to provide the correct amount of fuel for every operating condition, the engine control unit (ECU) has to receive signals from a number of input sensors.

- **Mass airflow sensor**, signals the ECU the mass of air entering the engine.
- **Oxygen sensor**, monitors the amount of oxygen in the exhaust so the ECU can determine how rich or lean the fuel mixture is and make adjustments accordingly.
- **Throttle position sensor**, monitors the throttle valve position (which determines how much air goes into the engine) so the ECU can respond quickly to changes, increasing or decreasing the fuel rate as necessary.
- **Coolant temperature sensor**, allows the ECU to determine when the engine has reached its proper operating temperature.
- **Voltage sensor**, monitors the system voltage in the vehicle so the ECU can raise the idle speed if voltage is dropping (which would indicate a high electrical load).
- **Manifold absolute pressure sensor**, monitors the pressure of the air in the intake manifold. The amount of air being drawn into the engine is a good indication of how much power it is producing. The more air that goes into the engine, the lower the manifold pressure, greater power is the result. Hence this reading is used to gauge how much power is being produced.
- **Engine speed sensor**, monitors engine speed, which is one of the factors used to calculate the pulse width.

Working principle of MPFI

In MPFI system, a high-pressure electric fuel pump is located in or near the fuel tank. The pump sends fuel from the tank through the fuel-pressure regulator to the injectors at constant high pressure all the time the engine runs. Injector is operated by a small electric solenoid. When the solenoid is connected to the battery, it pulls back on the nozzle needle, opening the valve. Now fuel can spray out of the valve. Various sensors located on the engine detect the operating mode of the engine and signal this condition to the electronic control unit (ECU). It processes the information. Then the ECU computes when and for how long the injectors should be opened. The ECU sends a voltage pulse to the injectors. The voltage pulse causes the injectors to remain open for the required time. To increase engine speed, the pulse width increases so that injectors stay open longer. If less fuel is needed, the pulse width decreases. Therefore the injectors close earlier and engine speed decreases.

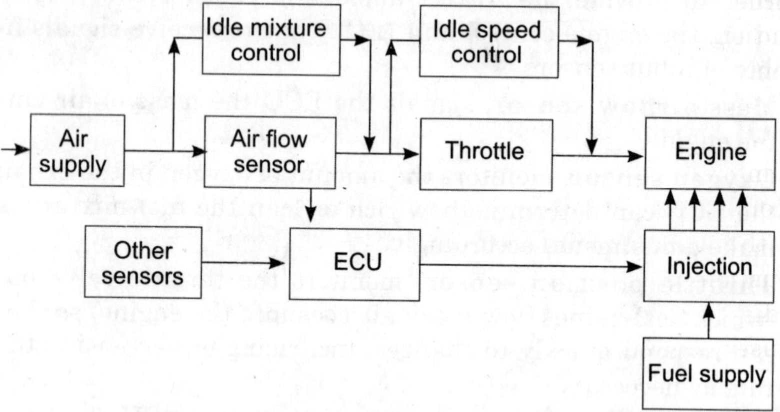

Fig.2.21: *MPFI system*

Types of MPFI

There are three types of MPFI systems: Sequential, Batched and Simultaneous.

Sequential injection

In sequential injection, injection is timed to coincide with each cylinder's intake stroke. Injectors are driven independently and sequentially by separate driver circuits. Injection is timed to deliver the entire fuel charge just prior to each intake valve opening. This timing method provides optimum engine performance, emissions and fuel economy.

Simultaneous injection

Fuel is injected at the same time to all the cylinders in simultaneous injection. All injectors are pulsed simultaneously by a common driver circuit. Injection occurs once per crankshaft revolution just prior to the crankshaft reaching TDC cylinder 1. This means that twice per engine cycle one half of the calculated fuel is delivered by the injectors. This is the simplest and most common injection timing method in use.

Grouped injection

In group injection, fuel is injected to the cylinders in groups, without precise synchronization to any particular cylinder's intake stroke. Injectors are grouped into pairs. The pairs consist of two consecutive cylinders in the firing order, each pair is driven by a separate driver circuit. Four cylinder engines use two groups, six cylinder engines three groups and the eight cylinder engine uses four groups of injectors.

This timing method ensures that fuel does not reluctant to leave behind the intake valve, thereby, reducing emissions, improving fuel economy and throttle response.

GASOLINE DIRECT INJECTION (GDI)

Gasoline direct-injection engines generate the air-fuel mixture in the combustion chamber. During the suction stroke, only the air flows through the intake valve. The fuel is injected directly into the combustion chamber at high pressure by special fuel injectors. As in a diesel engine, air-fuel mixture formation takes place inside the combustion chamber.

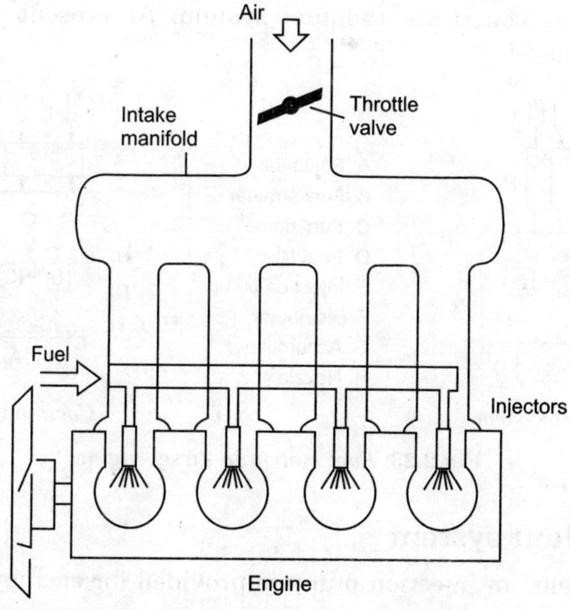

Fig.2.22: *Gasoline direct injection*

The electric fuel pump delivers fuel to the high-pressure pump at a pre-supply pressure of 3-5 bar. The highly pressurized fuel flows into and is stored in the fuel rail. The high-pressure fuel injectors are mounted on the fuel rail, also known as the "common rail". These injectors are actuated by the engine ECU and spray the fuel into the cylinder combustion chambers. For a GDI engine, fuel pressure is usually on the order of 100-500 bar.

Advantages of GDI

1. Fuel is injected from a tiny nozzle into a relatively large cylinder, so it has a high latent heat of vaporization, which efficiently cools the air within (in-cylinder cooling effect).
2. The air temperature in the cylinder decreases, which means:
   - more air may be charged into the combustion chamber, which produces increased torque.
   - the engine is less prone to knocking. This contributes to increased torque, and enables a higher compression ratio.
   - good fuel efficiency.

FUEL SYSTEM IN DIESEL ENGINES

The fuel from the fuel tank is compressed by the injection pump and injected into the cylinder through the nozzle. All parts concerned with this operation constitute the fuel system. At present two different systems are used.

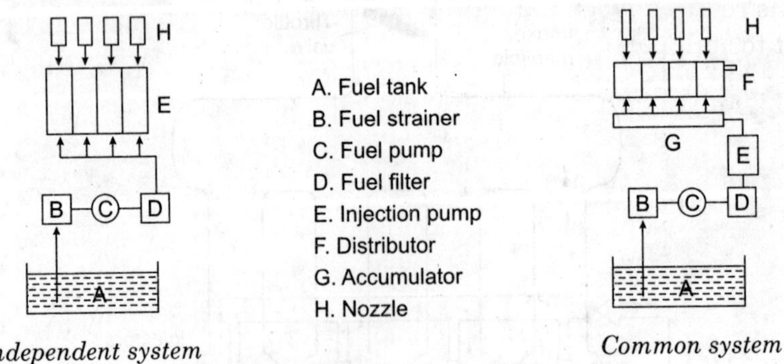

Fig.2.23: *Fuel system in diesel engines*

Independent system

In this system, an injection pump is provided for each cylinder and is high speed engines for passenger train service.

Common system

In this system, only one injection pump is used, which sends the oil under high pressure to the accumulator. From here, the distributor which controls the amount and timing of the high pressure oil, sends the oil to the various nozzles.

DIESEL INJECTION SYSTEMS

Precisely delivered fuel maintains timing that keeps the engine running smooth. Simultaneously, the pump also controls the amount of fuel needed to gain the desired power.

The main parts of this system are fuel filters, fuel transfer pump, fuel injection pump, fuel injector, and governor.

DIESEL INJECTION PUMP

The injection pump pressurises and supplies fuel sent by the feed pump to the nozzles according to the firing order. The injection pump is usually located on the side of the engine. It is driven by the crankshaft via timing gears or a timing belt.

There are two types of injection pumps. the in-line type and the distributor type.

INLINE-INJECTION (JERK) PUMPS

This type is widely used to supply fuel under high pressure in diesel engines. It has a plunger for each cylinder operated by a camshaft.

In this pump, the plunger stroke remains constant but the effective stroke is reduced by changing the position of helix on the plunger with respect to spill port.

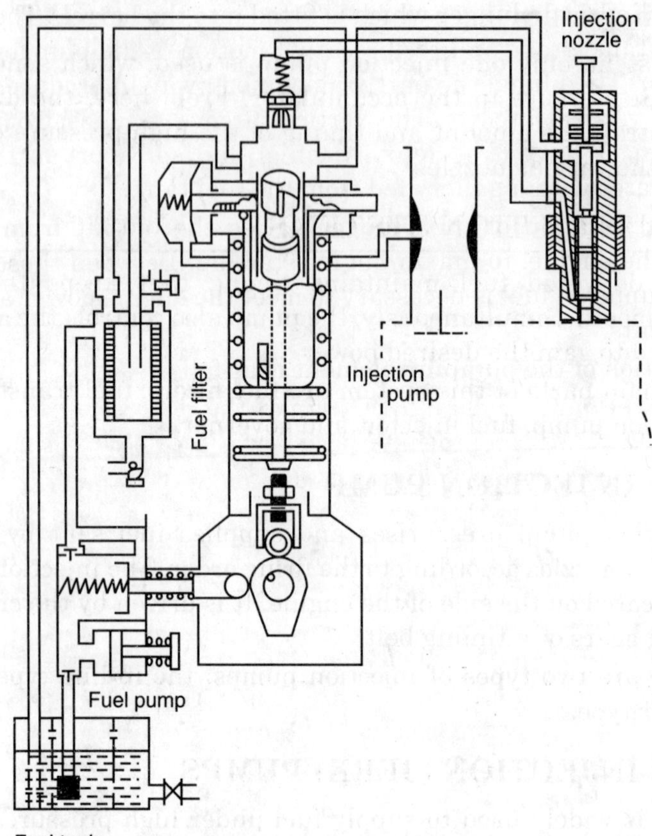

Fig.2.24: *Inline-injection (jerk) pump*

Pumping element

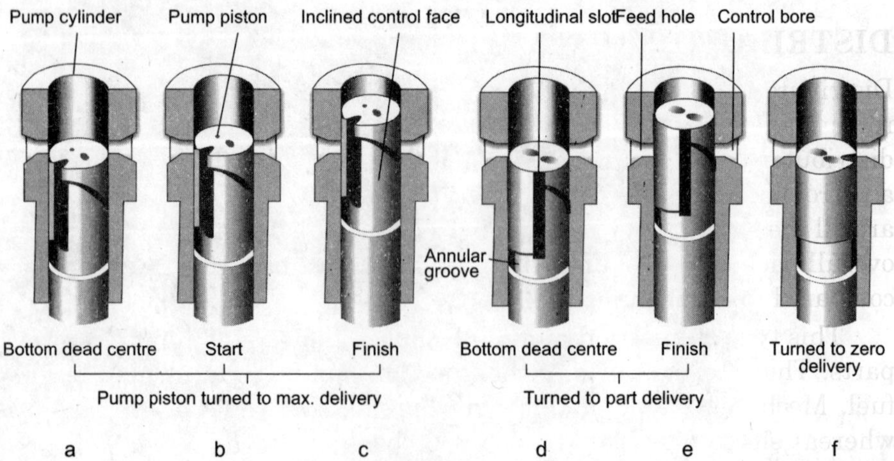

Fig.2.25: *Pump element*

It has a cylindrical plunger which is fitted into the barrel. The plunger has a constant stroke and reciprocating motion of the plunger is given by the cam. The plunger is partly machined away to produce a control helix and vertical groove. The vertical groove extends from the top of the plunger to control helix. An inlet and spill port in the barrel communicates with a gallery fed from the fuel tank.

Partial rotation of the plunger varies the output from zero for stopping the engine, to maximum for starting. Between these limits a variable supply of fuel is necessary to meet the engine power and speed requirements.

The operation of the pumping element is as follows:

- When the plunger is at BDC (Fig.a) the depression in the pump chamber causes fuel to enter both ports.
- This position (Fig.b) is known as the point of port closure (or spill cut-off), i.e. theoretical point of injection. Both ports have been covered and so the upward movement of the plunger raises the pressure of the fuel to produce injection.
- Injection stops when the edge of the helix uncovers the spill port (Fig.c). Pressure is relieved by fuel passing down the vertical groove, around the waist of the plunger and out of the spill port.
- Rotation of the plunger causes the helix to uncover the spill port either earlier or later, to give less or more fuel respectively.
- Moving the plunger to make the vertical groove coincide with the spill port means that the port will remain open. Therefore no fuel will be delivered and the engine will come to rest.

The injection pressures of in-line injection pumps generally varies in the range from 200 to 1250 bar.

DISTRIBUTOR (ROTARY) INJECTION PUMPS

Distributor pumps have only one fuel-metering plunger. A spinning rotor makes a hydraulic connection with the different ports on the distributor head, somewhat similar to the way a distributor works on a petrol engine. Benefits of a rotary-style pump with only one plunger are all the shots of fuel are exactly the same, and it makes for a smaller overall package. Also, distributor-style pumps have less moving parts compared to inline pumps.

This type of injection pump is compact, light weight and has fewer parts. The internal parts of the injection pump are lubricated by the fuel. Mechanical type of this pump produces 450-1200 bar pressure whereas electronic type produces 1600 bar.

Construction

The rotary distributor pump has a set of plungers located within a centrally mounted rotor shaft, which is driven by and timed to the engine. Behind the plungers is a set of rollers, which are mounted on a cam ring. The main purpose of the rotor is to transport fuel towards the plungers, and then to distribute the fuel to each of the injector outlet ports in the appropriate sequence. The cam ring has a number of cam lobes machined at equally spaced distances from each other. There are as many cam lobes as the engine has cylinders.

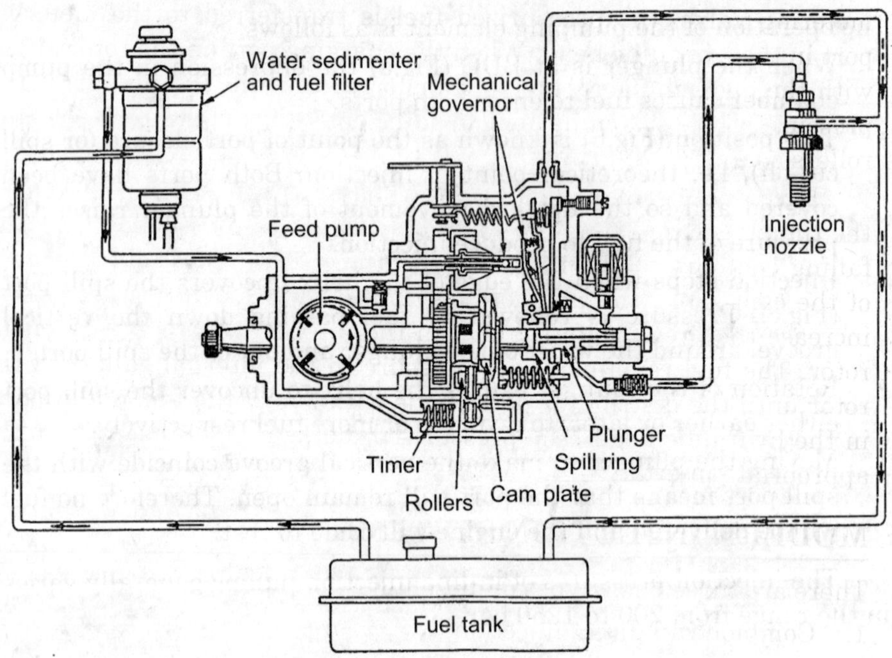

Fig.2.26: *Distributor pump system*

The main steel body of the distributor pump is called the hydraulic head. This contains a set of internal drillings, which allow the fuel into the pump. Then, in conjunction with the rotor, the pressurized fuel is distributed to each of the injector outlet pipes in the appropriate sequence. To control and regulate the pressure of the fuel accurately, a transfer pump and a regulator valve are fitted inside the injection pump. These two components ensure that the line pressure of the fuel is kept within specified tolerances.

In order to provide the pump with a means by which the fuel injection phase can be progressively advanced or retarded with engine speed and load changes, an automatic advancement mechanism is fitted.

Operation

When the fuel is delivered to the distributor injection pump by the fuel lift pump, a sliding vane-type transfer pump raises the fuel's pressure still further. The purpose of the transfer pump is to create a transfer pressure, so that the injection pump can supply all of the engine's fuel requirements under varying operating conditions. To prevent the transfer pressure inside the pump from exceeding safe limit, a spring-loaded regulator valve is fitted.

The position of the fuel-metering valve is controlled by a combination of throttle lever movement and the action of the governor. From the metering valve, the pressurized fuel is transferred to the metering port in the hydraulic head. When the charging port on the rotor aligns with the metering port, fuel is transferred towards the plungers. The pressure of the incoming fuel is sufficient to force the plungers and rollers against a cam ring.

Further movement of the rotor shaft, which is being driven by the timing gears, causes the rollers to rotate and follow the rising and falling contours of the cam ring. As the rollers approach the peaks of the cam lobes, the plungers are forced towards each other and so increase the pressure of the fuel in the internal centre gallery of the rotor. The fuel remains under pressure in the internal gallery of the rotor until the distributor port aligns with one of the fuel outlet ports in the hydraulic head. The pressurised fuel is then transferred to the appropriate injector.

MODERN INJECTION SYSTEMS

There are three basic types of diesel injection systems:

1. Common rail direct injection
2. Unit injector
3. Unit pump

The first two are used on the large diesel engines of sports utility vehicles, trucks, etc.

COMMON-RAIL DIRECT INJECTION

Conventional diesel engines (non-CRDi engines) are sluggish, noisy and poor in performance compared to a CRDi engine. This technology has turned diesel engines into fuel efficient, clean, high torque and quieter engines. CRDi is an intelligent way of controlling a diesel engine with use of modern computer systems. So it helps to improve the power, performance and reduce harmful emissions from a diesel engine.

Construction

A common rail diesel fuel injection system consists of a high pressure fuel pump driven by the engine of a vehicle, a thick-walled steel tube known as an accumulator or rail that runs alongside the cylinder head, and electrically operated injectors. The rail creates a reservoir of fuel "common" to all the fuel injectors, hence the name. Common rail diesel injectors are controlled by an Electronic Diesel Control (EDC) unit, which regulates fuel pressure and the timing and duration of fuel injection.

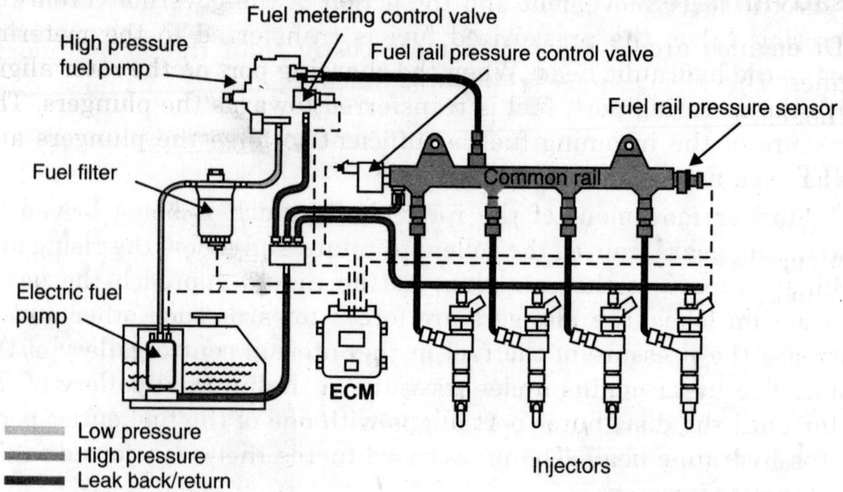

Fig.2.27: *Common rail direct injection*

A high-pressure pump increases the fuel pressure in the accumulator up to 1,800 bar. The pressure is set by the engine control unit and is independent of the engine speed and quantity of fuel being injected into any of the cylinders. The fuel is then transferred through rigid pipes to the fuel injectors, which inject the correct amount of fuel into the combustion chambers.

Operation

To run a CRDi system, the microprocessor works with input from multiple sensors which include throttle position sensor, crank position sensor, pressure sensor, lambda sensor, etc. Based on the input from these sensors, the microprocessor can calculate the precise amount of diesel and the timing when the diesel should be injected inside the cylinder. Using these calculations, the CRDi control system delivers the right amount of diesel at the right time to allow best possible output with least emissions and least possible wastage of fuel.

Advantages

The electronic diesel control unit precisely meters the amount of fuel injected, and improves atomization of the fuel by controlling the injector pulsations. This results in quieter, more fuel efficient engines, reduced exhaust emissions (upto 30%) and more power output.

CRDi engine fitted cars offer 25% more power than the normal direct injection engine with a superior pickup and torque offering sometimes up to 70% more power than the conventional diesel engines.

Disadvantages

CRDi engines are at least 25% more costly than the conventional engines. They also require a higher degree of maintenance and spares are not cheap also.

UNIT INJECTOR

The lines connecting the injection pump to the fuel injector produced problems for early diesel engineers. So in 1905, Carl Weidman got rid of them by combing the injection pump and the injector.

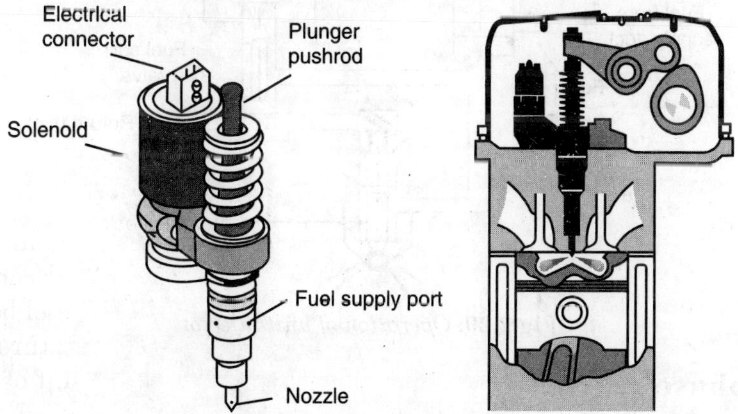

Fig.2.28: *Electronic type unit injector*

Construction

Unit Injector System (UIS) is an integrated direct fuel injection system for diesel engines, combining the injector nozzle and the injection pump in a single component. The pump is usually driven by a shared camshaft. The unit injector is fitted into the engine cylinder head, where the fuel is supplied via integral ducts machined directly into the cylinder head. Each injector has its own pumping element, and in the case of electronic control, a fuel solenoid valve as well.

Operation

The basic operation can be described as a sequence of four separate phases: the filling phase, the spill phase, the injection phase, and the pressure reduction phase.

The fuel system is divided into the low pressure fuel supply system and the high pressure injection system. A low pressure fuel delivery pump supplies filtered diesel fuel into the cylinder head fuel ducts, and into each injector fuel port of constant stroke pump plunger injector, which is overhead camshaft operated.

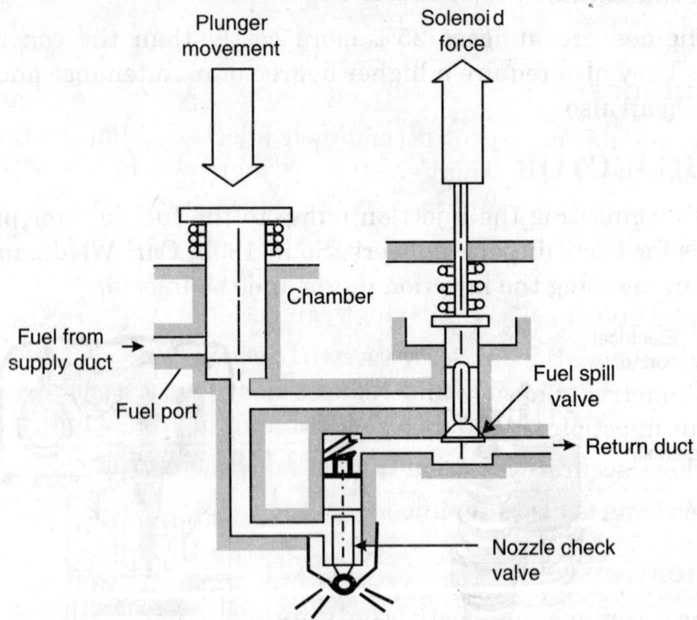

Fig.2.29: *Operation of unit injector*

Fill phase

The constant stroke pump element on the way up draws fuel from the supply duct into the chamber, and as long as electric solenoid valve remains de-energized fuel line is open.

Spill phase

The pump element is on the way down, and as long as solenoid valve remains de-energized the fuel line is open and fuel flows in through into the return duct.

Injection phase

The pump element is still on the way down, the solenoid is now energized and fuel line is now closed. The fuel cannot pass back into return duct, and is now compressed by the plunger until pressure exceeds specific opening pressure, and the injector nozzle needle lifts, allowing fuel to be injected into the combustion chamber.

Pressure reduction phase

The plunger is still on its way down, the engine ECU de-energizes the solenoid when required quantity of fuel is delivered, the fuel valve opens, fuel can flow back into return duct, causing pressure drop, which in turn causes the injector nozzle needle to shut, hence no more fuel is injected.

Advantages

1. As unit injection uses a combined injection pump and nozzle for each cylinder, it eliminates long, high pressure pipelines from pump to injector, and thereby the compressibility and wave effects associated with them.
2. Consequently, it is possible to attain very high injection pressure (up to 2000 bar) so that smoke can be reduced.
3. By reducing the degree of air swirl needed, high pressure improves volumetric efficiency and reduces heat loss to coolant. Therefore, unit injection also offers a reduction in fuel consumption.
4. Allows accurate injection timing, and amount control
5. Servicing is also simplified.

Application

The most common application for unit injectors are Volkswagens and large diesel engines.

UNIT PUMP SYSTEM (UPS)

The UPSs use an individual injection pump mounted on the engine block for each cylinder so it is primarily designed for OHV or cam in the engine block. The fuel pumps are driven by an extra camshaft lobe where each pump unit is connected to the injector via a short precise length high-pressure fuel line.

This system also feature electronically controlled fuel solenoids for precise timing and the injection of fuel quantity is variably adjusted for each cylinder.

Benefits

- High pressure capability

- Robust design
- Proven reliability
- Low emissions
- Low fuel consumption and less CO_2 output
- High torque even at low engine speeds

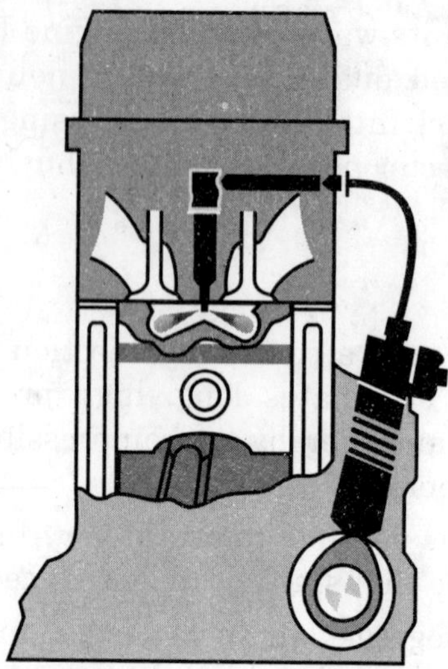

Fig.2.30: *Unit pump system*

Applications

The Delphi Diesel Electronic Unit Pump is suitable for heavy duty diesel engines for on- and off-highway applications and premium medium duty diesel engine programs.

GOVERNOR

Governors control engine speeds by varying (controlling) the quantity of fuel injected into the engine.

Higher gas pressures in the cylinder of a diesel engine usually demand stronger components than those used in a petrol engine. Strength is normally improved by increasing the component dimensions, but this also increases weight and leads to engine damage if the speed exceeds a given value which is governed by the component's strength/weight ratio.

The pneumatic and centrifugal governors are explained below:

CENTRIFUGAL GOVERNOR

The Fig.2.34 shows a centrifugal governor, also called mechanical governor which is often used on heavy vehicle engines.

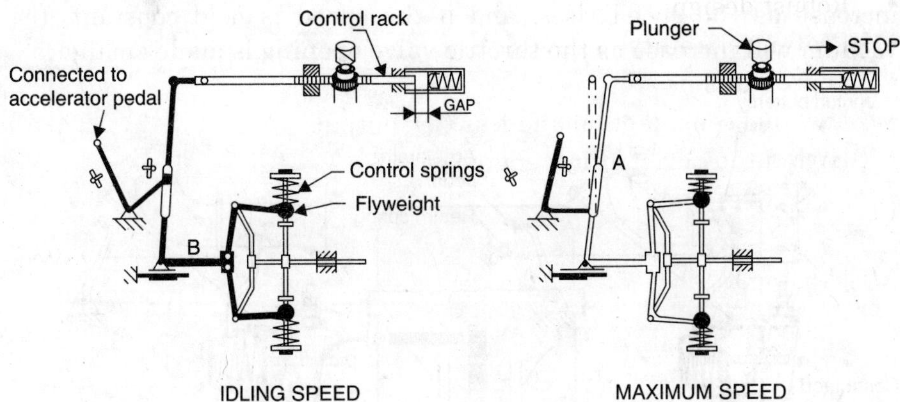

Fig.2.31: *Centrifugal governor operation*

Construction

A mechanical governor consists of a flyweight and various levers and springs. To control the injection volume, a mechanical governor uses the centrifugal force that is generated by a rotating flyweight. The flyweights are rotated by the pump camshaft.

Operation

When the engine is running close to its maximum speed or exceeds a prescribed maximum speed, the centrifugal force that is generated in the flyweight becomes greater than the spring force pushing the control block (B). Thus, control block (B) moves to the right, and together with this movement, the guide lever (A) also moves to the right. This movement is transmitted via the control lever to pull the control block towards the "fuel decrease" direction, thus decreasing the injection volume. As a result, the engine speed decreases to prevent the engine from overrunning.

Thus, to control engine output at maximum engine speed range, conventional mechanical governors regulate the fuel injection volume by using the centrifugal force that is generated by the flyweight and in the medium-speed range, it relies on the movement of the accelerator pedal.

PNEUMATIC GOVERNOR (VARIABLE SPEED GOVERNOR)

This pneumatic governor is an all-speed type and is actuated by the vacuum developed at the venturi part of the engine air intake. As a result, the governor is divided into the venturi part and the governor part. If the throttle valve is set at a given position, the vacuum will increase as the speed rises, and if the speed is held constant, the vacuum will increase as the throttle valve opening is made smaller.

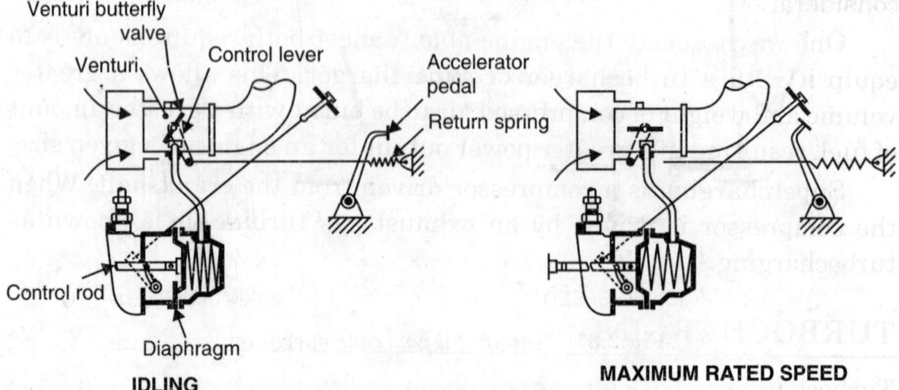

Fig.2.32: *Pneumatic governor operation*

Though the changes in the degree of vacuum and the tension of governor spring, the control rack is shifted to the required position to control the amount of fuel injected.

This type of governor is mainly used with passenger vehicles, tractors and small stationary diesel engines.

TURBOCHARGING AND SUPERCHARGING

NECESSITY OF TURBOCHARGER AND SUPERCHARGER

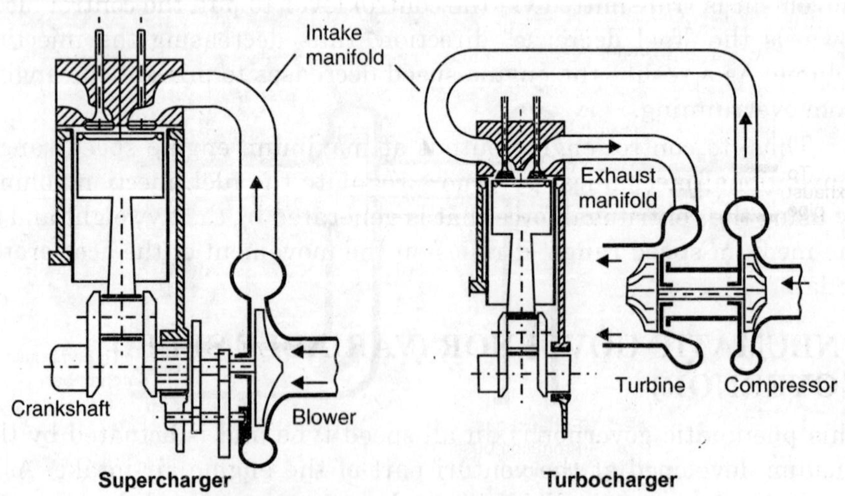

Fig.2.33: *Supercharger and turbocharger*

In designing an engine, the balance between higher power on the one hand and lower weight and smaller size on the other must be taken into consideration.

One way to make the engine able to meet both requirements is to equip it with a turbocharger or supercharger. This allows a greater volume and weight of compressed air to be burnt with a greater amount of fuel, resulting in a greater power output for an engine of a given size.

Supercharging is a compressor driven from the crankshaft. When the compressor is driven by an exhaust gas turbine, it is known as turbocharging.

TURBOCHARGING

Turbocharger is basically an air pump *driven by the exhaust gas* from the engine.

"More fuel + More air = Bigger explosion = Greater power"

PURPOSE

Turbocharger forces an additional amount of air or air-fuel mixture into the engine cylinder. With more air-fuel to burn, the engine produces more combustion pressure and hence more engine power. On many turbocharged engines, *power output is boosted 30% or more* (for example, a 100 horsepower engine could produce 130 horsepower with a turbocharger).

CONSTRUCTION

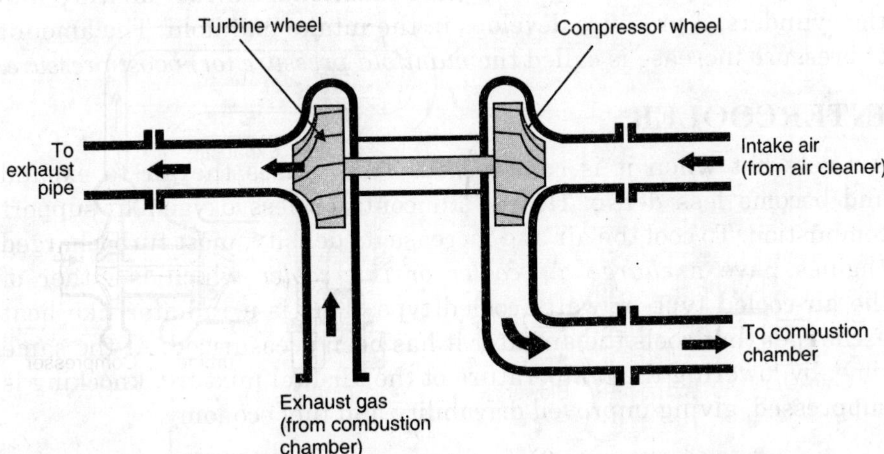

Fig.2.34: *Turbocharger*

The turbocharger consists of the turbine housing, compressor housing, center housing, turbine wheel, compressor wheel, full floating bearings, waste gate valve, actuator, etc.

The turbine wheel and the compressor wheel are mounted on the same shaft. When the engine is running, the exhaust gases flow from the exhaust manifold to the turbine wheel. They strike the blades of the turbine, spinning it up to 100,000 rpm or higher. Since the compressor is on the same shaft, it also spins at high speed. As the compressor spins, it forces the intake air into the engine cylinders.

Since the turbine and compressor wheels turn at very high speed, full floating bearings are used to ensure the absorption of vibrations from the shaft and lubrication of the shaft and bearings.

OPERATION

The turbocharger has two operating phases, the atmospheric phase and the boost phase. During light-load and cruising conditions, the compressor is almost idling. It is rotating too slowly to pressurise the intake air. The engine operates about the same as an engine of similar size without a turbocharger. As the engine is accelerated for more power, the combustion of more air-fuel mixture, thereby increasing the flow of exhaust gas. This speeds up the turbine and compressor. Now additional air-fuel mixture is delivered to the cylinders and the engine delivers more power. A turbocharged engine operates in the atmospheric phase most of the time. The turbocharger does not add power during cruising at normal highway speeds. It usually provides boost only when acceleration or full power is required.

As the compressor begins to force additional air-fuel mixture into the cylinders, a pressure develops in the intake manifold. The amount of pressure increase is called the *manifold pressure (or) boost pressure.*

INTERCOOLER

Air gets hot when it is compressed. This causes the air to expand and become less dense. Heated air contains less oxygen to support combustion. To cool the air and increase its density, most turbocharged engines have a *charge air cooler or intercooler* which is either of the air-cooled type or water-cooled type. This is a radiator like heat exchanger that cools the air after it has been pressurized. At the same time, by lowering the temperature of the air-fuel mixture, knocking is suppressed, giving improved drivability and fuel economy.

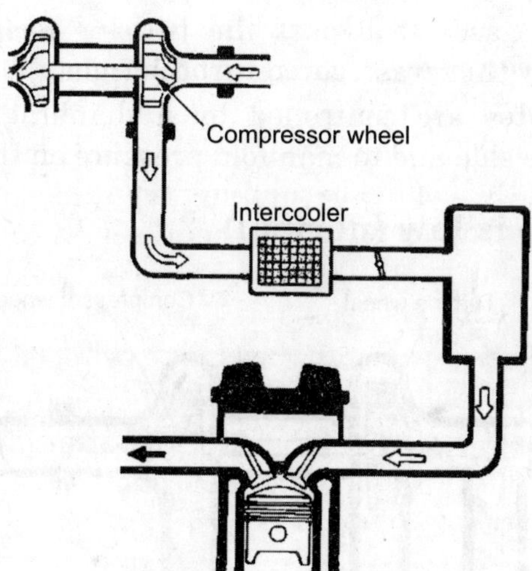

Fig.2.35: *Turbo intercooler*

Turbo Boost

The amount of air density a turbo can provide is known as **boost pressure**, or *boost*. The point where boost begins is called the **boost threshold**. This might be something like 1800 rpm, for instance.

Turbo Lag

When a turbo is spinning at low speed, little or no boost is produced. The time required to bring the turbo up to a speed where it can function effectively is called **turbo lag**—hesitation in throttle response when coming off idle.

TURBOCHARGER CONTROL

The turbocharger can raise boost pressure high so that detonation and engine damage occur. To limit boost pressure and prevent over boost, most turbochargers have control systems.

Common control designs are turbocharger with waste gate, and variable geometry turbochargers (VGT). A wastegate is a simple way of varying turbocharger geometry. A more sophisticated way is to use a variable geometry turbine (VGT), also called variable nozzle turbine (VNT) or variable displacement turbine.

WASTE GATE

These turbochargers have an exhaust bypass valve allowing exhaust gases to bypass the turbine leaving some exhaust energy unused.

Allowing exhaust gases to bypass the turbine wastes some of the exhaust gas energy thus wastegated turbocharger.

Most wastegates are controlled by a diaphragm, open to the atmosphere on one side and to manifold pressure on the other.

Boost pressure is low (normal)

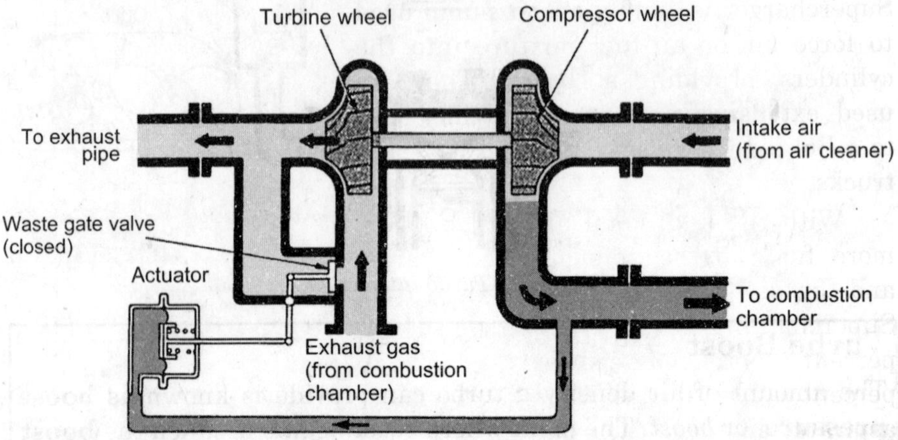

Fig.2.36: *Turbocharger operation: Low boost pressure*

As long as the boost pressure inside the intake manifold is normal, the actuator does not operate and the waste gate valve remains closed. All exhaust gas is therefore routed into the turbine housing.

Boost pressure is high

As the accelerator pedal is depressed (fuel volume increases), the exhaust gas pressure increases, thus increasing the boost pressure.

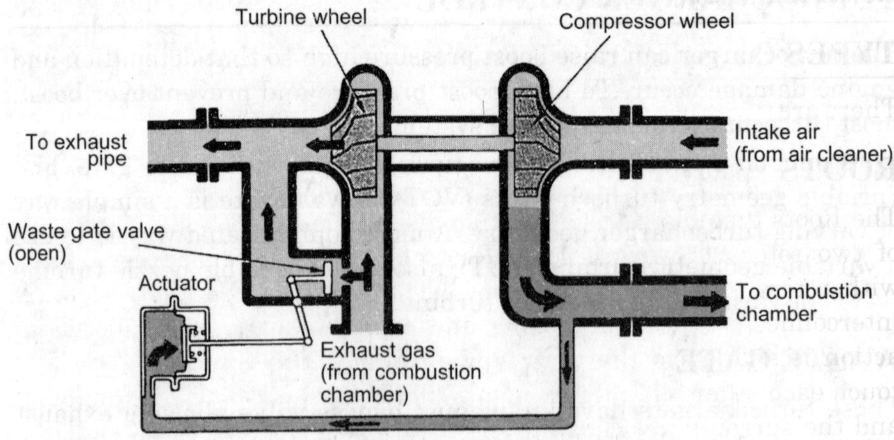

Fig.2.37: *Turbocharger operation: High boost pressure*

When the boost pressure reaches high, the waste gate valve is opened by the actuator and part of the exhaust gas is diverted around the turbine wheel. As a result, the turbine speed is kept within the optimal range to prevent the boost pressure from rising excessively.

SUPERCHARGER

Supercharger is simply an air pump used to force air or air-fuel mixture into the cylinders, providing a "boost." They are used extensively on racing engines and to a limited extent on passenger cars and trucks.

With the additional air in the boost, more fuel can be added to the charge, and the power of the engine is increased. Supercharging adds an average of 50 percent more horsepower and 30-40 percent more torque than a normally aspirated engine of the same size.

The crankshaft usually drives the supercharger with a belt, but it is sometimes driven by a chain or gears. To pressurise the air, a supercharger must spin more rapidly than the engine itself. Making the drive gear larger than the compressor gear causes the compressor to spin faster. Superchargers can spin at speeds as high as 50,000 to 65,000 rpm which is much slower than turbochargers.

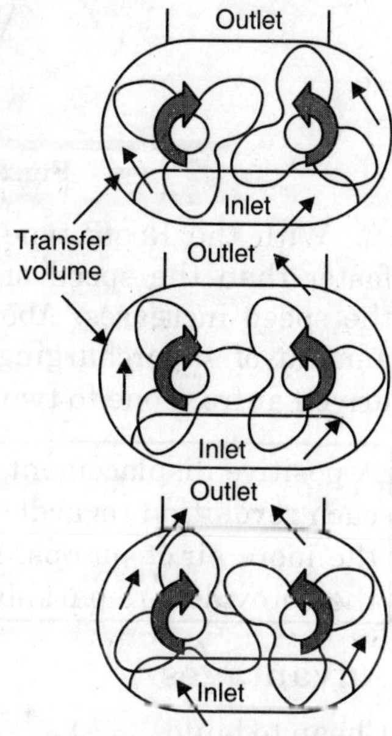

Fig.2.38: *Turbocharger*

TYPES OF SUPERCHARGERS

There are two types of superchargers, the Roots type and the centrifugal.

ROOTS BLOWER

The Roots type blower is a positive displacement compressor, consists of two rotors. However, some Roots type of superchargers is fitted with rotors of three or more lobes. The shafts of the two rotors are interconnected through gearing and operate at the same speed. Its action is similar to the gear type oil pump. The rotors do not quite touch each other and there is also a slight clearance between the rotors and the surrounding housing. In operation, air enters the housing by the action of the rotors, and passes between the lobes of the rotors and the housing. The air is then forced out through the outlet opening of the unit.

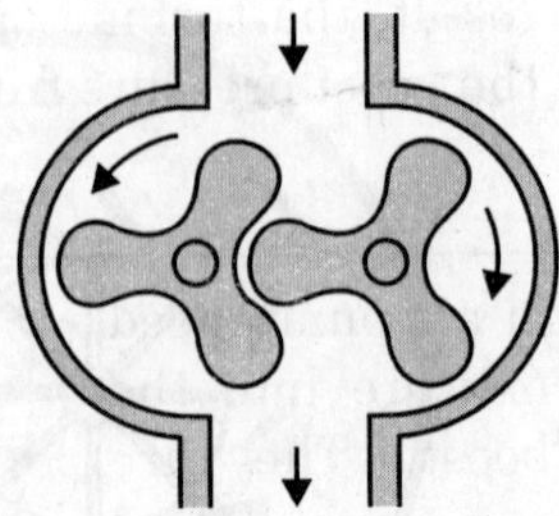

Fig.2.39: *Roots type supercharger*

With the Roots type blower, the rate of delivery varies slightly faster than the speed of rotation, because the leakage decreases as the speed increases. Above a certain minimum speed, therefore, the amount of supercharging is almost constant. Roots type blowers are driven at from one to two times engine speed.

A positive displacement pump delivers the same amount of air with each revolution regardless of the speed. Therefore, the faster it turns the more air it pumps. Because there is no lag with these blowers, they provide a remarkable increase in torque right off idle.

Advantages

Cheap to build

Disadvantages

Low efficiency, low boost, consumes lots of power at high revolutions, ugly noise.

Application

Mercedes M111 and M271 4-cylinder, GM 3800 V6.

CENTRIFUGAL TYPE

The centrifugal type supercharger consists of an impeller rotating at a high speed inside a housing. Clearance between the blades of the impeller and the housing is kept at a minimum. Speed of rotation on racing engines is approximately 5 times crankshaft speed. On an engine rotating at 5000 rpm, the centrifugal supercharger will therefore be rotating at 25000 rpm. Because of the terrific speed, it is absolutely essential that the impeller blades must be made strong enough so that the centrifugal force at the high speeds will not cause them to stretch and possibly strike the housing. Impellers are made of alloy steel, duraluminum or magnesium. Housings are generally made of aluminium with external fins to provide cooling.

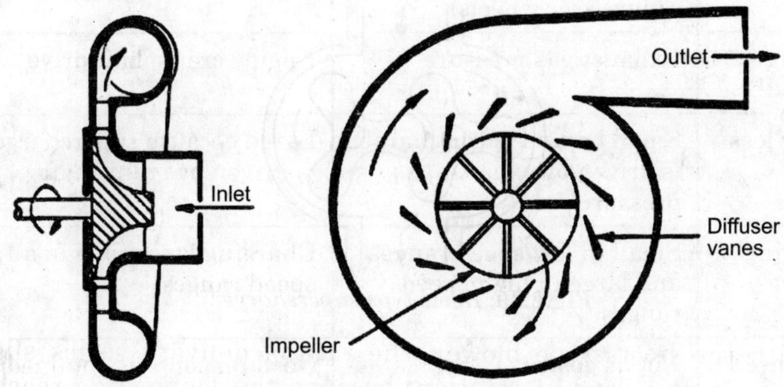

Fig.2.40: *Turbocharger*

On racing car, the air from the impeller first passes to a diffuser, where the force of the moving air is converted to static energy. The diffuser consists of ring-shaped housing containing blades or vanes. Coolers are also used to reduce the temperature of the air. This is important as the act of compressing the air will increase its temperature. The warm air entering would reduce the efficiency of the engine. The rate of delivery of the centrifugal supercharger increases as the square of the speed of rotation. That is, if the speed is doubled the rate of air delivery will be four times as great. As a result very little supercharging is obtained at lower speeds and the variation between different speeds is large. Carburetion is therefore more difficult with a centrifugal supercharger than with a Roots type unit.

Advantages

Highest efficiency among mechanical superchargers, high boost pressure, small and light.

Disadvantages

Weak boost at low to medium revolutions

Application

Koenigsegg, Farbio

COMPARISON OF TURBOCHARGER AND SUPERCHARGER

The features of turbocharger and supercharger are summarised in the following table:

	Turbocharger	Supercharger
Type of charger	Turbine (turbine and compressor wheels)	Mechanical (Roots type)
Driving method	Exhaust gas pressure	Engine crankshaft drive
Power loss	Small because turbocharger is driven by exhaust gas pressure	Large because supercharger is driven by crankshaft
Charging effect	Small in low-speed ranges and large in high-speed ranges	Charging is possible in all speed ranges
Response	Not as good as supercharger	Good response is obtained because supercharger is directly driven by crankshaft.

COOLING SYSTEM

The burning of fuel in an internal combustion engine produces heat, which is sufficient to melt the metal of the cylinder. If the engine operating temperature is allowed to go unchecked, it would burn and dry up the lubricating oil film, so that the pistons would seize in their cylinders and distortion would result from over-expansion of metals. It is the function of the cooling system to prevent the engine overheating but it must also allow it to operate at a temperature high enough to assist in effective combustion.

NEED FOR COOLING

The engine burns fuel in the cylinder to transform its thermal energy (heat energy) into a driving power. But the thermal energy of the fuel is not converted into power entirely. It can be stated that approximately 25 per cent of the energy contained in the fuel is converted into useful work; 30 per cent is absorbed by the engine itself, and the remaining 45 per cent passes away in the hot exhaust gases, frictional losses and pumping losses.

The temperature attained during combustion in the cylinder of an engine may be as high as 2000°C. These temperatures are considerably higher than the melting points of the materials used in the construction of the engine. Therefore, it is essential that excessive heat is removed quickly and efficiently to avoid serious damage to the engine. Therefore a cooling system is built into the engine to cool it and prevent overheating.

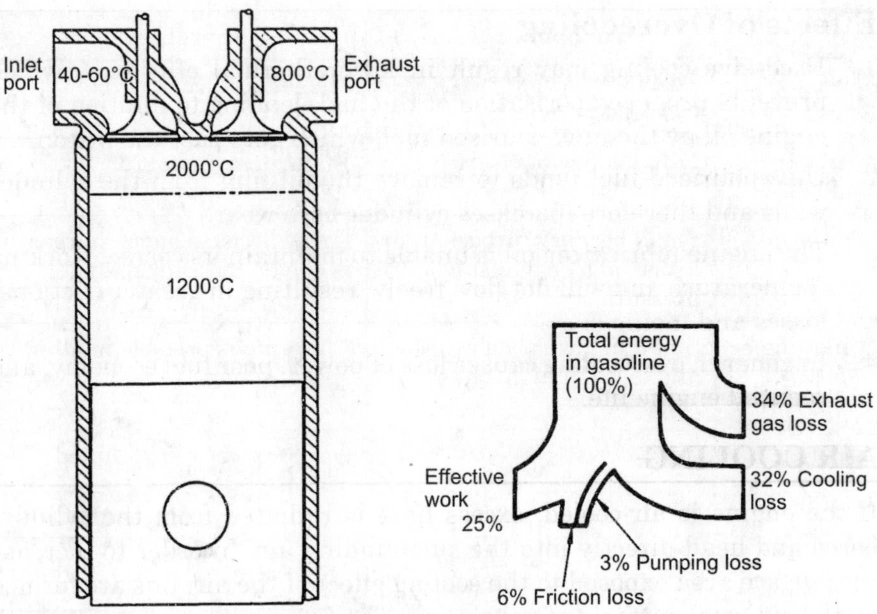

Fig.2.41: *Turbocharger*

To ensure that the engine operates efficiently it must be kept within a given range of temperatures. In the case of a normal water-cooled engine this would be in the region of 75°C-90°C. For air-cooled engines this range would be higher.

Effects of Overheating

1. Operation at high temperatures, high loads, or extremely high speeds without adequate cooling, results in oxidation of the lubricating oil. This may result in high temperature varnish and sludge deposits being formed, also piston ring sticking. Ring scuffing may follow ring sticking.
2. If working temperatures are too high, excessive expansion will take place in the pistons and the crankshaft journals, causing them to exceed their running tolerances and, as a result, they may seize into their bearings.
3. Areas within the combustion chamber such as the exhaust valve seats, spark plugs, etc., may become overheated and ignite the fresh charge of mixture before the spark occurs. This condition is known as 'pre-ignition' and may result in permanent damage to the engine if allowed to continue.
4. If the fresh charge of mixture entering the cylinder is overheated, its density will be reduced and a loss of power will result.

Effects of Overcooling

1. Excessive cooling may result in lower thermal efficiency, which prevents proper vaporisation of the fuel, leading to dilution of the engine oil by the unvapourised fuel, which gets past the piston.
2. Unvapourised fuel tends to remove the oil film from the cylinder walls and therefore increases cylinder bore wear.
3. The engine lubricating oil is unable to maintain its correct working temperature and will not flow freely, resulting in greater frictional losses and inefficiency.
4. In general, overcooling causes loss of power, poor fuel economy, and reduced engine life.

AIR COOLING

If the engine is air-cooled, excess heat is radiated from the cylinder bores and head directly into the surrounding air. In order to increase the surface area exposed to the cooling effect of the air, fins are formed on the external surfaces of the engine. The fins are cast integrally with the cylinder and cylinder head and act as a means of increasing the amount of radiation surface presented to the cooling air stream. Heat produced by the combustion process passes through the walls of the cylinder and cylinder head to the fins, where it is dissipated into the passing air.

DIRECT COOLING

It is the most straightforward method and normally used on *motorcycles* which have an engine exposed to the air stream on its forward motion and no surrounding bodywork. To improve the thermal conductivity of cylinders and cylinder heads, they are usually alloy castings with cooling fins (in some cases the crankcase also) to maximize the cooling effect.

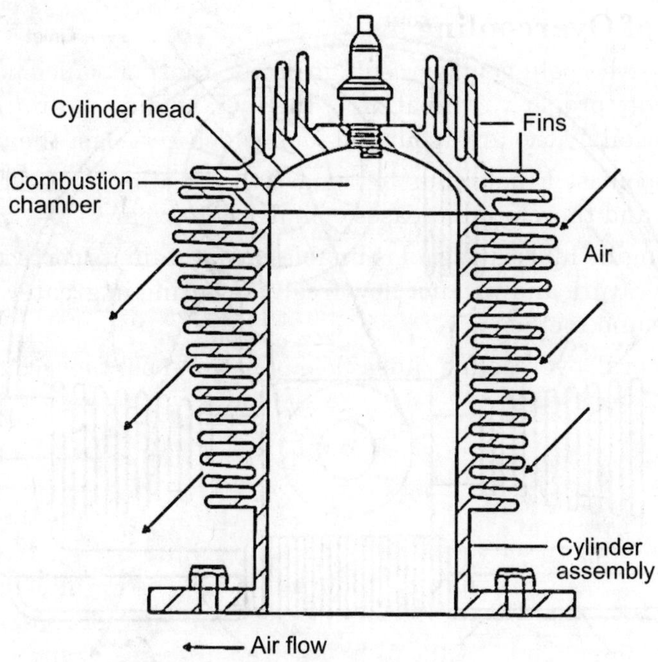

Fig.2.42: *Direct cooling*

Drawback

- Cooling effect depends on vehicle speed and outside temperature.

FAN (OR) BLOWER COOLING

Fan cooling enables engines enclosed within bodywork, for instance in scooters and autorickshaws, to be adequately cooled. The fan can be mounted directly on the crankshaft or driven by a V-belt.

The cowling (enclosed bodywork) often encloses the engine to control the flow of air over the hotter sections of the engine. In many arrangements, baffles or deflector plates are located near or between the cylinders to deflect the air through the finning and around the rear of the engine cylinders. A thermostat may be fitted on an air-cooled engine to prevent the fan from forcing air past the engine cylinders until they have reached their normal working temperature. This will take the form of a flap or shutter assembly which is controlled by a thermostatically operated valve.

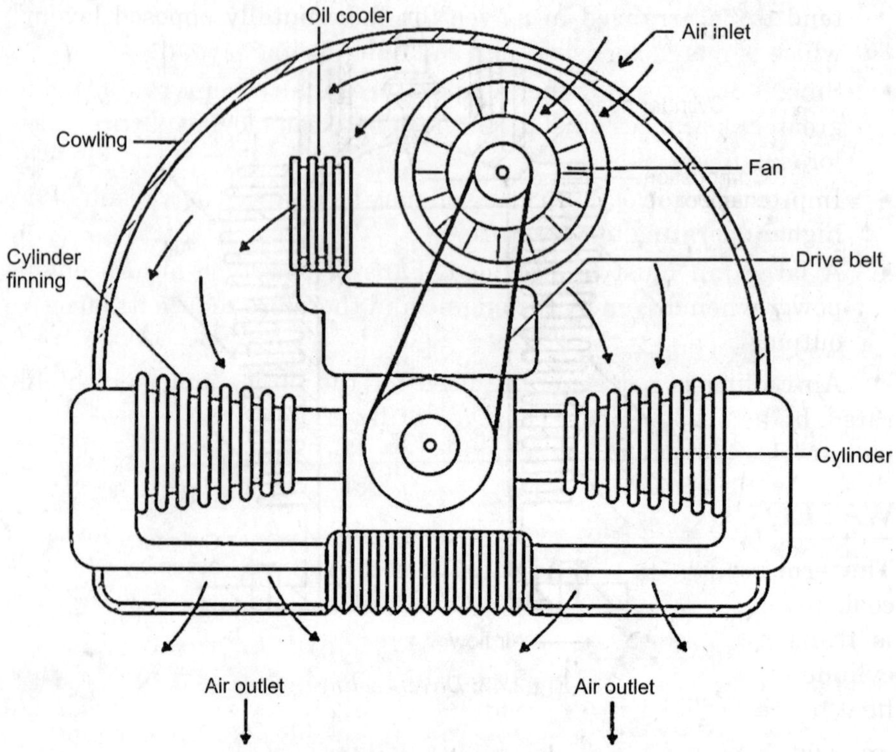

Fig.2.43: *Blower type cooling*

Advantages of air-cooling system

- Simplicity, with no problems associated with water leaks, radiators, water pumps, etc. so that maintenance is minimum.
- Aluminium alloys are generally used for the construction of engine components as they are better heat conductors than cast iron or steel. This ensures that the air-cooled engines are lighter than water-cooled engines.
- Can reach its normal working temperature in a shorter period than a water-cooled engine.
- Able to operate at a higher temperature than a water-cooled engine.

Disadvantages of air-cooling system

- Absence of a sound-muffling water jacket results in noisy engine operation.
- In most cases it is necessary to cast each cylinder separately to ensure proper formation of the fins. The cylinders must also be a considerable distance apart to accommodate the depth of fins between each cylinder. Therefore, multi-cylinder air-cooled engines

tend to be arranged in a "vee" or "horizontally opposed layout", which is more expensive than a simple "in-line layout".

- Since it operates at higher temperatures than other types of engine, greater clearances must be provided between moving parts to allow for expansion.
- Improved grade of lubricating oil may be required as a result of the higher operating temperature.
- A large fan must be provided. Fans of this type absorb engine power when driven by the engine and therefore reduce its effective output.

Air-cooling is not generally used for the larger and more highly rated, high-speed engines. There are still very few vehicles and all the motorcycles preferred to employ this type of cooling medium.

WATER COOLING

The term 'indirect air cooling' refers to the conventional liquid-or water-cooled engine. In a water-cooled system the heat from the cylinders is transferred to the coolant contained in jackets surrounding the cylinders. This coolant then passes through a radiator which transfers heat to the cool air flowing around it.

The simplest method of coolant circulation is the 'thermo-syphon' system which depends upon the natural circulation of heated water. An improvement on this basic method of circulation can be found in the modern 'forced circulation (or) pump assisted' cooling system which ensures a more positive circulation of the coolant.

THERMO-SYPHON COOLING SYSTEM

It this simple cooling system, which is rarely used in modern vehicle engines, movement of the coolant depends upon the natural circulation. The water surrounds each of the cylinders, combustion chambers, valves, etc. As the water in these areas absorbs heat it expands, becomes less dense, and so rises to the top of the radiator, thus making room for cold water from the bottom of the radiator to take its place and so maintain circulation by natural means.

The cooling fan is essential when the vehicle is moving at *low speeds or is stationary (engine idling).*

Although the thermo-syphon system of coolant circulation has the advantage of being simple and relatively efficient in small, low-speed engines, it is seldom used in modern vehicle design, because of the above mentioned disadvantages.

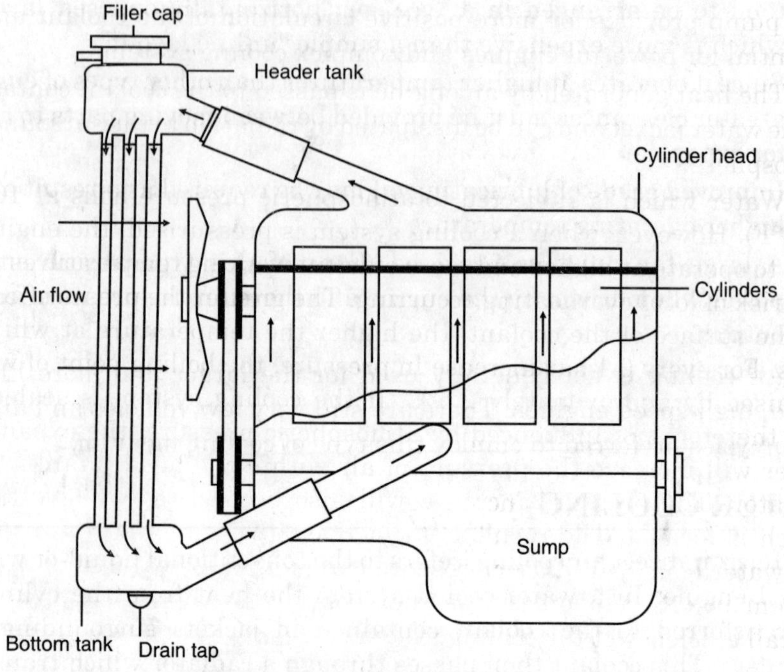

Fig.2.44: *Thermo-syphon cooling system*

FORCED CIRCULATION COOLING SYSTEM

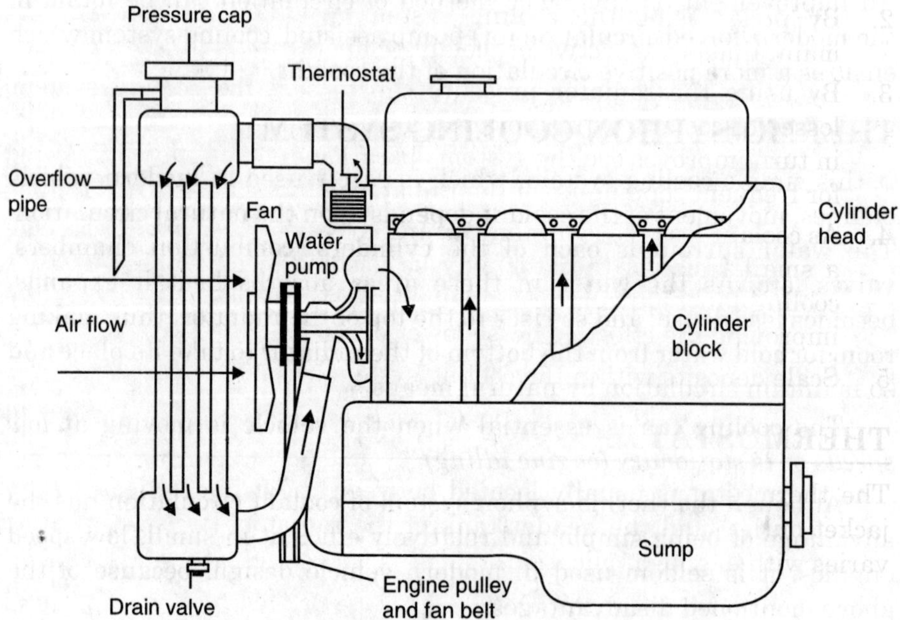

Fig.2.45: *Pressurised cooling system*

The pump provides for more positive circulation of the coolant and is essential for powerful engines and complex cooling systems.

The heat generated by an engine is passed to the coolant contained in the water jacket and can be dissipated by radiation to the surrounding atmosphere.

Water which is subjected to atmospheric pressure boils at 100°C (212°F). However, when a cooling system is pressurised, the engine is able to operate at higher and more efficient working temperatures with less risk of local overheating occurring. The greater the pressure acting on the surface of the coolant, the higher the temperature at which it boils. For every 0.1 bar increase in pressure, the boiling point of water is raised by approximately 2.5°C. If the cooling system is sealed off and therefore not influenced by atmospheric pressure, the expanding water will increase the pressure of air within the header tank of the radiator. The pressure increase will also be assisted by any steam which is formed. The result is to increase the temperature at which the water boils. A radiator pressure cap is used to seal off the cooling system in order to pressurise the coolant and thereby improve the overall efficiency of the cooling system.

Advantages

1. The engine is able to operate at a higher, more efficient temperature without the coolant boiling and producing dangerous steam pockets.
2. By pressurising the cooling system, water pump efficiency is maintained irrespective of coolant temperature.
3. By using the radiator pressure cap to seal the cooling system, losses of coolant by evaporation and surging are prevented. This in turn improves cooling system operation and eliminates the need for frequent topping-up of the radiator.
4. As coolant losses are minimised and cooling efficiency is increased, a smaller size of radiator may be used. Alternatively, a smaller cooling fan requiring less engine power may be fitted, therefore improving the effective horsepower of the engine.
5. Scale deposits within the cooling system are slow in forming.

THERMOSTAT

The thermostat is usually located between the cylinder head (water jacket outlet) and the header tank of the radiator. Its exact location varies with engine application.

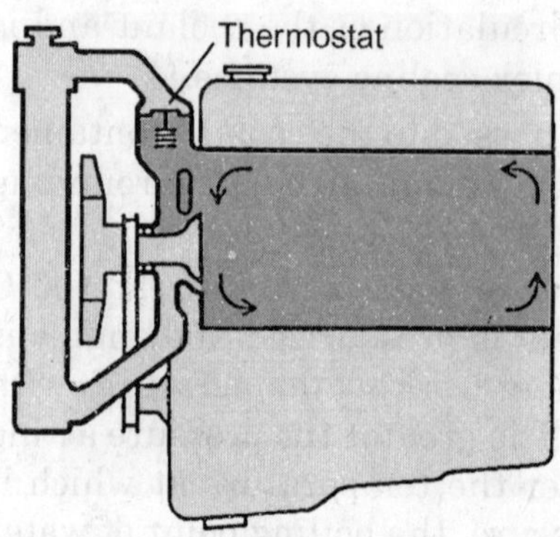

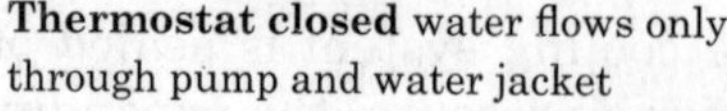

Thermostat closed water flows only through pump and water jacket

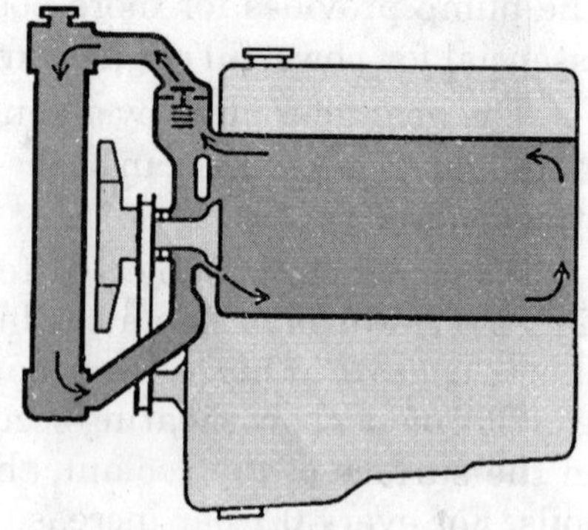

Thermostat open water can pass through the radiator for cooling

Fig.2.46: *Thermostat function*

Internal combustion engines are designed to operate most efficiently at a particular temperature (operating or working temperature) and it is most important that this temperature should be attained as soon as possible after starting the engine, and then maintained. The thermostat continues to control the flow of coolant to prevent overheating or overcooling and ensure that the engine will operate at its most efficient temperature. Therefore the function of a thermostat is to maintain ideal engine working temperature, which it does by controlling the rate of coolant flow through the radiator according to engine operating conditions.

The opening and closing of the thermostat is controlled by the temperature of the coolant in the system. When the coolant is cold and during the 'warm-up' period, the thermostat remains closed and the water pump circulates the coolant through the cylinder block and cylinder head only. As the normal operating temperature is reached, the valve opens to allow the coolant to circulate throughout the entire system. The thermostat is usually marked with the temperature at which the valve begins to open. A standard temperature range used in modern engines may be in the region of 80°C to 90°C and the valve should be fully opened at 93°C. However, this range will be recommended by the manufacturers. Thermostats are of the bellows type and wax type.

BELLOWS TYPE

The unit comprises a disc valve and seat with the valve attached to a brass bellows. These bellows contain a volatile liquid, which is usually a mixture of methyl alcohol and water, in a partial vacuum.

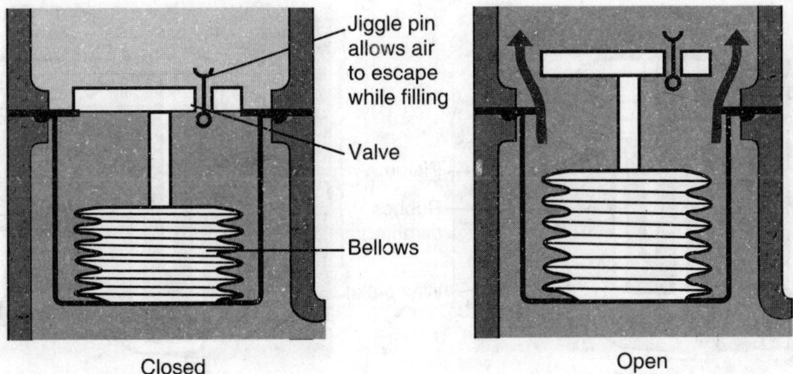

Fig.2.47: *Bellows type thermostat*

Operation

The thermostat is a heat-sensitive valve, which opens and closes according to the temperature of the coolant acting upon it.

Engine cold

When the engine is cold, the disc valve of the thermostat is held on its seating by the contraction of the bellows. In this situation there can be no flow of coolant between the cylinder head and the radiator.

Engine hot

As the engine warms up, the coolant in the jacket is heated, which in turn heats the volatile liquid contained within the bellows. This increase in pressure causes the bellows to expand and lift the valve off its seat. The coolant is then free to flow through the open valve and circulate through the radiator to be cooled.

WAX TYPE

In most engines the wax thermostat has almost completely superseded the bellows type. The wax type thermostat has a stronger action and is more reliable in service. It is not affected by normal coolant pressures and is particularly suitable for cooling systems where the operating pressures are in excess of 0.48 bar.

The thermostat consists of a temperature sensitive wax compound within a copper capsule. This wax compound also contains a copper powder which improves thermal conductivity. As the coolant temperature increases the wax melts and expands rapidly and uniformly, conversely as the temperature of the coolant is reduced, the wax contracts and solidifies again.

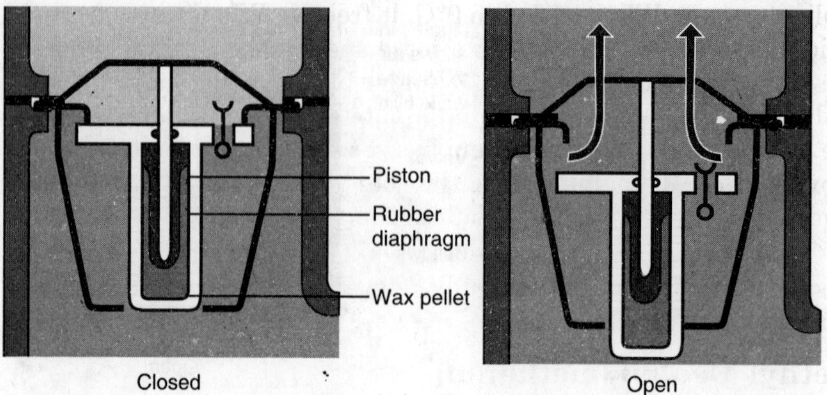

Fig.2.48: *Bellows type thermostat*

Operation

The wax type thermostat relies on expansion for its functioning. At an operating temperature, the wax expands rapidly to operate the thermostat valve.

Engine cold

At lower temperatures the thermostat valve is held in the closed position by spring pressure so that the coolant may only circulate around the engine cylinder block. There is no flow of coolant between the radiator and engine.

Engine hot

As the coolant in the engine begins to absorb heat and reaches operating temperature, the small quantity of wax contained in the capsule melts and begins to expand rapidly. Due to the construction of the capsule, the expansion force of the wax exerts pressure through the synthetic rubber sleeve on to the thrust rod which is located at the top by the bridge section. As the rod is retained by the bridge section and cannot move, the capsule is gradually forced downwards against spring pressure, opening the valve which is attached to it, and allowing the flow of coolant between the engine and radiator. Once the coolant temperature is reduced, the wax contracts and solidifies and the spring closes, preventing the flow of coolant.

ANTI-FREEZE AND INHIBITING SOLUTIONS

ANTI-FREEZE

There is a major disadvantage in using water as coolant in a water cooling system. When cooled to 0°C, it freezes. When water frozen, the solid mass expands and exerts a force against the internal surfaces of the cooling system. So that the cylinder block casting may be damaged and cracked. Furthermore, radiator tubes and connections may split if the water contained within them freezes. Adequate protection must be provided for the engine during the cold winter months to ensure that damage does not occur as a result of the effects of frost.

Anti-freeze lowers the freezing point of water and gives a margin of safety in cold operating conditions. There are two anti-freezes in common use:

Methyl alcohol (methanol)

When mixed with water in correct proportions, methanol will provide adequate protection at very low temperatures. It is least favoured because of the following reasons:

- Lower boiling point (approx. 82°C) which causes it to evaporate after long engine operating periods.
- Inflammable which increases the risk of fire in the engine compartment.

Ethylene glycol

Most modern anti-freeze solutions are based upon either ethylene glycol or propylene glycol. The following table shows coolant mixture for different freezing points.

5 liters of coolant		Ratio of water to ethylene glycol	Freezing point (in degree centigrade)
Quantity of water	Quantity of ethylene glycol		
3.75	1.25	3:1	–10
3.3	1.9	2:1	–20
3.0	2.0	3:2	–30
2.5	2.5	1:1	–35

INHIBITING SOLUTIONS

- Corrosion inhibitors such as benzotriazole or tolytriazole are added which prevent corrosive attack on the internal components of the cooling system.
- Other compounds such as phosphoric acid and sodium tetraborate may be added to improve the quality of the anti-freeze solution.
- Dye stuff like fluorescein may be used to give the solution an identifying colour.

LUBRICATION SYSTEM

The engine has many moving parts. All the moving parts are operating at high temperatures and at high speeds and most of these components are subjected to very high pressures. If the parts rub against one another, they will wear out quickly due to friction. It becomes essential to reduce this friction. The lubricating system performs this job by supplying oil to prevent metal-to-metal contact. The lubricant actually replaces dry friction with a fluid friction.

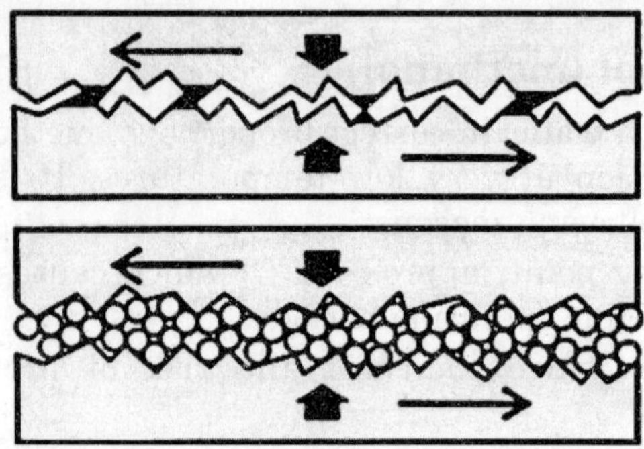

Fig.2.49: *Friction between the metal parts*

It also performs the following tasks:

- **Act as a seal** and prevent leakage between the parts such as pistons, rings and cylinders.
- **Remove the heat** generated by friction. This keeps down the temperature of the moving parts and thus prevents seizure.
- **Act as a cleaning agent**. Wash away dirt, carbon particles and abrasive metal worn from the friction surfaces.
- **Absorbs shock** between bearings and other engine parts, thus reducing engine noise and extended engine life.

The lubrication system consists mainly of the **oil pan** which stores the oil; the **oil pump** which pumps the oil to each part of the engine; the **relief valve** which controls the maximum oil pressure; the **oil filter** which filters out the impurities which get into the oil; the **pressure sender gauge (oil pressure switch)** which detects the oil pressure and the **oil level gauge** which measures the oil level in the oil pan.

ENGINE LUBRICATION

The principal friction surfaces in an engine are piston, cylinder, crankshaft, camshaft, main bearings, valves and valve operating mechanism. Friction between moving surfaces can be reduced by interposing a lubricant between them.

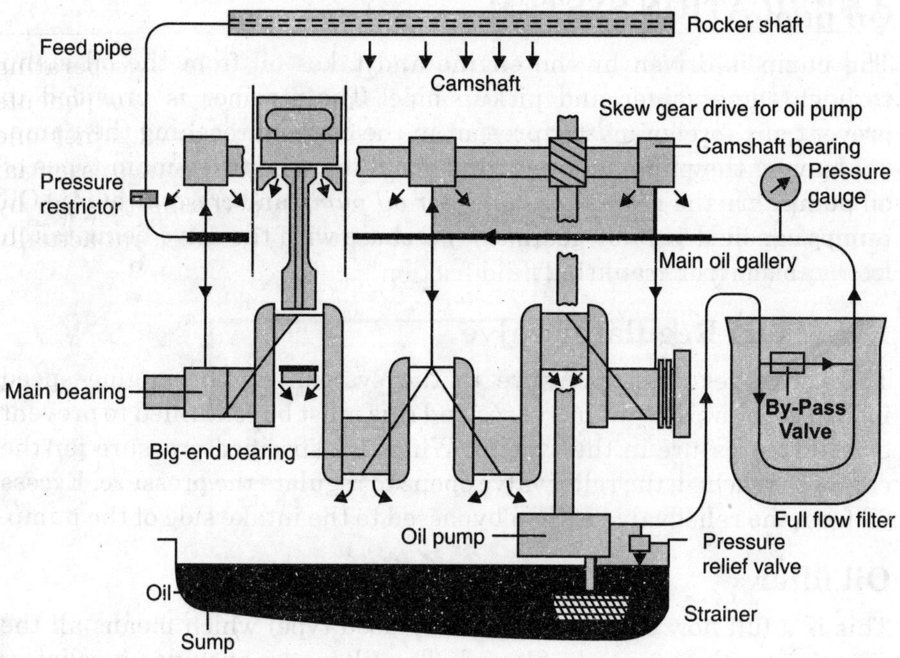

Fig.2.50: *Bellows type thermostat*

The various frictional parts of the engine are lubricated by oil, which is pumped to them from the oil pan (or) sump. The oil reaches thosc parts through smaller pipes, passages, drilled holes and grooves. These together with oil pump, oil filter, oil pressure controls and oil sump form a lubricating system. The main parts of the engine such as crankshaft bearings are provided with positive lubrication, receiving oil directly under pressure (pressure feed). The other parts that are not heavily loaded are lubricated with oil that is sprayed on to the moving parts through a hole or jet. The cylinder walls and pistons are lubricated in this way. Some parts of the engine do not require pressure lubrication, but must still be lubricated in this way. These parts receive lubrication oil on its way back to the oil sump. The push rods, cam followers and cams on some engines are lubricated in this manner. Some parts depend on the oil mist that is present in the crankcase when the engine is operating. Oil thrown from the other parts (for example connecting rod bearing) creates a general spray or mist of oil within the crankcase.

In the lubrication system, oil is delivered to the various parts of the engine by the oil pump and after performing its functions, returns to the oil pan. Oil circulates through the systems as follows:

Oil pump

The pump is driven by the engine and takes oil from the oil sump through the strainer and pickup pipe. The strainer is provided to prevent any foreign matter present in the oil from reaching the pump. Oil leaving the pump is directed to the filter. The three main types of oil pumps are the *gear oil pump, rotor oil pump* and *crescent pump*. Oil pumps can be driven by gears or by a chain with the drive being taken from camshaft or crankshaft.

Relief (or) Regulator valve

This valve regulates pressure in the system. As the engine speed increases, pump output increases and this must be controlled to prevent excessive pressure in the system. When the specified pressure for the engine is reached, the relief valve opens to regulate the pressure. Excess oil from the relief valve is then bypassed to the intake side of the pump.

Oil filter

This is a full flow oil filter (commonly used type) which means all the oil entering the engine is filtered. The filter also includes a relief or bypass valve which will open if the filter becomes blocked. This will allow unfiltered oil to be circulated in the system, thus preventing the damage to the engine that would occur if were being starved of oil.

Main oil gallery

The main oil gallery is a passage that runs full length of the crankcase. A number of drillings and passages from the gallery to other parts of the lubrication system. Fig.6.17 shows two passages from the gallery, one to the crankshaft and its associated parts, and the other to the camshaft and its associated parts.

Crankshaft main bearings

Each of the crankshaft main bearings is supplied with oil through a separate drilling from the main oil gallery.

Connecting rod bearings

Oil from the drillings in the crankshaft lubricates the connecting rod bearings.

Oil jet

An oil jet is used to lubricate the cylinder walls and pistons. A small hole in the side of the connecting rod is arranged to line up with the oil drilling in the crankpin journal. As this occurs, a stream of oil is directed onto the cylinder walls as a mist to provide lubrication. This happens once in each crankshaft revolution.

Camshaft bearings

The camshaft receives its oil supply from the main oil gallery. The oil to lubricate the camshaft bearings is carried through the hollow rocker shaft to the camshaft bearing caps where drillings take it to the camshaft bearings.

Rocker shafts

Drillings in the rocker shafts provide oil to the rocker arms and these inturn have a small drilling to provide lubrication to the cams.

Cams

After receiving oil directly from the rocker arms, the cams and the other parts are lubricated by splash. The cylinder head under the camshaft is shaped to form an oil reservoir, from which oil is splashed by the cams to lubricate the cam surface, valve stems and valve springs.

Timing chain

Where a timing chain is fitted, oil is provided from the end of camshaft bearing or similar source to that the chain and sprockets are lubricated.

Return oil

After lubricating the various components, oil from the cylinder head drains back to the oil sump through oil drain holes provided for this purpose. Oil from the camshaft drops directly back to the oil sump.

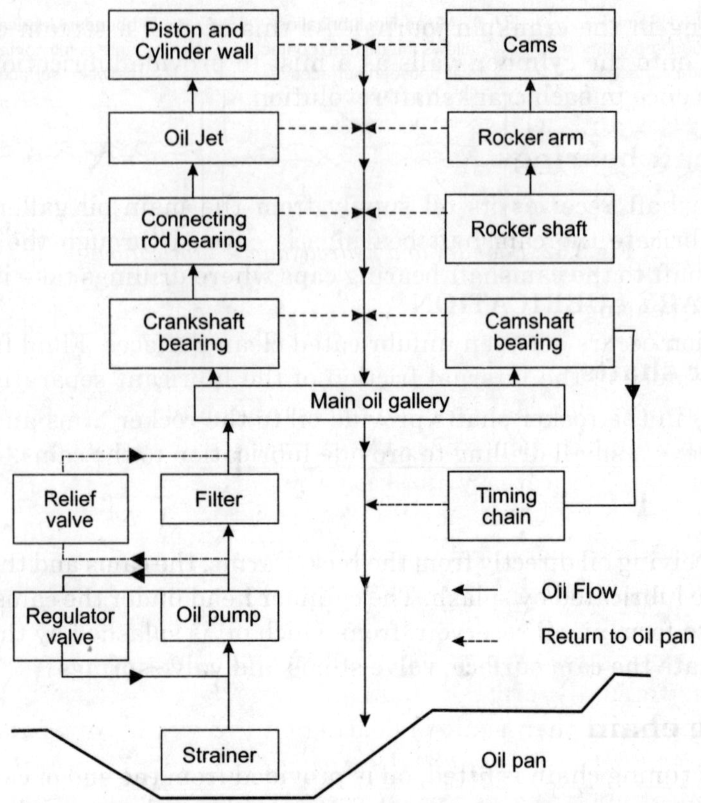

Fig.2.51: *Engine lubrication*

BOUNDARY LUBRICATION AND HYDRODYNAMIC LUBRICATION

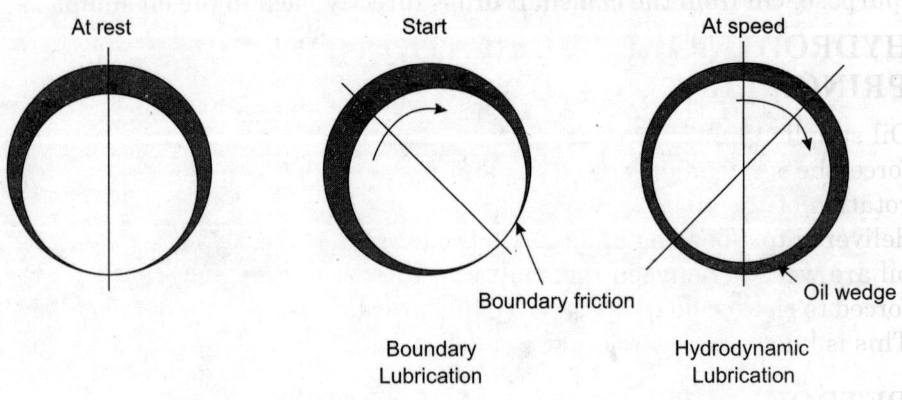

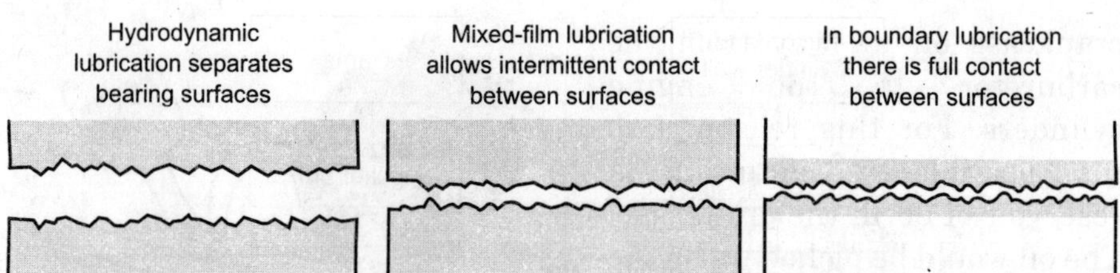

Fig.2.52: *Boundry and hydrodynamic lubrication*

BOUNDARY LUBRICATION

Dry friction occurs between unlubricated clean surfaces. Fluid friction occurs because of the internal friction of the lubricant separating the moving surfaces. Boundary friction refers to a condition between the dry friction and fluid friction where the surfaces are almost unprotected, being coated only by a greasy lubricant film. These surfaces are therefore actually in contact and not separated by a layer of lubricant oil.

If we consider a crankshaft bearing, when the shaft comes to rest any lubricant that is in the bearing will be forced out due to the load of the shaft. This is particularly true if the shaft remains stationary for quite some time. When the shaft starts to rotate, then several revolutions occur before the lubricant circulates completely around the bearings (and then shaft is floating on the oil). During this time, *boundary friction occurs and bearing wear takes place.*

Chemical additives in the oil can assist in protecting the parts against the effects of boundary friction. The term boundary lubrication may be used in relation to very small amount of lubricant (oil film) on the shaft and bearing under these conditions. All engine parts rely on boundary lubrication alone when the engine is first started. This is the critical period that is responsible for most engine parts wear.

HYDRODYNAMIC LUBRICATION (OIL-WEDGE PRINCIPLE)

Oil supplied under pressure from a pump to bearings does not merely force the shaft off the bearing and cause it to float on the oil. The rotation of the shaft is responsible for this. It collects layers of the oil delivered to a bearing and carries them around with it. These layers of oil are wedged between the shaft and the bearing, so that the shaft is forced to rise (or float) on the oil. Thus prevents metal-to-metal contact. This is known as hydrodynamic lubrication.

PETROIL LUBRICATION (MIST LUBRICATION)

In the small two-stroke engine, the air-fuel mixture passes through the crankcase on its way from the carburetor to the engine cylinders. For this reason it is not possible to maintain a reservoir of oil in the crankcase. The oil would be picked up by the passing air-fuel mixture, carried to the engine cylinders and burned. Therefore, to provide lubrication of two-stroke engine parts, the oil is mixed with the fuel (petroil). As the air and oil-fuel mixture enters the crankcase, the fuel, being more volatile, evaporates and passes on to the engine as an air-fuel mixture. Some of the oil is carried along with the air-fuel mixture and is burned, but enough oil is left behind to keep the moving parts coated with oil and thus adequately lubricated. The oil and fuel are mixed together in proportion by volume. Different proportions are recommended by different engine manufactures which vary from 16:1 to 50:1.

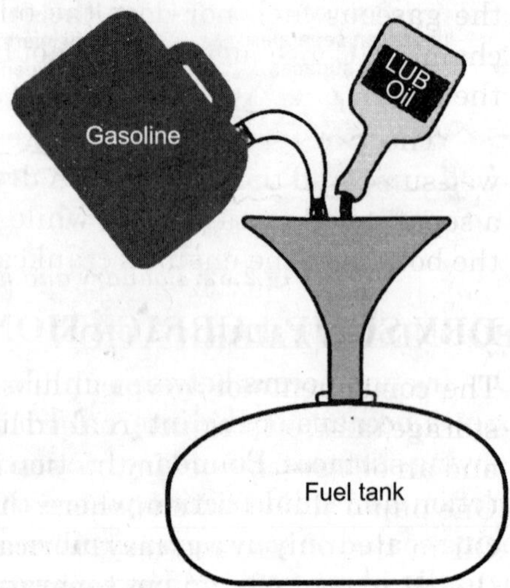

Fig.2.53: *Petroil lubrication*

SPLASH LUBRICATION

Small four-stroke engines of the type used in motor bikes and for small stationary engines, use a splash system of lubrication, although a small pump may be used as well. Oil is held in the crankcase and a dipper on the bottom of the connecting rod bearing cap is used to pick up oil for the bearings. The dipper also splashes oil into the upper parts of the engine to lubricate the valve mechanism, piston, cylinder wall, etc.

Fig.2.54: *Splash lubrication*

PRESSURE LUBRICATION SYSTEMS

Unlike the two-stroke engine, the lubrication system for four-stroke engines requires the engine oil and fuel to be kept separate from each other. Therefore, the four-stroke engine does not mix the oil with the gaseous fuel, nor does the oil intentionally enter the combustion chamber. Consequently, oil is not burned but is recirculated throughout the engine.

The two types of four-stroke engine lubrication systems are the wet-sump and the dry-sump. A dry-sump engine stores its oil supply in a separate oil storage tank, while the wet-sump system stores its oil in the bottom of the engine's crankcase.

DRY SUMP LUBRICATION SYSTEM

The components of a typical dry-sump lubrication system are the oil storage tank, oil strainer, oil feed line, oil pumps, engine oil passageways and an oil return line. In this system, two oil pumps are typically used. One pump acts as an oil pressure feed. The other is an oil return pump. Oil in the oil storage tank is gravity-fed to the pressure-feed side of the oil pump. The oil strainer keeps relatively large pieces of dirt and debris from entering the feed line. The pump then forces oil, under pressure, through oil passes in the engine. This lubricates the moving internal engine components. Oil that is thrown off the pressure-fed parts lubricates other internal engine components using the splash lubrication method. Finally, the excess oil collects in the sump and is returned to the oil storage tank by the oil return pump.

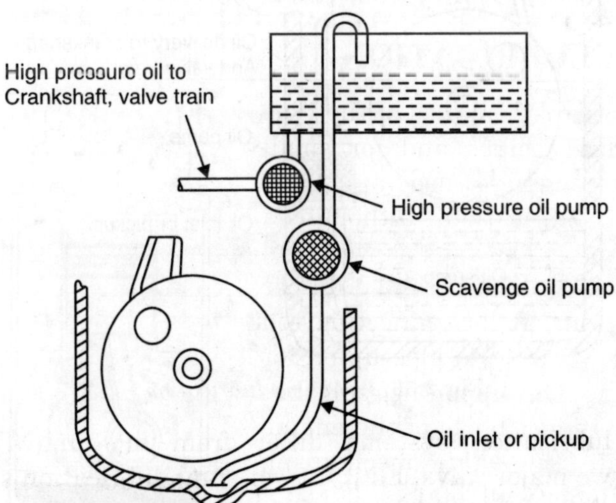

Fig.2.55: *Dry sump lubrication*

This system in four-stroke engines has a better place to cool by being stored in a separate storage tank away from the hot engine. A one-way check valve prevents the oil in the storage tank from leaking back into the sump.

Applications

They are used on race cars and high performance production cars like the Audi A8, Corvette ZR1 and Porsche 911. Dry sumps are also used in airplanes and in large diesel engines found in bulldozers and earthmovers.

Advantages

1. Protecting against oil sloshing
2. Dry sump oil tank can be made as large as necessary
3. Can be located anywhere on the vehicle.
4. Oil pan is no longer a reservoir for oil so it can be made smaller, lowering a race car's center of gravity.

Disadvantages

More complex so they cost more to produce

WET SUMP LUBRICATION SYSTEM

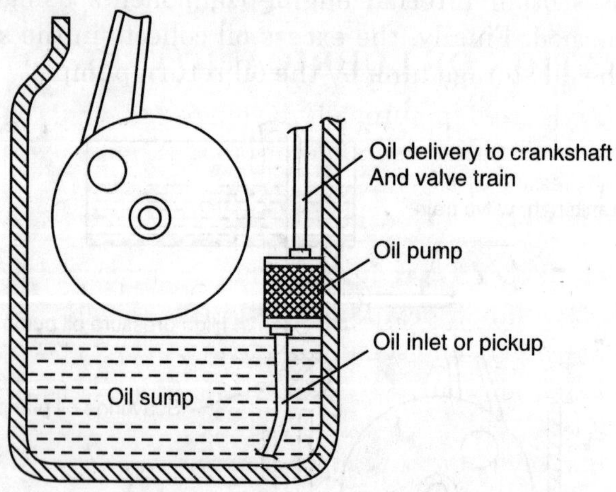

Fig.2.56: *Wet sump lubrication*

Wet-sump lubrication systems differ from dry-sump lubrication systems in two major ways. First, in wet-sump lubrication systems, oil is stored in the engine crankcase, not in a separate oil tank. Second, only one oil pump is used in wet-sump four-stroke engines. Similarly to dry-sump lubrication systems, oil is pressure-fed to all areas in need of lubrication such as cam bearing areas, pistons and cylinders and main crankshaft bearing areas. The oil that is pressure-fed to these areas is thrown off the rotating components and drained back to the sump, where the oil pump picks it up once again and recirculates it to those high-friction areas that need lubrication.

Advantages

1. Has fewer components
2. No external oil lines carrying the major supply of engine oil.
3. Less prone to oil leaks
4. Lower center of gravity, because the engine oil is contained in the bottom of the engine. This helps to improve the handling of the vehicles.
5. No second (scavenge pump) required.
6. Can be operated in much cooler temperatures.

Disadvantages

1. Oil supply limited by sump
2. Cooling is difficult
3. Higher operating temperatures
4. Not adaptable to inverted flying

CLASSIFICATION OF LUBRICATION OIL

The lubricating oils used in automotive vehicles are usually petroleum base and a variety of additives. Some are made from a synthetic base.

REQUIREMENTS

The following properties are required of engine lubricating oil:

- It must have an appropriate viscosity. If it is too low, the oil film will be broken easily and seizure of parts will occur. If it is too high, it will create additional resistance in parts movement and cause hard engine starting and loss of power.
- The viscosity level must remain relatively stable regardless of changes in temperature.
- Engine oil must be suitable for use with metals
- It must not corrode or rust parts
- It must not foam

LIQUID LUBRICANTS

Lubricating oils form a highly adhesive, pressure resistant film of lubricant between sliding metal surfaces. If this function is performed correctly, friction occurs only within the actual lubricant. Since this is a liquid, the level of friction is extremely low.

Classification by viscosity

To simplify the overall range of engine oil viscosities, the American Society of Automotive Engineers (SAE) has drawn up a table of SAE viscosities classes. SAE indexes are usually determined according to the temperature at which a particular oil may be used. But engine oil must be selected carefully not only according to the expected ambient temperatures but also to the expected operating conditions of the vehicle.

Viscosity of an oil is expressed by a number called the viscosity index. The lower the index, the thinner the oil. The higher the index, the thicker the oil.

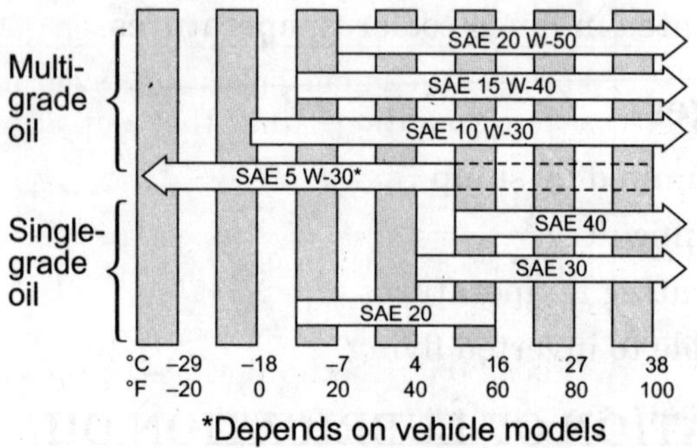

Fig.2.57: *Dry sump lubrication*

Oil whose viscosity indexes are indicated as a range (10 W-30, 15 W-40, etc) are called "multi-grade" oils. Since their viscosity is not affected much by a change in temperature, they are usually used throught the year. These oils are easier starting in cold weather and stability at high temperatures.

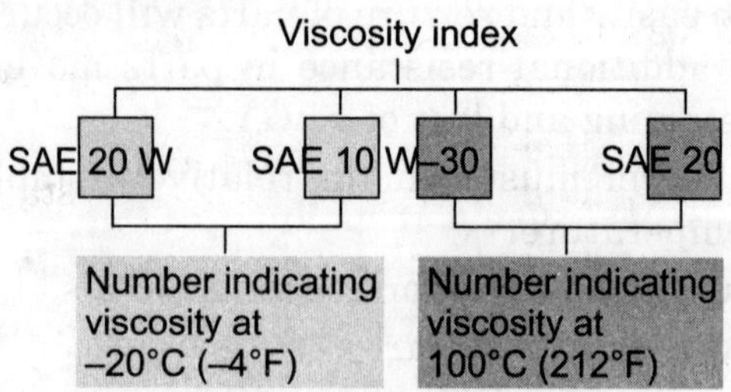

Fig.2.58: *Viscosity index*

Viscosity indexes followed by the letter E (10 W, etc) indicate oil's viscosity measured at -20°C (-4°F). Using a low viscosity oil helps to start the engine more easily in cold climates. Viscosity numbers after a "W" indicate the viscosity at 100°C (212°F). For example, "SAE 10

W-30" means that the engine oil meets standards for SAE10 oil at -20°C (-4°F) and standards for SAE30 oil at 100°C (212°F).

Classification by quality

Engine oil is classified by quality according to standards set by American Petroleum Institute (API) and by test methods also set by API.

API classification	Service and oil description
SD	For use in engines operating under extreme temperatures or other severe conditions; contains detergent-dispersant, wear-resisting agent, anti-oxidant, etc.
SE	For use in engines operating undermore extreme or severe conditions than those in SD; contains large amounts of detergent-dispersant, wear-resisting agent, anti-oxidant, etc.
SF	Superior to SE, the SF type oil offers outstanding wear resistance and durability.
SG	This oil offers better performance thin SF to prevent oxidation, wear and rust.
SH	The top-most categry, this oil is superior to SG.

SEMI-SOLID LUBRICANTS

Greases are semi-solid (plastic) lubricants. Greases have a high viscosity or waxy consistency (density). The principal automobile grease categories are calcium soap greases (for general lubrication), sodium soap greases (for anti-friction bearings) and lithium soap greases (multipurpose greases).

SOLID LUBRICANTS

Solid lubricants are extremely resistant to temperature change and possess a high load capacity when exposed to high pressures. The most frequently used solid lubricants are molybdenum sulphide and graphite.

Frame

The frame is that part of the vehicle to which the body, engine, and other main components are mounted. It also maintains the correct relationship between the other related parts in order to secure their normal function and freedom from strain and wear caused by operation in misaligned condition.

FUNCTIONS

1. To support the engine, body, power train, suspension system, steering system, brake system, etc.
2. To maintain the correct spacing and relationship between the other units.
3. To withstand all the dynamic and static loads without undue twist and deflection.

LOADS ON THE FRAME

1. Impact loads due to sudden collision of the vehicle.
2. Inertia loads due to applying brakes of the vehicle.
3. Vertical loads due to a crossing the bump or a hollow, by the vehicle.
4. Loads due to taking turn and applying the brakes at a time.

MATERIALS FOR FRAMES

Frames should be constructed by the materials of good quality so that they bear all the uneven twists, shocks and other loads in different conditions. Generally steels used for pressed frames are mild steels, carbon steel and nickel alloy steel sheets. An aluminium alloy is also used for frame material. The composition of alloy steel used for frames is carbon 0.25 to 0.35, Nickel up to 3, Manganese 0.3 to 0.7, Silicon up to 0.3, Sulphur up to 0.5 and Phosphorus up to 0.05 percents.

FRAME FOR PASSENGER VEHICLE

Frame consists of two longitudinal members connected by cross-members by bolting, riveting or welding. The longitudinal-members are brought closer together at the front to give an adequate steering lock, and at the rear are upswept to allow the axle movement. The depths of section of the longitudinal-members are usually higher at the centre where the greatest bending loads occur.

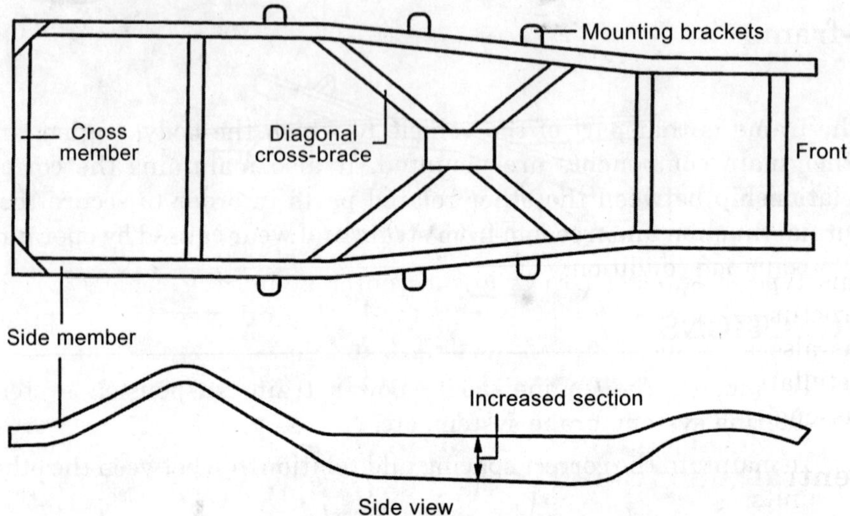

Fig.3.1: *Passenger car chassis frame*

On the sides of the frame some brackets and dumb irons are provided to support the body, springs, bumpers and bearings for spring shackles.

The engine is mounted at the one end of the frame, usually on the front end, and supported on the frame at the three places with the help of rubber blocks to avoid the effects of road-shocks on the engine and engine vibrations on the body.

TYPES OF PASSENGER CAR FRAME

Platform-type frame

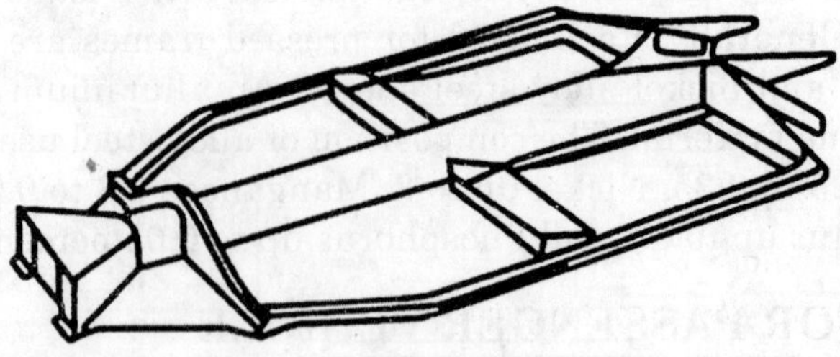

Fig.3.2: *Platform-type frame*

This is a separate chassis frame which consists of side and cross members welded to floor panels together with a center section shaped to act as a transmission tunnel. This type of frame possesses high torsional rigidity, relatively low weight and has low center of gravity. It is used for passenger cars.

X-frame

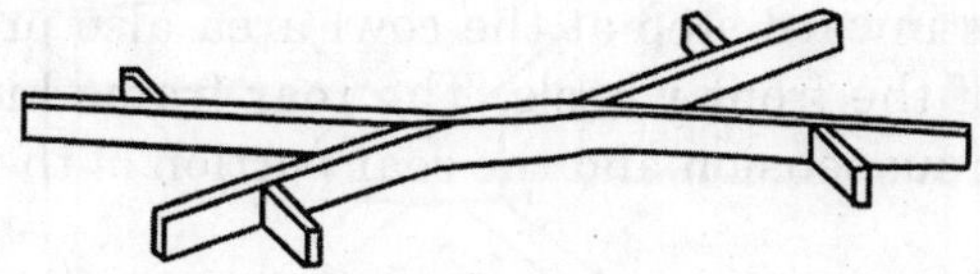

Fig.3.3: *X-frame*

This type of frame narrows in the center giving the vehicle a rigid structure that is designed to withstand a high degree of twisting. It has also high resistance to bending loads, light in weight, and easy installation of attachments. This type of frame is mainly used in passenger cars.

Central tube framc

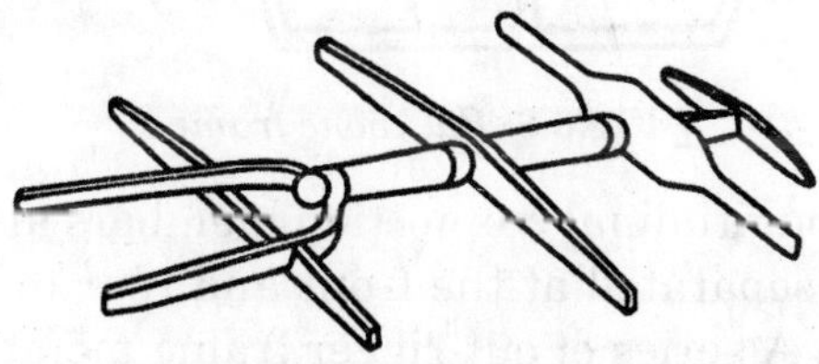

Fig.3.4: *Central-tube frame*

It has a central member which may have a tubular or rectangular cross section. Several cross members and forked ends are connected to the central section. These are essential for attachments. This type of frame is highly resistant to bending and twisting loads, light in weight and easy installation of attachments. The propeller shaft can be placed inside the tubular central member. It is applicable for passenger cars.

Perimeter frame

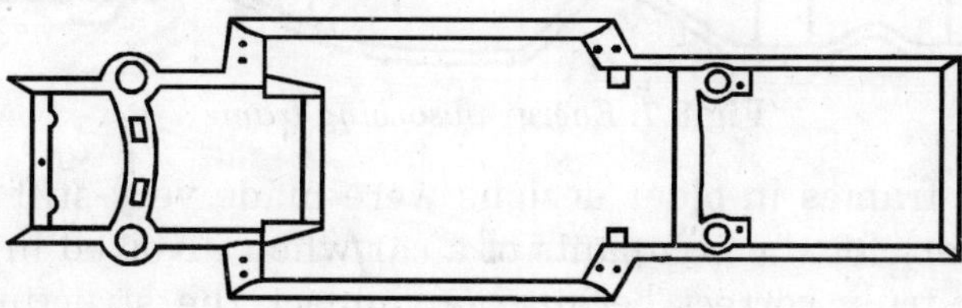

Fig.3.5: *Perimeter frame*

The perimeter frame is used in some rear wheel drive cars. A perimeter frame consists of welded or riveted frame members around the entire perimeter of the body. This means that the frame members provide support underneath the sides as well as for the suspension and suspension components.

The frame rails are stepped inward at the cowl area to provide increased strength, which supports the engine mounts and front suspension. This inward step at the cowl area also provides room for the movement of the front wheels. The rear frame kick up side rails support the rear suspension and the rear portion of the body weight.

Backbone frame

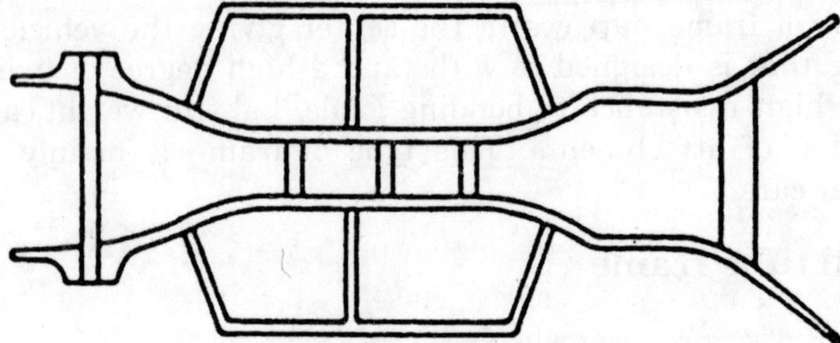

Fig.3.6: *Backbone frame*

In this type, two longitudinal box section members are welded together at the centre and separated at the front and rear to accommodate the main components. A series of out-rigger frame members are welded to the spine to support the floor of the body.

Energy-absorbing frame

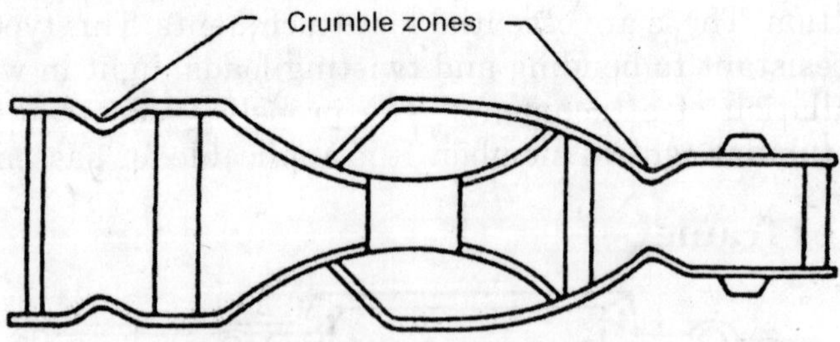

Fig.3.7: *Energy-absorbing frame*

The chassis frames in older designs were made very stiff in order to improve safety for the occupants of a car when involved in a collision. This is not truly correct because on impact the structure provides the occupants an extremely high deceleration and the force acting on the human body as it dashes against a hard surface is likely to cause serious injury or death.

This problem is overcome by constructing the front-and-rear-end of the frame in a manner so that it crumples in a concertina manner during collision and absorb the main shock of the impact. Actually the body panel in the vicinity of these crumple zones are generally damaged beyond repair, but this is a small price to pay to minimize the injury to the occupants.

DESIGN OF FRAME FOR COMMERCIAL VEHICLE

The arrangement of the commercial vehicle chassis frame is similar to that of the SUVs/passenger cars but it must be capable of carrying heavier and more varied loads. The various parts of the chassis are usually manufactured from relatively heavy gauge steel plate and are of much deeper section than those in the SUVs/passenger cars chassis. Diagonal cross-bracing is rarely used. Instead a series of channel or tubular section cross members are welded or riveted to the side members, as illustrated in Fig.3.8. In most designs the side members are not swept up at the front and rear to allow for the vertical movement of the axles. The front end of the chassis is not swept in to allow for the movement of the steered wheels. This is because the engine is usually mounted higher in the chassis than a SUV/passenger car engine and therefore there is more space available for the front road wheels to turn on the larger vehicle and the ground clearance is greater.

Ladder frame

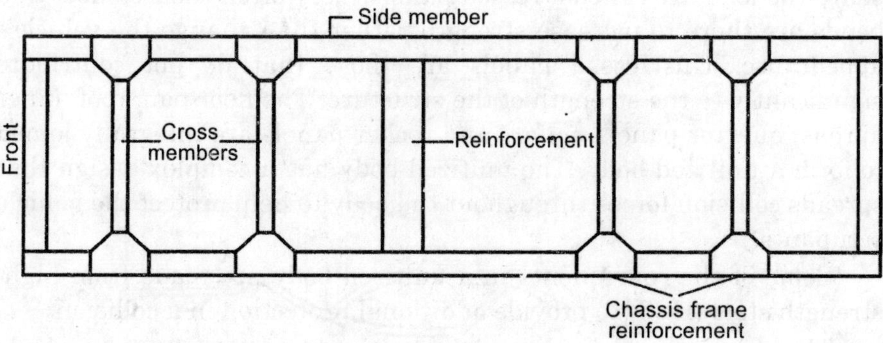

Fig.3.8: *Ladder frame*

This type of frame is highly resistant to bending and torsional (twisting) loads, and provides the high load bearing capacity in many cases on trucks, commercial vehicles, etc.

Box-type frame

It has side and cross members with box shaped or rectangular cross section. This type is mostly applied on buses but generally it is applicable for all types of vehicles.

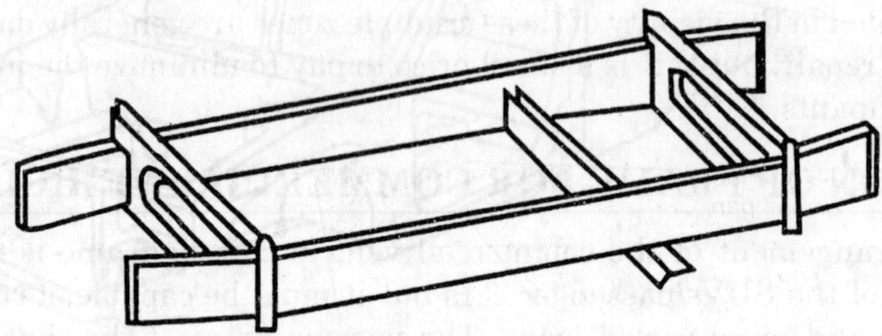

Fig.3.9: *Box-type frame*

UNITIZED BODY DESIGN

Today, almost all passenger cars, most vanes and even some buses are of integral construction, also known as frameless or monocoque construction.

In this type of construction heavy side members used in conventional construction are eliminated and the floor is strengthened by cross members and the body, all welded.

Most of the modern cars have this integral construction. In this body design, the frame and body are combined as one unit. All the members of a unitized body are load carrying members. Panels or members that share the load are called stressed panels. Very often their creases and bends are there to increase strength rather than change the vehicle's appearance. Unstressed panels are those that do not contribute significantly to the strength of the structure. The floor pan, roof, inner aprons, quarter panels, pillars and rocker panels are integrally joined to form a unitized body. The unitized body has a complex design that spreads collision forces throughout the body to help protect the vehicle occupants.

Some of the components in a unitized body are made from high-strength steel (HSS) to provide additional protection in a collision.

The strength and rigidity of the unitized body is achieved by body design rather than by having a heavy steel frame to support the body. In this body design, body sheet metal is fabricated into a box design which provides good resistance to bending and torsion.

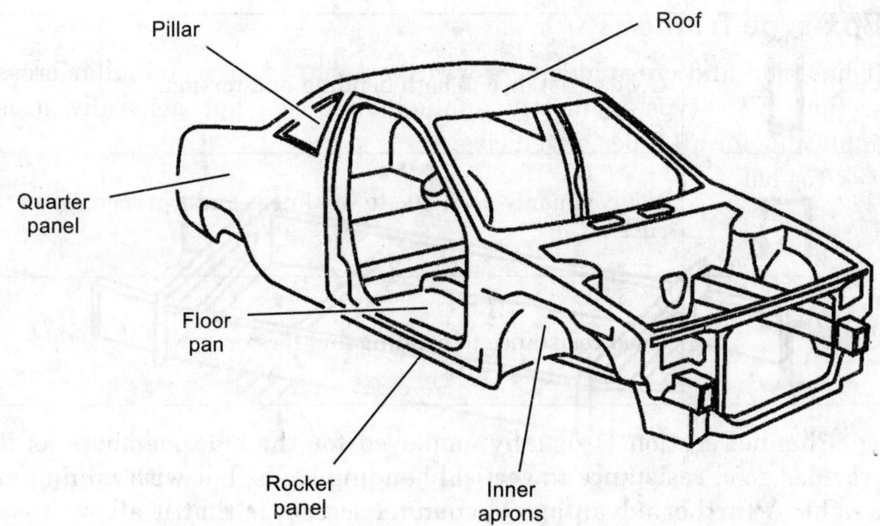

Fig.3.10: *Parts of an unitized body of a car*

The advantages of this construction are low weight and the ability to style the outer body panels as desired.

Design and production costs are very high, which means that integral construction is used only for mass-produced vehicles. A further disadvantage of this type of construction is that the resulting box-like shape tends to amplify road and engine noise for the occupants. Special sound-deadening materials are used to minimize the transmission of noise.

FRAME SECTIONS

In order to provide a good resistance to bending and torsional effect, the frame sections are made of proper forms.

Table 3.1: Frame sections and behaviours

Frame Section	Behaviour
Flat	Offers little resistance to bending and twisting.
Channel	Excellent resistance to bending. Resistance increases as depth of section is increased.
Tubular	Excellent resistance to torsion. Resistance increases as diameter is increased.

Section	Properties
Box	Good resistance to both bending and torsion.
Top-hat 	Not commonly used due to weakness and a propensity to rust.
I 	Good resistance to bending and torsion.

Channel section is usually employed for the side members as it provides good resistance to vertical bending loads, but with minimum weight. A further advantage of channel section is that it allows easy attachment of the cross members and mounting of other components. However, channel section provides only a limited resistance to torsion forces (i.e. poor in torsion) and is therefore unsuitable for the construction of cross members. As the cross members must be capable of resisting torsional effects, they may be formed from tubular or circular section, although for convenience a box section is sometimes employed.

Various other material sections may be used for the construction of cross members. These include the top-hat section and the I-section which is manufactured from two channel sections placed back-to-back. Provided the correct material section is used, the chassis frame can be made rigid enough to withstand the loads placed upon it.

SUB-FRAMES

Generally the different units are directly bolted to the main frame of the vehicle. But sometimes these units or components are mounted on the separate frame which is known as sub-frame. The sub-frame carrying the components is further supported on the main frame at some points. By the use of the sub-frame the units are isolated from the effects of flexing and twisting of the main frame. The main advantages of the sub-frame are as given under:

1. Mechanically some units are difficult to mount directly on the main frame even harm the main frame but providing sub-frame simplifies production on the assembly line as well over-hauling and repairing of the units.
2. The vibrations of the engine or the units and shocks of the springs transmitted to the main frame are observed and damped by the sub-frame.

3. Any alteration in the unit systems can easily be performed by making changes in the sub-frame instead of mainframe.

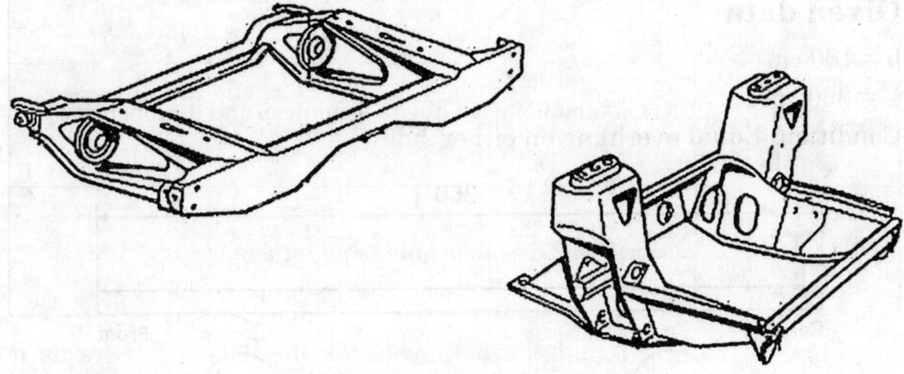

Fig.3.11: *Sub frame*

DEFECTS IN FRAMES

The main defect occurs in the frames is the alignment fault. The main causes of this fault are under :

1. Side frame members bent due to accident.
2. Bent dumb irons.
3. Engine mountings damaged.
4. Cross-bracing members buckled.
5. Rivets loosened or sheared.
6. Welds broken.
7. Defective front axle, steering system, suspensions, wheels, hubs and tyres.

PROBLEMS

Problem 3.1: Calculate the maximum bending moment and maximum section modulus for the side member of a vehicle frame having channel section and the following particulars are:

Wheel base = 180 cm
Overall length = 360 cm
Equal overhang on either side

2700 N acting at CG of load 45 cm in front of front axle; 1800 N acting at CG of load 45 cm behind front axle; 1800 N acting of CG of load 45 cm in front of rear axle; 675 N acting at CG of load 45 cm behind the rear axle.

In addition, there is a uniformly distributed load of 1.75 N/mm run over entire length of the chassis.

Given data

b = 180 cm

L = 360 cm

Condition: Equal overhang on either side

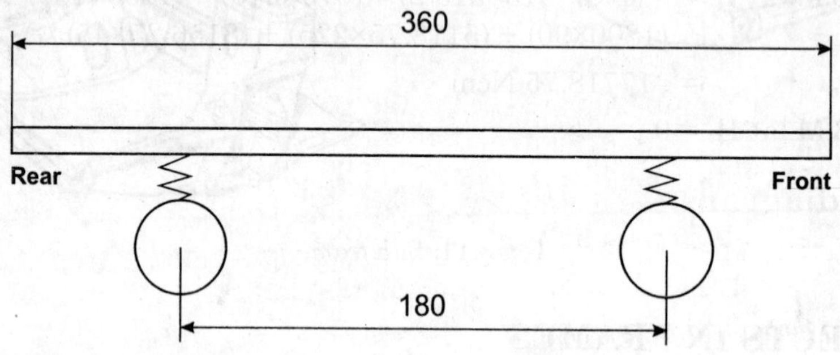

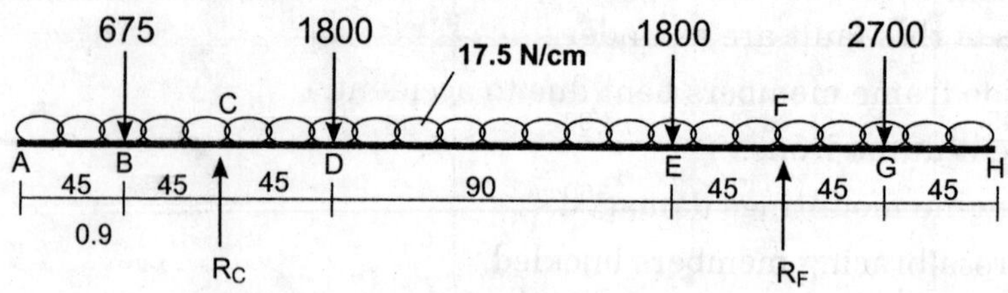

$R_C + R_F = 675 + (2\times1800) + 2700 + (17.5\times360)$

$R_C + R_F = 13275$ N ....(1)

Take moment about C,

$(R_F\times180) + (675\times45) + (17.5\times90\times90/2) = (1800\times45) + (1800\times135) + (2700\times225) + (17.5\times270\times270/2)$

$R_F = 8156.25$ N ....(2)

From (1), $R_C + 8156.25 = 13275$

$R_C = 5118.75$ N

B.M. at A = 0

B.M. at B = – (17.5×45×45/2) = –17718.75 Ncm

B.M. at C = – (17.5×90×90/2) – (675×45)

= –101250 Ncm

B.M. at D = – (17.5×135×135/2) – (675×90) + (5118.75×45)

= + 10125 Ncm

B.M. at E = – (17.5×225×225/2) – (675×180) – (1800×90) + (5118.75×135)

= –35437.5 Ncm

B.M. at F = – (17.5×270×270/2) – (675×225) – (1800×135) – (1800×45) + (5118.75×180)

= –192375 Ncm

B.M. at G = – (17.5×315×315/2) – (675×270) – (1800×180) – (1800×90) + (5118.75×225) + (8156.25×45)

= –17718.75 Ncm

B.M. at H = 0

B.M. diagram

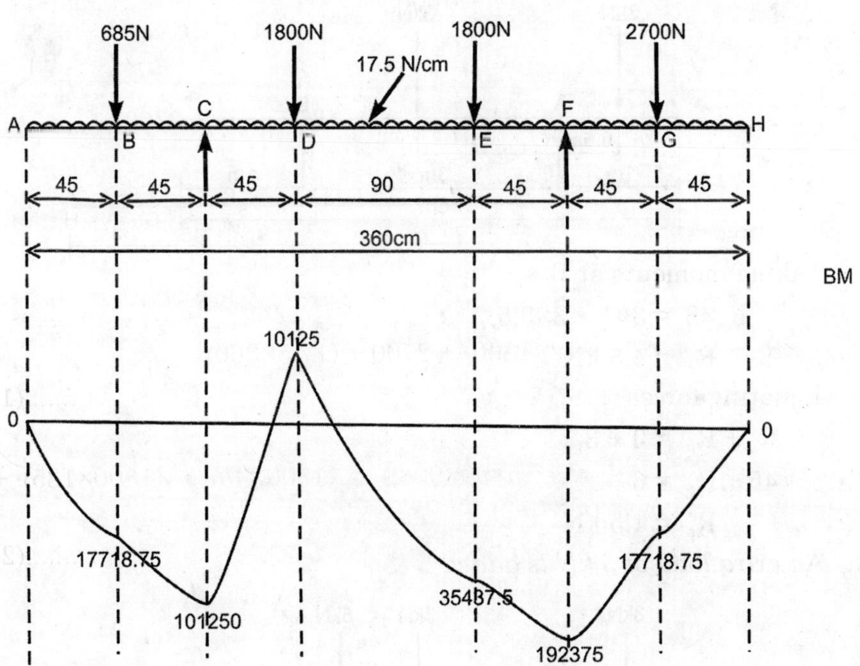

(a) Maximum B.M.

From the diagram, Maximum B.M. = 192375 Ncm

We know that, maximum B.M. due to dynamic load is twice that due to static load.

M_D = 2 × 192375 = 384750 Ncm

(b) Maximum section modulus

Assume allowable stress in the material (f_b) as 6000 N/cm².

$$\text{Section modulus } Z = \frac{M_D}{f_b} = \frac{384750}{6000} = 64.12 \text{ cm}^3$$

Problem 3.2: A vertical chassis can be considered as a overhang supported beam of 5 m long and is supported at A and B each being 1 m from its front and rear ends. Loads of 3 kN are carried 0.5 m in front of A and 2 m behind A respectively.

i. Find the magnitude of the reactions A and B
ii. If an extra load of 5 kN is to be added to the beam in such a position that the reactions at A and B are to be equal, what will these reactions can be, and at what distance from A must the 5 kN load be situated.

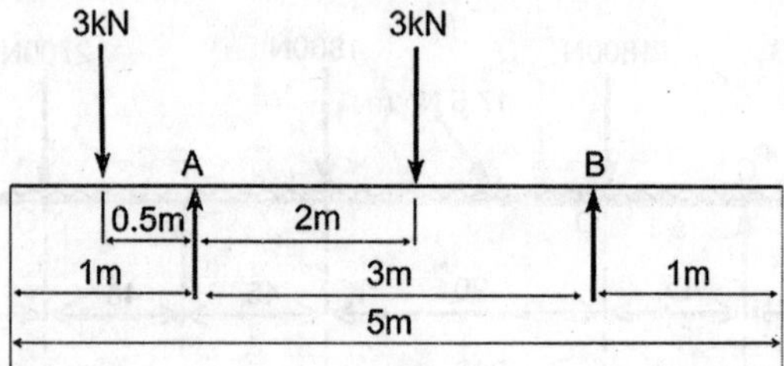

i. Taking moments at B

$$R_A \times 3 = 3\times1 + 3\times3.5$$

$$R_A = 4.5 \text{ kN}$$

Equating forces

$$R_A + R_B = 3 + 3$$

$$45 + R_B = 6$$

$$R_B = 1.5 \text{ kN}$$

ii. *An extra load of 5 kN is added*

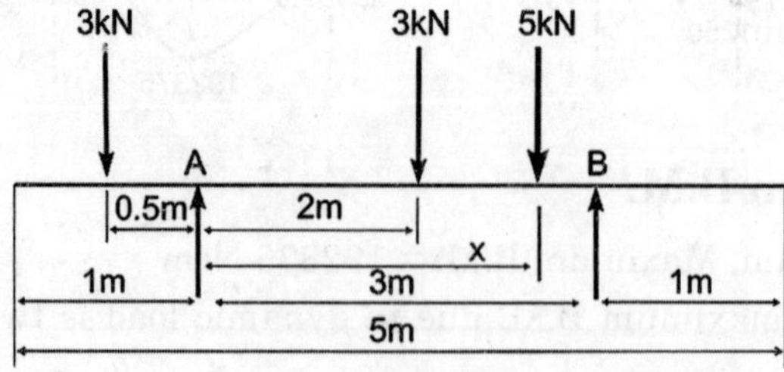

Equating forces

$$R_A + R_B = 3 + 3 + 5$$

Given condition is reactions at A and B are to be equal.

$$R_A = R_B$$

$$\therefore \quad R_A + R_A = 3 + 3 + 5$$

$$2R_A = 11$$

$$R_A = 5.5 \text{ kN}$$

$$\therefore \quad R_B = 5.5 \text{ kN}$$

Taking moment along A

$$R_B \times 3 = 5 \times x + 3 \times 2 - 3 \times 0.5$$

$$5.5 \times 3 = 5 \times x + 3 \times 2 - 3 \times 0.5$$

$$x = 2.4 \text{ m}$$

Problem 3.3: A bus chassis 5.4 m long consists of two longitudinal members and many cross members. The distance between front and rear axle is 3.6 m. Front axle is at 0.9 m from the front of the chassis frame. The details of the load and distances are given below.

Load	Magnitude, kN	Distance from the front of the frame, m
Engine weight (at front)	2	0.6
Engine weight (at rear)	2.5	1.8
Gearbox weight	0.5	2.4
Vehicle body weight	W	3

If the reaction by front axle is 8.5 kN, determine the magnitude of vehicle body weight W, and the reaction at rear axle.

Solution

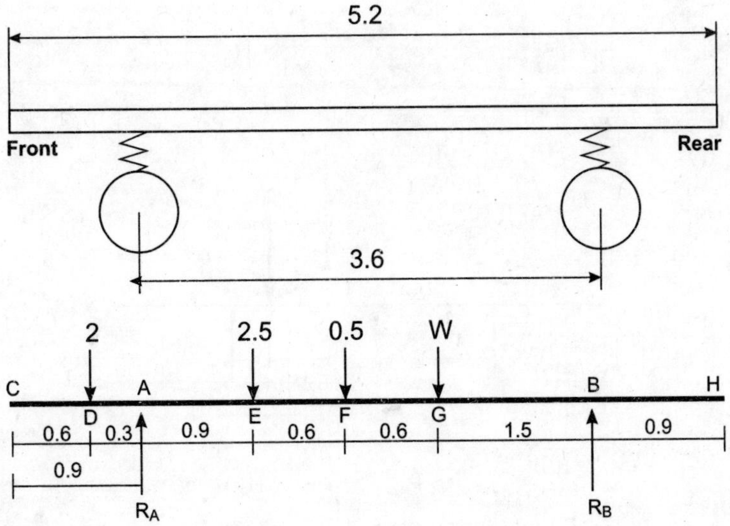

Taking moment about B

$$8.5\times3.6 = 2\times3.9 + 2.5\times2.7 + 0.5\times2.1 + W\times1.5$$

$$W = 10 \text{ kN}$$

For equilibrium, upward forces must be equal to total downward forces

$$R_A + R_B = 2 + 2.5 + 0.5 + W$$

$$8.5 + R_B = 2 + 2.5 + 0.5 + 10$$

$$R_B = 6.5 \text{ kN}$$

Clutch

The clutch is a device used to connect and disconnect engine torque to the gearbox. The clutch is usually disengaged,

- when starting the engine
- when shifting the gears
- when stopping the vehicle, and
- when idling the engine

REQUIREMENTS OF CLUTCH

1. Gradual engagement

The clutch should be engaged gradually so that occurrence of sudden jerks avoided.

2. Compact size

The size of the clutch should be so small that should occupy minimum space.

3. Torque transmission

The clutch should be so designed that the maximum torque of the engine can be transmitted through it.

4. Heat dissipation

Due to friction occurring in this system a large amount of heat is generated. Therefore the design of the clutch should ensure adequate dissipation of the heat.

5. Dynamic balancing

For the high speed clutches, the dynamic balancing is necessary.

6. Clutch free pedal play

To take the full efficiency and to avoid the wearing of clutch there should be slight free pedal play.

7. Ease of operation

Engaging and disengaging of the clutch should not be tiresome to the operator.

8. Vibration damping

Clutch system should be so designed that noise produced in the transmission can be eliminated easily.

TYPES OF CLUTCHES

There are many types of clutches including friction clutch, magnetic clutch, fluid clutch, and centrifugal clutch, but those most commonly used are the friction type and the fluid type clutches.

The main types of clutches are as given under:

(a) Friction clutches

(b) Fluid flywheel

When one rotating disc is brought in contact with the other forcibly, then due to friction between two, the power is transmitted from former to the later, this system is known as 'friction clutches". When power is transmitted from one vane to other due to centrifugal force by means of fluid this mechanism of transfer of energy from one rotor to the other is called "fluid flywheel".

FRICTION CLUTCHES

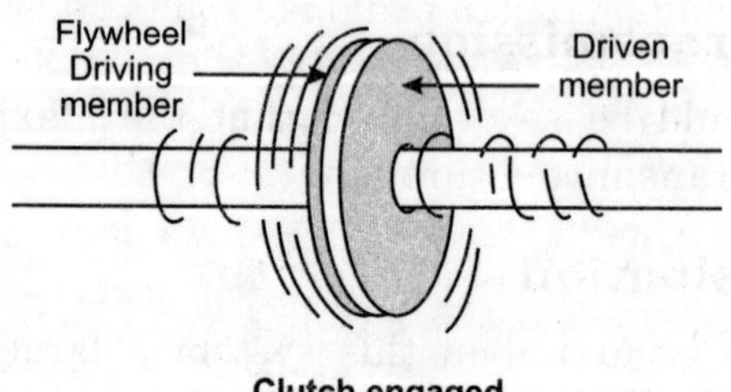

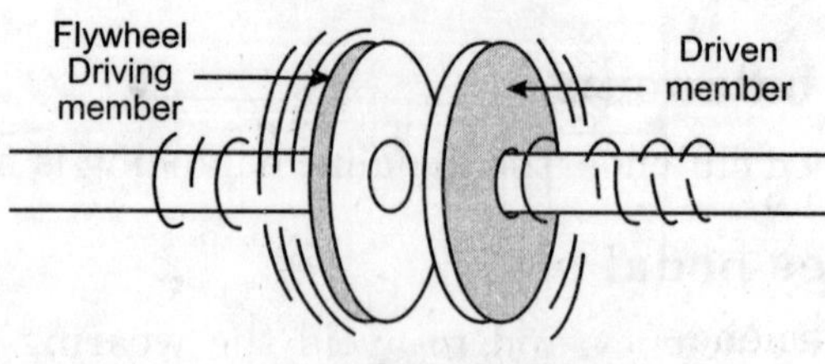

Fig.4.1: *Principle of a friction clutch*

There are different types of friction clutches, either running dry or in oil. The clutches discussed here are dry friction types, which are common practice in today's vehicles.

- Cone clutch
- Single plate clutch
- Multiplate clutch
- Semi-centrifugal clutch
- Centrifugal clutch
- Electromagnetic clutch

CONE CLUTCH

The cone clutch was used in the early days of vehicles because of its mechanical simplicity.

CONSTRUCTION

As shown in the line diagram of cone clutch in Fig.4.2, both the driving and driven components are made of cone shaped. While fully engaged the male cone remains fully inside the female cone so that the friction surface may remain in complete contact. This engagement of the cone remains all the time by means of pressure of springs.

The cone clutch is so called because each of its two friction faces is a frustrum of a cone, one male and the other female. The clutch uses these two conical surfaces to transmit torque by friction. The female cone is keyed to the crankshaft, whereas the male cone is free to axially slide on the output shaft. A clutch pedal is used to engage and disengage the clutch, by bringing the two conical surfaces in contact. Leather, cork, asbestos or brass is generally used as friction material.

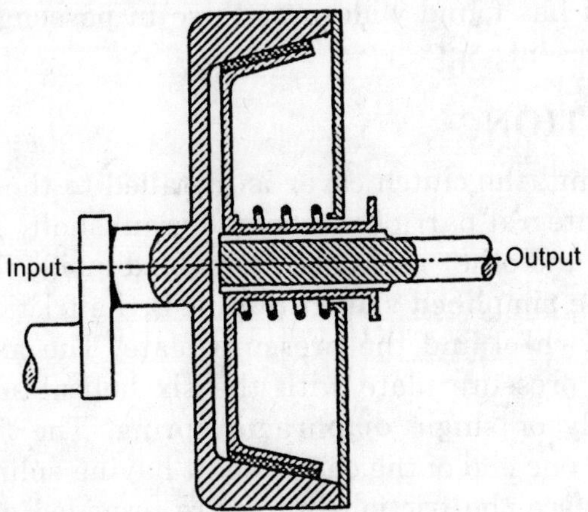

Fig.4.2: *Cone clutch*

OPERATION

Torque is transmitted due to the frictional forces set up at the contact. With a correct choice of cone angle the clutch is free from slip as long as the friction linings remain in good condition. The axial force required to engage the clutch is provided by a helical compression spring. A short rearward displacement of the male member against helical compression spring is needed to disengage the clutch completely. This is done by the driver through clutch operating mechanism.

Advantages

The advantages of the cone clutch system is that in this case the normal force acting on the friction surface is larger than the axial force in case of simple single plate clutch.

Disadvantages

This type of clutch is less practicable due to some disadvantages given under.

1. When the angles of the cones remain smaller than 20° then male cone tends to bind or jam in the female cone. In this way it becomes very difficult to disengage the clutch.
2. Continuous rubbings and wearing of the cone surfaces result in the need of axial movement of the male cone which is difficult to be allowed after a limit.

SINGLE-PLATE CLUTCH

This clutch is relatively simple in construction, allows engaging and disengaging of power, and comparatively easy to repair. Because of these merits, it has found widespread use in passenger cars, trucks, and buses at present.

CONSTRUCTION

In the clutch unit, the clutch cover is installed to the flywheel which rotates as an integral part of the engine crankshaft. The clutch disc, pressure plate, and other parts are assembled inside the clutch cover. As shown in the simplified sketch (Fig. 3.4) the friction plate is held between the flywheel and the pressure plate. The axial pressure is applied on the pressure plate with the six helical springs mounted circumferentially or single diaphragm spring. The friction plate is mounted on the one end of the clutch shaft having splines. To increase the friction surface the friction lining are provided on both sides of the clutch plate. For disengaging the clutch a pedal is provided to pull the pressure plate against the springs to make loose the friction plate from both sides. While there is no effort on the clutch pedal, the clutch remains always in engaged position.

OPERATION

Coil spring type

When the clutch pedal is depressed, the pressure plate moves out against the pressure springs. The gap between pressure plate and flywheel increases. Now, the clutch plate is free to rotate. Thus power is not transmitted. This position is called *clutch release or declutching.*

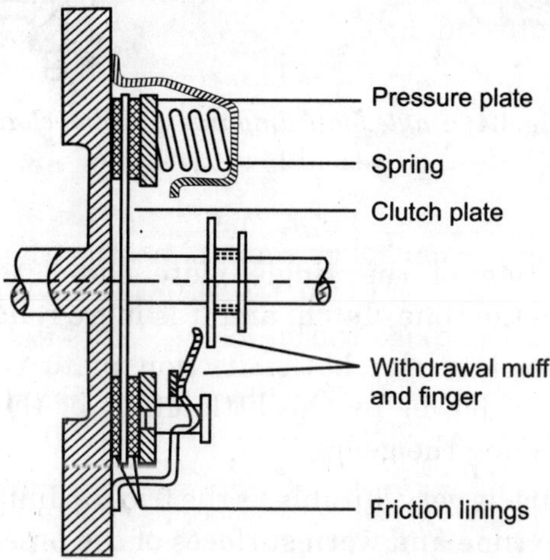

Fig.4.3: *Single plate coil spring clutch*

Diaphragm spring type

In the diaphragm type, a diaphragm spring is used instead of a number of coil springs. When the clutch is engaged the spring is almost flat and its outer rim exerts pressure against the clutch plate. To disengage the clutch, the clutch pedal is depressed. This causes the linkage to move the throwout bearing against the inner portion of the diaphragm spring. This in turn causes the rim of the diaphragm spring away from the flywheel. The gap between pressure plate and flywheel increases. Now, the clutch plate is free to rotate. Thus power is not transmitted.

As soon as the pedal is released, the pressure spring forces the pressure plate towards the flywheel. The clutch plate is gripped between the flywheel and pressure plate. This is called clutch engaging. At this position, the power is transmitted to the gearbox.

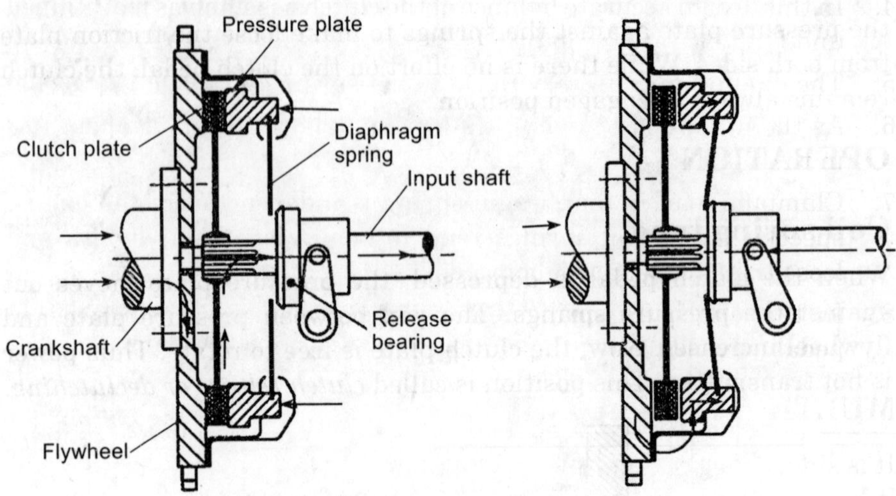

Fig.4.4: *Single plate diaphragm spring clutch*

Advantages

1. The construction of the single plate clutch is very simple in comparison to the cone clutch, and it is more reliable.
2. In the single plate clutch the friction plate is made free from transmission of power by the little effo.'t of the driver and it is easier in changing the gears.
3. The plate clutch is very durable as the friction lining can be changed easily time to time and worn surfaces of the pressure plate as well fly wheel can be smoothen easily when required.

Disadvantages

Space required to accommodate the clutch is more as compared to multi plate clutch.

MERITS OF THE DIAPHRAGM SPRING OVER MULTI-COIL SPRING

1. The diaphragm spring is compact permitting the use of a shallow clutch bell-housing to enclose the clutch unit.
2. Due to fewer moving parts squeaks, rattles, and wear are eliminated in diaphragm spring.
3. This system does not require initial adjustment of the pressure-plate unit unlike the multi-coil spring clutch units, where a small clearance is necessary between the release-lever plate and the thrust bearing.

4. In this design accurate balance of the clutch assembly is maintained under all operating conditions.
5. The diaphragm acts as both clamping spring and release-finger.
6. As the driven-plate wears, the spring axial load self-compensates in this clutch.
7. Clamping load in diaphragm-spring is independent of the engine speed whereas coil springs tend to bend along their length and loose their thrust at high speeds.
8. In this clutch disengagement pressures reduce with increase in pedal movement.

MULTIPLATE CLUTCH

It is similar to the single plate clutch but only the difference is this that it has more number of clutch plates. Due to increasing the number of frictional plates the frictional surface is also increased which increases the capacity of the clutch to transmit torque. To transmit the same torque the diameters of plates are reduced in comparison to the single plate clutch.

It is most commonly used in motorcycles, scooters and sports cars.

CONSTRUCTION

The construction of the muliplate clutch is clear from the simplified diagram Fig. Almost construction and system is the same as that of single plate clutch except the number of frictional plates. It contains two sets of frictional plates one set of plates slides on splines of the pressure plate and the other set slides in grooves on the flywheel. The both types of frictional plates are arranged in alternate manner such so that one lies on the other. The construction of the frictional plates is made clear from the Fig.4.5 (a) and (b).

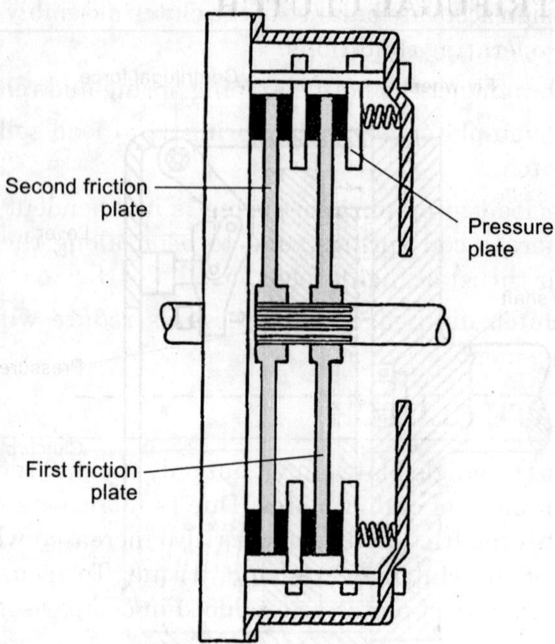

Fig.4.5: *Multiplate clutch*

OPERATION

A separator plate is used in these assemblies to separate the clutch plates and it is fitted to the flywheel by a driving pin. When the clutch is engaged, the first clutch plate is held between the clutch pressure plate and separator plate and the second clutch plate is held between the separator plate and the flywheel. When disengaged, the separator plate, flywheel and pressure plate assembly rotate as a unit while the clutch plates which are not in contact with the pressure plates do not transmit power to the transmission.

Advantages

1. High torque can be transmitted
2. Compact
3. Better increase in acceleration
4. Better clutch control

Disadvantages

1. Heavy
2. Expensive
3. Too many components

SEMI-CENTRIFUGAL CLUTCH

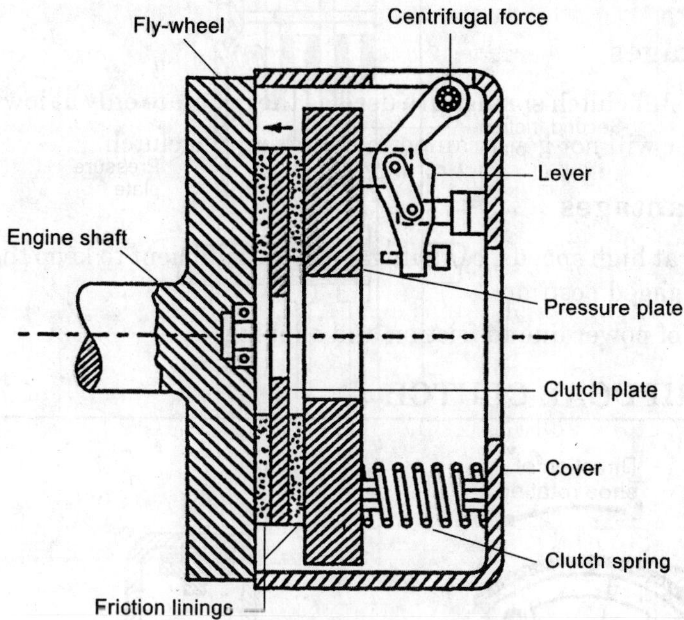

Fig.4.6: *Semi-centrifugal clutch*

In this system the springs are designed to transmit the power at normal speeds and for high speeds the assistance of the centrifugal force is taken to reduce driver fatigue. Such types of clutches are called "semi-centrifugal clutches". This type of clutch is used in some models from Vauxhall motors, British automotive manufacturing and distribution company.

CONSTRUCTION

One of the hinged and weighted lever is shown in the Fig.4.6. The lever is having fulcrum at B and is hinged to pressure plate at C. The upper end of the lever is weighted at A. At the tail of the lever an adjusting screw is provided, by means of which the centrifugal force on the pressure plate can be adjusted. To reduce the friction the needle roller bearing is used at the hinge C.

OPERATION

At the normal speeds the pressure of the springs is sufficient to transmit the small amount of the required torque and the assistance of the centrifugal force is not needed. But when the high speed occur the weight A moves about B as fulcrum due to the centrifugal force of the weight. In this way pressure plate moves towards left side by the pressure of the lever transmitted through hinge C.

Advantages

1. Less stiff clutch springs are used as they operate only at low speeds.
2. Driver will not get strained in operating the clutch.

Disadvantages

1. Only at high speeds, centrifugal force is sufficient to keep the clutch in engaged position
2. Loss of power due to friction and slipping.

CENTRIFUGAL CLUTCH

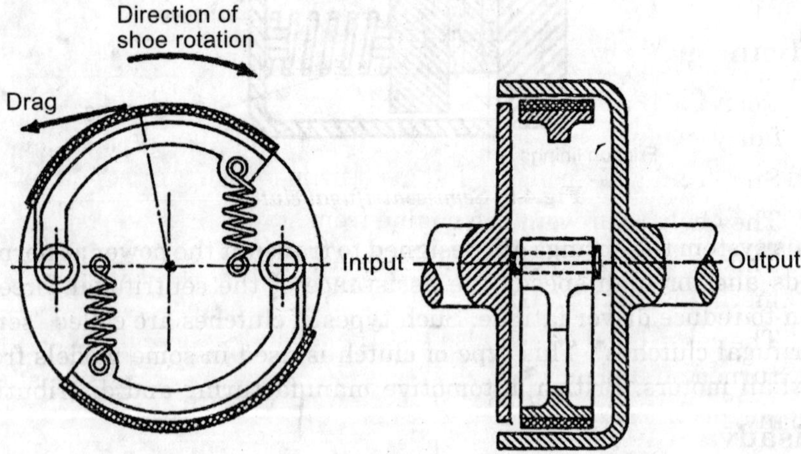

Fig.4.7: *Centrifugal clutch*

Centrifugal clutches are often used in mopeds, lawn movers, go-karts, chainsaws, and mini bikes. Centrifugal clutch is the common name given to a clutch which uses centrifugal forces to engage. This type of mechanical clutch is **fully automatic**, and works depending on the speed of the engine. If the engine is turning over very slowly, then the centrifugal clutch is disengaged, and the vehicle comes to a stop. If the engine speeds up then the automatic clutch will engage, and the vehicle will move forward.

Centrifugal clutches are often used to:

- keep the engine from stalling when the vehicle is stopped abruptly; and,
- disengage loads automatically when starting and idling.

CONSTRUCTION

The two shoes are carried on a back plate which is driven by the engine. Springs hold the shoes out of contact with the drum at very low speeds, but once the spring tension is overcome by centrifugal force the shoes rub on the drum, friction builds up and as the speed increases the drum is dragged round by the shoes. Ultimately rotation takes place without slipping.

OPERATION

Engagement takes place smoothly as the throttle is opened gradually. The springs expand outward when engine rpm goes above a certain point, usually about 800 to 900 rpm. As the speed of the driven member drops with reduction of road speed, disengagement takes place when the engine reaches the critical speed at which centrifugal force is no longer sufficient to overcome the spring tension.

Advantages

1. Compared to other types of clutches, a clutch centrifugal type is fairly cheap.
2. Simple and inexpensive and need little maintenance.
3. They help to prevent the engine from stalling.
4. This type of mechanical clutch is automatic, so no kind of control mechanism is necessary.
5. The centrifugal clutch also has the advantage of not needing to be turned off and on.

Disadvantages

1. Automatically disengaging at low engine speeds
2. Only at high speeds, centrifugal force is sufficient to keep the clutch in engaged position
3. Loss of power due to friction and slipping.
4. This type of clutch not appropriate for the high amount of torque, the shoes will slip at the heavy loaded condition.
5. They engage at full or near-full power, shoes get heated very quickly may cause overheating.

Note: Centrifugal force usually refers to a fictional force that seems to give an outward "center-fleeing" acceleration to an object traveling in a circle. Centrifugal forces come into effect when an object is continuously going around in a circle.

ELECTROMAGNETIC CLUTCH

An electromagnetic clutch is a clutch that is engaged and disengaged by an electromagnetic actuator. It is used in few models of Renault cars.

CONSTRUCTION

The flywheel incorporating the winding which is supplied electric current from battery or dynamo. Clutch plate is lined with friction surfaces and is free to slide over the splines on the clutch shaft.

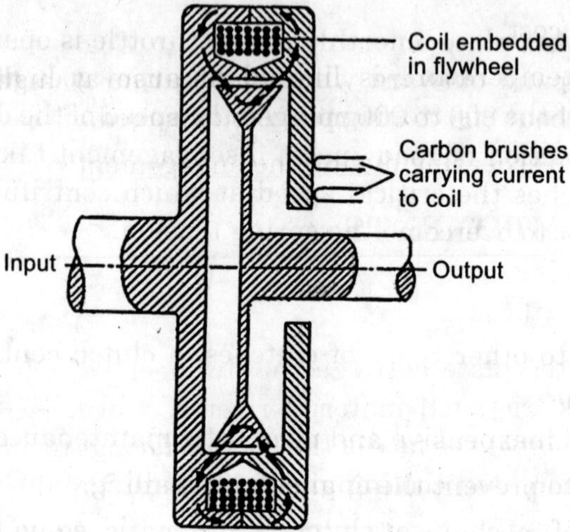

Fig.4.8: *Electromagnetic clutch*

OPERATION

When electric current is supplied to the winding it is energized and attracts the pressure plate thus pressing the clutch plate between two surfaces and thereby engaging the clutch. When the supply to the winding is cut off, the clutch is disengaged by releasing the pressure plate by means of the helical springs.

The gear lever of the system is provided with clutch release switch, which is operated as soon as the driver holds the gear lever to change the gear position, cutting off current to winding and then the clutch is disengaged.

When the vehicle is stalling, the engine speed is lower and dynamo output is also low in this way force between winding and pressure plate is very small which can be balanced by the helical springs to disengage the clutch.

The forces of the electromagnet can be regulated by means of an electrical resistance provided with acceleration system and controlled by the accelerator pedal. When the speed is increased the accelerator pedal is pressed and the resistance is gradually cut off and thus in this way force of electromagnet is increased and clutch transmission becomes more rigid.

Advantages

The complicated linkages are not required to control clutch.

Disadvantage

1. High initial cost.
2. Operating temperature is limited because at high-temperature insulation of the electromagnet gets damaged.
3. The risk of overheating during the engagement.

CLUTCH COMPONENTS

CLUTCH PLATE

Clutch plate is the plate between the flywheel and pressure plate. A clutch plate has frictional material riveted or bonded to both sides. Grooves are cut across the face of the friction facings to allow for smooth clutch action, increased cooling and a place for the dust to go as the clutch plate wears.

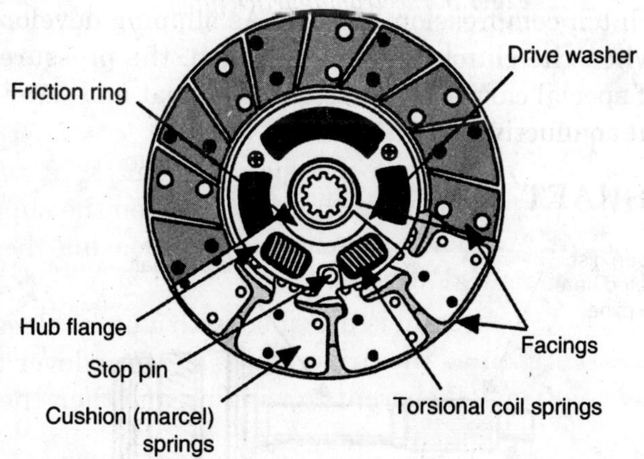

Fig.4.9: *Clutch plate*

The clutch facings are made of materials such as paper based, ceramic materials and asbestos that are highly heat and wear resistant. Some of the facings used lately contain metal pieces to increase the strength and improve heat conductivity.

The disc center (clutch hub) is provided with splines. These mesh with the splines on the transmission input shaft so as to allow the disc to slide back and forth along the shaft.

The clutch disc is also provided with torsion springs, the purpose being to cushion the clutch engagement so as to eliminate the danger of the clutch disc warping or breaking.

PRESSURE PLATE

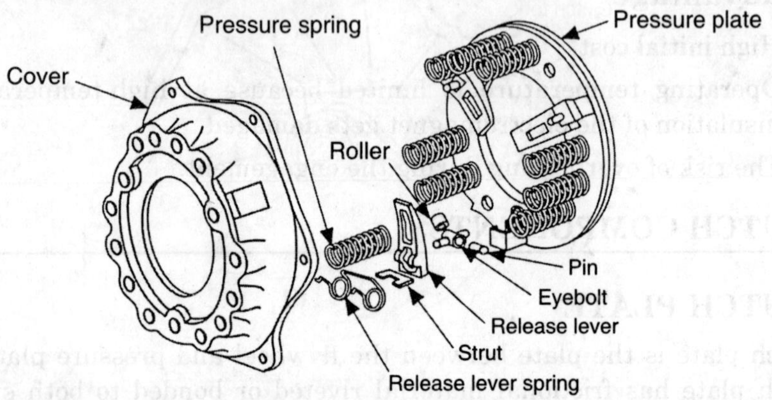

Fig.4.10: *Pressure plate*

The pressure plate is a cast iron ring with its front face finished accurately flat. It serves to press the clutch disc against the flywheel by the force of clutch compression springs. As slipping develops with the clutch disc when the clutch is being engaged, the pressure plate has been made of special cast iron to assure good heat and wear resistance and high heat conductivity.

CLUTCH SHAFT

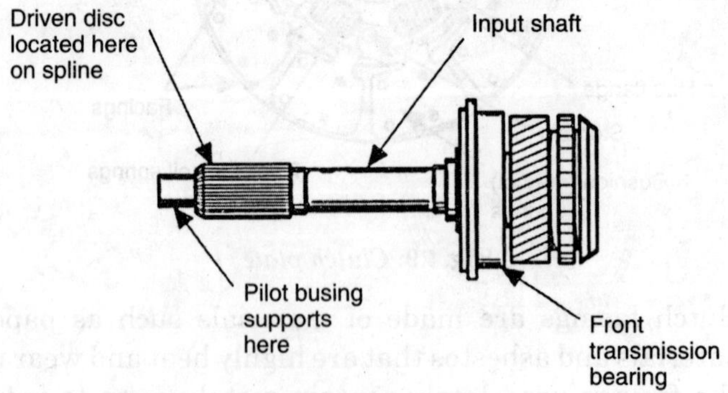

Fig.4.11: *Clutch shaft*

The clutch shaft projects from the front of the transmission. The clutch plate will be splined to the clutch shaft. When the clutch is engaged, the clutch plate drives the transmission's input shaft through these splines.

RELEASE BEARING

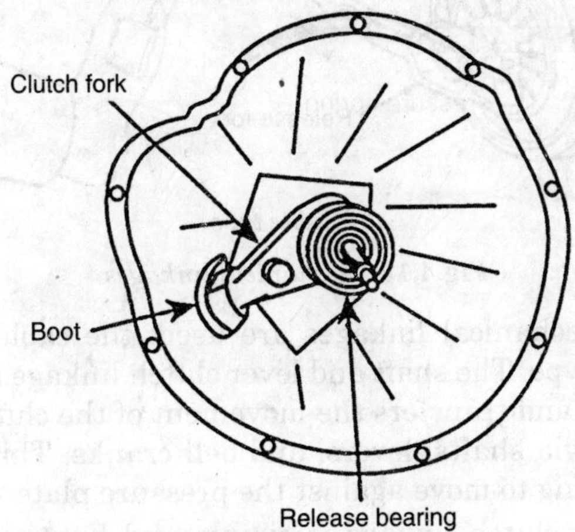

Fig.4.12: *Release bearing*

The release bearing slides back and forth over the guide inside the clutch housing when moved by the clutch release fork. When shifted ahead, it pushes the revolving pressure levers to disengage the clutch.

OPERATING MECHANISM

Clutches are normally operated by either mechanical or hydraulic linkages.

Mechanical linkage

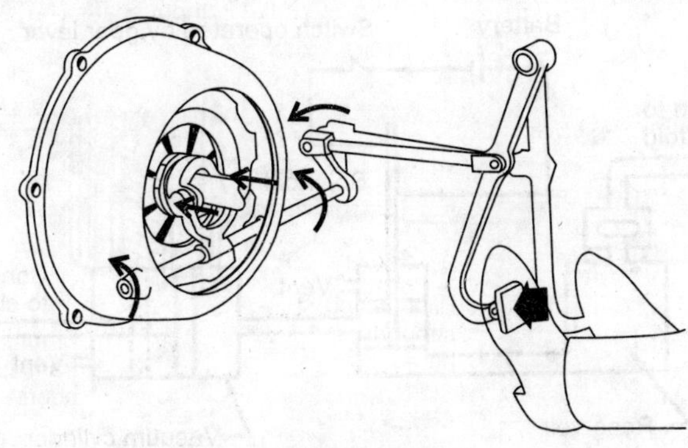

(a) Lever type

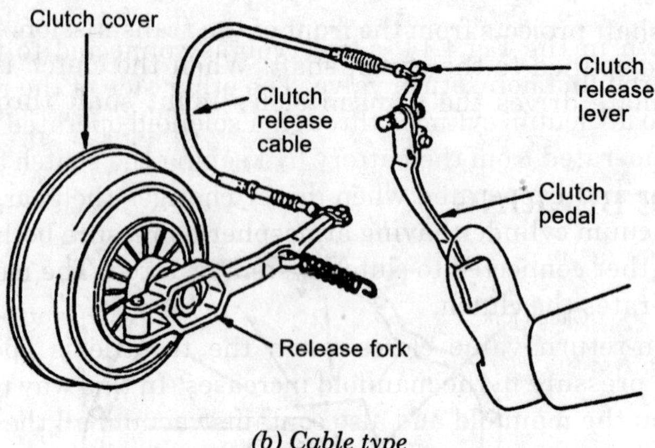

(b) Cable type

Fig.4.13: *Mechanical linkages*

Two types of mechanical linkages are used: the cable type, and the shaft and lever type. The shaft and lever clutch linkage has many parts and pivot points and transfers the movement of the clutch pedal to the release bearing via shafts, levers, and bell cranks. This action causes the release bearing to move against the pressure plate.

A cable-type clutch linkage is simple and lightweight. However, cables will gradually stretch and can break due to electrolysis. When the clutch pedal is depressed, the cable pulls on the clutch fork, which causes the release bearing to move against the pressure plate.

Vacuum operated

In this unit the systems of the clutch plate as well pressure plate are the same as of the ordinary clutch. Pressure is exerted on the clutch plate by means of the helical springs. The only difference is this that the clutch disengagement and engagement rod is moved by the vacuum system.

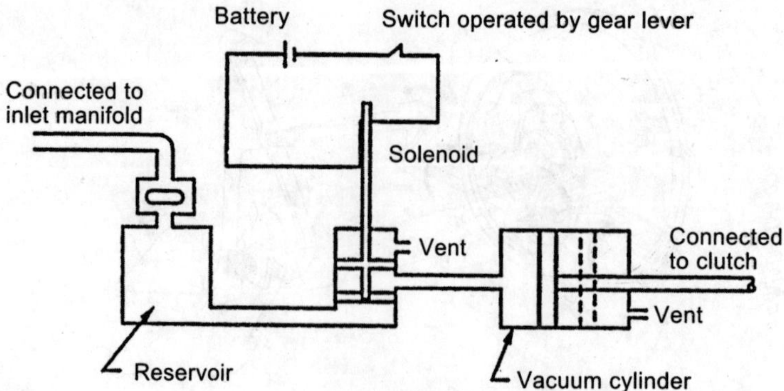

Fig.4.14: *Vacuum clutch*

As shown in the Fig.4.14, a reservoir is connected to the engine manifold through a non-return valve. The other side of the reservoir is connected to a vacuum cylinder through a solenoid operated valve. The solenoid is operated from the battery by keeping the switch at the gear change lever and is operated when driver changes the gears. There is piston in vacuum cylinder having atmospheric pressure both sides and which is further connected to clutch system by a rod. The movement of this rod operates the clutch.

The non-return value closes when the throttle is opened wide because the pressure in the manifold increases. In this way reservoir is isolated from the manifold and also contains vacuum all the time.

When the piston of the vacuum cylinder and valve of the solenoid are in the dotted position, there is atmospheric pressure on both the sides of the piston that means there is no movement in the piston. When the driver is to change gears with gear lever this action closes the switch and the solenoid is energized which pulls the valve up and the vacuum cylinder is connected to the vacuum in the reservoir. Then due to pull of the vacuum and the push of the atmospheric pressure on the piston causes it to move. This movement is transmitted to clutch through clutch rod to disengage. After changing the gear the driver releases the lever and then switch is also opened thus the clutch is again engaged.

Hydraulic linkage

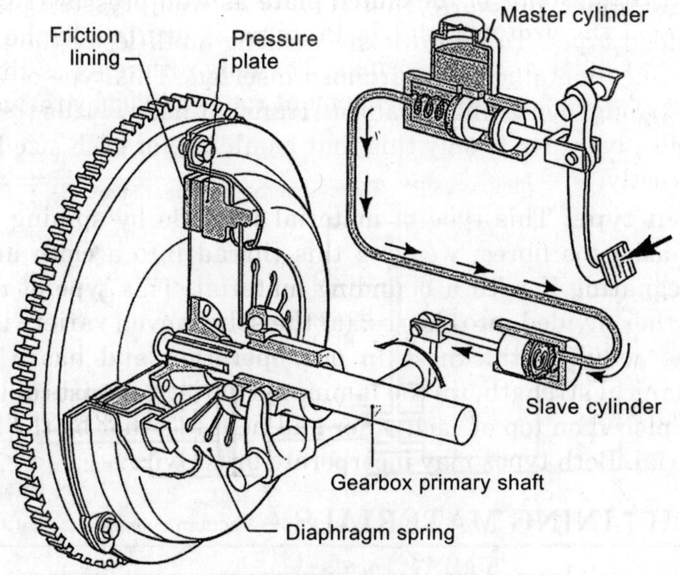

Fig.4.15: *Hydraulic linkage*

In this type of clutch, the movement of the clutch pedal is converted by the master cylinder into a hydraulic pressure which is then transmitted to the clutch release fork via the slave cylinder. It consists of a master cylinder, a slave cylinder, and an oil reservoir. When the clutch pedal is depressed, the fluid under pressure from the master cylinder reaches the slave cylinder. The slave cylinder is mounted on the clutch itself. The fluid actuates slave cylinder push rod which further operates the clutch release fork to disengage the clutch.

The hydraulic system is designed to multiply the driver's efforts so that a light force applied to the foot pedal produces a much greater force on the clutch fork. A small piston in the master cylinder travels a relatively long distance with only a low input force. This moves larger piston in the slave cylinder a relatively small distance but provides a greater force to disengage the clutch.

FRICTION MATERIALS

The various types of friction materials are as given under:

(a) **Millboard type:** This material is in the form of sheets made of asbestos and treated with certain impregnants. From the standard sheet we can cut facing discs according to different size requirements. This is the cheapest material and gives satisfactory results.

(b) **Moulded type:** When asbestos fibres are mixed with a suitable binder and heated to a definite temperature and then moulded in dies under pressure, this type of friction material is called "moulded type". To improve the wearing qualities of the material sometimes metallic wires are also inserted. This type of lining is more strong, dense and capable to transmit heavy loads through it. The disadvantage is only this that moulding for each size lining is very costly.

(c) **Woven type:** This type of material is made by spining threads from asbestos fibres, weaving this thread into a cloth and then impregnating it with a bounding material. This type of material is further divided into classes; (a) the solid woven variety is woven to the required thickness in one operation and has a greater mechanical strength, (b) the laminated variety consists of layers of cloth placed on top of each other and held together by the bonding material. Both types may incorporate brass wire.

CLUTCH LINING MATERIALS

Common materials used for clutch lining are as given under:

(a) **Asbestos:** It can be used at high temperature also and linings of this material have coefficient of friction of about 0.2.

(b) **Reybestos and Ferodo:** These materials are most suitable and generally used for clutch friction linings and have a coefficient of friction of about 0.35.

(c) **Leather:** The co-efficient of friction between dry leather and iron is 0.27.

(d) **Cork:** The co-efficient of friction between cork and steel or iron is 0.32.

(e) **Fabric:** This material has co-efficient of friction about 0.4 but it cannot be used at high temperature.

BONDING MATERIALS

The bonding materials roughly classified as given under:

(a) **Vegetable gums:** The coefficient of friction of this material is from 0.35 to 0.45 up to 250°C temperature. It has good wearing properties.

(b) **Rubber:** The co-efficient of friction of this material can be made up to 0.6 by the incorporation of filling materials. The bond of this material can be produced either flexible or rigid but both types are more affected by water than the other types of bond.

(c) **Asphaltic basis with additions of natural gums and oils:** The coefficient of friction of this material is from 0.3 to 0.4 and temperature up to 250°C. The higher co-efficient can be obtained by using excessive bonding materials but due to rise of temperature the excess of bond has been driven off thereby such coefficient are not maintained much long.

(d) **Synthetic resins, alcohol soluble:** The co-efficient of friction of this material is from 0.4 to 0.5 at any temperature up to 230°C but at higher temperatures the coefficient fall of up to 0.1. This type of bond has a quality that it is not affected by lubricating oil and can withstand heavy pressures.

(e) **Synthetic resins oil soluble:** The coefficient of friction of this material is from 0.35 to 0.38 which is maintained at quite high temperatures. Asphaltic basis and vegetable gums are often used in conjunction with this type of bond. This material also has excellent wearing quality.

FLUID CLUTCHES

Fluid clutch automatically disconnects the drive when the engine speed is low and gradually connects it as the vehicle is moved from a stationary position. It is either a fluid coupling or torque converter. The fluid clutch utilizes oil to transmit the power. As it is able to transmit the power continuously and smoothly, it has started to find greater popularity lately. This clutch is used in the automatic transmission together with the planetary gear transmission.

FLUID COUPLING

The fluid coupling serves to transmit the engine torque to the gearbox, same as the friction clutch, but does it automatically and more smoothly.

Construction

Fluid coupling is comprised essentially of the pump impeller (drive side) and turbine runner (driven side). These parts have the appearance of a hollow doughnut sliced vertically into two equal halves.

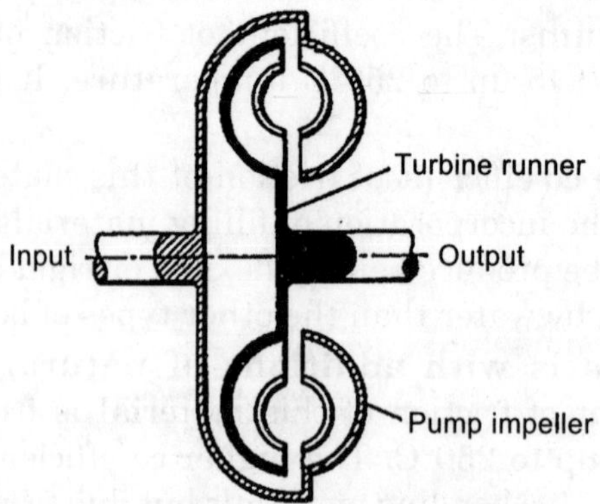

Fig.4.16: *Fluid coupling*

Both of these parts are provided with vanes radiating out from the center. These parts are contained in a housing filled with oil and installed facing each other, with very little clearance provided between them. The pump impeller is installed on the engine crankshaft and the turbine runner coupled to transmission input shaft.

Operation

When the engine rotates, the pump impeller provides motion to the oil in between the blades to turn the turbine runner and transmit the engine power to the transmission. When the engine speed drops, the turbine runner will turn faster than the pump impeller so that the turbine runner will now be working as a pump impeller and exert a braking action to the engine.

At idling, the pump impeller is rotating slowly and since the oil pressure is low, the turbine runner receiving the oil does not turn.

The oil merely circulates between the opposing blades and there is no transmission of engine power to the propeller shaft. This is because the oil pressure generated by the pump impeller is not large enough to overcome the resistance of the drive wheels.

Advantages of the fluid coupling

The principle merits of this type of fluid flywheel or coupling are as given under:

1. Gives a smoother power take-up.
2. The fluid of the coupling reduces or absorbs transmission shocks as where braking or coasting down an incline.
3. The fluid in the coupling acts a kind of buffer between the transmission and the engine to reduce shocks.
4. The rotating unit acts as a flywheel on the crankshaft to minimise the torque variations.
5. No wear on the moving parts.
6. No adjustment to be made.
7. Simple design and no maintenance necessary except oil level.
8. The vehicle can be stopped in gear and moved off also by using accelerator only.
9. No skill required for operating.

CLUTCH TROUBLESHOOTING

Slipping, grabbing, dragging, and abnormal noises are clutch problems. It is important to know the symptoms produced by these problems and the parts that might be the cause.

Slipping

Slipping occurs when the driven disc fails to rotate at the same speed as the driving members when the clutch is fully engaged. This condition is due to worn clutch disc facings or grease or oil on the facings resulting in the disc frictional coefficient to drop so that the clutch is unable to transmit the engine power properly. When slipping develops the car is unable to attain high speeds and fuel consumption becomes higher.

Grabbing

A grabbing or chattering clutch will produce a very severe vibration or jerking motion when the vehicle is accelerated from a standstill. Even when the operator slowly releases the clutch pedal, it will seem like the clutch pedal is being pumped rapidly up and down. A loud bang or chattering may be heard, as the vehicle body vibrates. Clutch grabbing and chatter is caused by problems with components inside the clutch housing (friction disc, flywheel, or pressure plate).

Dragging

This trouble is due to the pressure plate not returning completely when the clutch pedal is depressed and causing the clutch to remain engaged. When the gears are shifted, a clashing sound is produced.

Abnormal noises

Faulty clutch parts can make various noises.

DESIGN AND PROBLEMS

DESIGN CALCULATION FOR SINGLE PLATE CLUTCH

The design of a single plate friction clutch is given below.

Dimensions of friction facing

Rated torque capacity of a clutch = Maximum engine torque

$$= \frac{\pi \mu p \eta_m (d_1^3 - d_2^3)}{600 R_f} \text{ kgfm}$$

Where

m = Co-efficient of friction = 0.3 to 0.4

p = Permissible unit pressure = $\frac{11.17}{\sqrt{d_1}}$ kgf/cm^2

d_1 = Outside diameter of friction facing, cm

d_2 = Inside diameter of friction facing, cm

= $0.6d_1$

η_m = Mechanical efficiency

= 0.8 when spring force is transmitted to pressure plate through a system of levers

= 0.85 when spring force is transmitted directly acts on pressure plate

R_f = Rate factor

= 1.25 if spring force is adjusted

$= 1.35 - 1.5$ for passenger cars

$= 1.1$ in semi-centrifugal clutch

$= 2$ for tractor clutch

Calculate d_1 and d_2

Select the nearest standard sizes.

Dimensions of pressure plate

Outside diameter of pressure plate $D_{p1} = d_1 + 0.3$

Inside diameter of pressure plate $D_{p2} = d_2 - 0.3$

Thickness of pressure plate t = 18 mm (up to d_1=18.5 cm)

= 22 mm (up to d_1=23.0 cm)

Design of pressure spring

(a) Coil spring

Spring force exerted by each spring is given by

$$P = \frac{\pi}{4} \times \frac{(D_{p1}^2 - D_{p2}^2)p}{N_o}$$

$$= \frac{\pi}{4} \times \frac{(D_{p1}^2 - D_{p2}^2)}{N_c} \times \frac{11.77}{\sqrt{d_1}} \text{ kgf}$$

Where

N_c = Number of springs used

= 6 (up to d_1= 20 cm)

= 9 (up to d_1= 23 cm, 26 cm, etc)

Stress equation $P = \dfrac{\pi \tau_s d^3}{8D}$

Where

τ_s = Permissible shear stress for the spring material

= 5250 kgf/cm²

D = Mean coil diameter = 1.8 to 2.4 cm

d = Spring wire diameter

Find d and select the nearest standard wire gauge of the wire.

Deflection of spring $y = \dfrac{8PD^3 n}{Gd^4}$

Where

G = Modulus of elasticity = 8.82×10^5 kgf/cm^2

n = Number of active coils = 6 (assume if not given)

Free length of the spring $L = (n+0.5)d+y_{max}+0.15y_{max}$

(b) Diaphragm spring

Axial load applied over the periphery of the diaphragm spring is given by

$$P = \frac{\pi}{4}(D_{p1}^{\ 2} - D_{p2}^{\ 2})p$$

$$= \frac{\pi}{4}(D_{p1}^{\ 2} - D_{p2}^{\ 2})\frac{11.77}{\sqrt{d_1}} \text{ kgf}$$

Almen-Lassle formula

$$P = \frac{E\delta}{(1-\sigma)^2 Ma^2}\left[(h-\frac{\delta}{2})(h-\delta)t + t^3\right] \text{ kgf}$$

Where

E = Young's modulus = 2.1×10^6 kgf/cm^2

δ = Axial deflection

= h at the time of complete disengagement of clutch

σ = Poisson's ratio = 0.3

M = Radial moment of tangential forces in the section about a neutral point.

$$= \frac{6}{\pi \log_e \alpha}\left(\frac{\alpha-1}{\alpha}\right)^2$$

$$\alpha = \frac{\text{Outside diameter of diaphragm spring}}{\text{Inside diameter of diaphragm spring}} = \frac{D_{do}}{D_{di}} = 1.2 \text{ to } 1.3$$

D_{do} = 0.8 to $0.9D_{p1}$

$$\frac{h}{t} = 1.75$$

h = 1.75t

h = Free height of cone

t = Thickness of disc

$$a = \frac{D_{do}}{2}$$

Calculate 't' and correct to the nearest sheet gauge

Checking

Maximum load in the upper edge of spring

$$P_m = \frac{E\delta}{(1-\sigma)^2 Ma^2}\left[h - \frac{\delta'}{2}\right]C_1 \pm C_2 t \quad \text{kgf}$$

Where

$$d'C_2 = h + 0.028\sqrt{d_1}$$

$$C_1 = \frac{6}{\pi \log_e \alpha}\left[\frac{\alpha - 1}{\log \alpha} - 1\right]$$

$$C_2 = 3(a-1) - \pi \log_e \alpha$$

Permissible value of P_m = 25000 kgf/cm²

Clutch shaft diameter

Clutch shaft diameter $d_c = 10.6\sqrt[3]{T}$ mm

Where

T = Maximum engine torque, kgfm

Depth of the spline = $0.095d_c$

Width of the spline = $0.156d_c$

Number of splines = 10

Type-straight sided

Axial length of hub = d_c

Problem 4.1: A single dry plate clutch is to be designed to transmit 10 hp at 900 rpm. Find:

(a) Diameter of the shaft

(b) Mean radius and face width of the friction lining assuming the ratio of mean radius to the face width as 4.

(c) Outer and inner diameters of the pressure plate, and

(d) Dimensions of the spring assuming that number of spring=6, spring index=6, allowable shear stress for the material of the spring wire=4300 kg/cm².

Given data

$P = 10$ hp

$N = 900$ rpm

$\frac{R}{b} = 4$

$N_c = 6$

$\frac{D}{d} = 6$

$\tau_s = 64300$ kg/cm^2

Solution

(a) Clutch shaft diameter

Clutch shaft diameter $d_c = 10.6\sqrt[3]{T}$ mm

Where

T = Maximum engine torque, kgfm

$$P = \frac{2\pi NT}{4500}$$

$$10 = \frac{2\pi \times 900 \times T}{4500}$$

$T = 7.95$ kgfm

$d_c = 10.6\sqrt[3]{7.95}$ mm

$= 21.15$

@ 22 mm

Depth of the spline = $0.095d_c = 0.095\times22 = 1.98$ mm

Width of the spline = $0.156d_c = 0.156\times22 = 3.43$ mm

Number of splines = 10

Axial length of hub = $d_c = 22$ mm

(b) Mean radius and face width of the friction lining

$\frac{R}{b} = 4$

$R = 4b$

$$\frac{r_1 + r_2}{2} = 4(r_1 - r_2)$$

$r_1 + r_2 = 8(r_1 - r_2)$

$r_1 + r_2 = 8r_1 - 8r_2$

$7r_1 = 9r_2$

$r_1 = 1.28r_2$

or

$d_1 = 1.28d_2$

Rated torque capacity of a clutch = Maximum engine torque

$$= \frac{\pi \mu p \eta_m (d_1^3 - d_2^3)}{600 R_f} \text{ kgfm}$$

$$7.95 = \frac{\pi \times 0.3 \times 0.85 (d_1^3 - d_2^3)}{600 \times 1.35} \frac{11.77}{\sqrt{d_1}} \text{ kgfm}$$

$$7.95 = \frac{\pi \times 0.3 \times 0.85 ((1.28d_2)^3 - d_2^3)}{600 \times 1.35} \frac{11.77}{\sqrt{1.28d_2}}$$

$$\frac{2.09d_2^3 - d_2^3}{1.13\sqrt{d_2}} = 682.94$$

$$\frac{1.09d_2^3}{1.13\sqrt{d_2}} = 682.94$$

$$\frac{d_2^3}{\sqrt{d_2}} = 708$$

$d_2 = 13.8$ @ 14 cm

$d_1 = 1.28 \times 14 = 17.92$ @ 18 cm

$\therefore r_1 = 7$ cm and $r_2 = 9$ cm

$b = (r_1 - r_2)$

$= (9-7)$

$= 2$ cm

$R = 4b$

$= 4 \times 2$

$= 8$ cm

(c) Outer and inner diameters of the pressure plate

Outside diameter of pressure plate $D_{p1} = d_1 + 0.3$

$= 18 + 0.3$

$= 18.3$ cm

Inside diameter of pressure plate $D_{p2} = d_2 - 0.3$

$= 14–0.3$

$= 13.7$ cm

Thickness of pressure plate t = 18 mm (up to d_1=18.5 cm)

(d) Dimensions of the spring

$N = 6$

$$C = \frac{D}{d} = 6$$

$$d = \frac{D}{6} = \frac{2R}{6} = \frac{2\times8}{6} = 2.66\text{cm}$$

Spring force exerted by each spring is given by

$$P = \frac{\pi}{4}\frac{(D_{p1}^2 - D_{p2}^2)}{N_c}\frac{11.77}{\sqrt{d_1}} \quad \text{kgf}$$

$$= \frac{\pi}{4}\frac{(18.3^2 - 13.7^2)}{6}\frac{11.77}{\sqrt{18}}$$

$= 53.45$ kgf

Stress equation $P = \dfrac{\pi\tau_s d^3}{8D}$

$$\frac{D}{d} = 6$$

$D = 6d$

$\tau_s = 4300$ kg/cm²

$$53.45 = \frac{\pi\times4300\times d^3}{8(6d)}$$

$d = 0.435 \cong 0.44$ cm

$\therefore D = 6\times0.44 = 2.64$ cm

Deflection of spring $y = \dfrac{8PD^3 n}{Gd^4}$

Where

G = Modulus of elasticity = 8.82×10^5 kgf/cm²

n = Number of active coils = 6 (assume if not given)

$$= \frac{8 \times 53.45 \times 2.64^3 \times 6}{8.82 \times 10^5 \times 0.44^4}$$

= 1.42 cm

Free length of the spring

$L = (n+0.5)d+y_{max} + 0.15y_{max}$

$= (6+0.5)0.44+1.42+0.15(1.42)$

= 4.47 cm

DESIGN OF SINGLE AND MULTI CLUTCH PLATES

r_1 = Outer radius of friction surface

r_2 = Inner radius of friction surface

μ = Co-efficient of friction

n = number of pair of contact surfaces

= 2 for single-plate clutch

= n_1+n_2-1 for multi-plate clutch

n_1 = Number of driving plates

n_2 = Number of driven plates

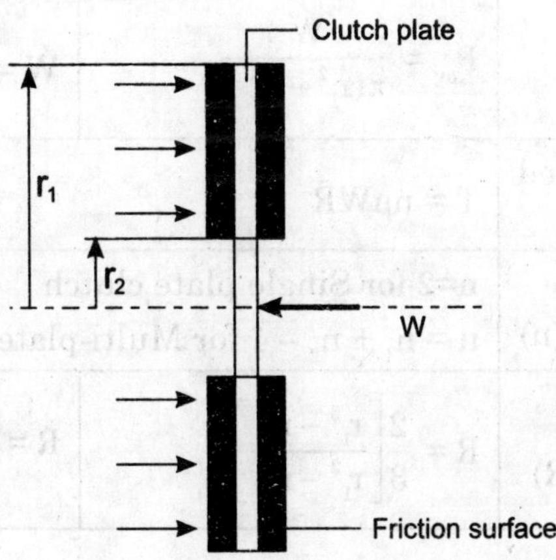

Case 1: Considering uniform pressure

Total axial spring force (axial load) required

$$Pav = \frac{W}{\pi(r_1^2 - r_2^2)}$$

Torque to be transmitted

$$T = n \mu W R$$

$$R = \frac{2}{3}\left[\frac{r_1^3 - r_2^3}{r_1^2 - r_2^2}\right]$$

Case 2: Considering uniform wear

Total axial spring force (axial load) required

$$W = 2\pi C (r_1 - r_2)$$

$$C = p_{max} \times r_2 \text{ (or) } p_{min} \times r_1$$

Torque to be transmitted

$$T = n \mu W R$$

$$R = \frac{r_1 + r_2}{2}$$

Power transmitted by the clutch

$$P = T \omega \text{ (or) } P = \frac{2\pi NT}{60} \text{ Watts}$$

	Case 1: Uniform pressure	Case 2: Uniform wear
Axial force (load) required (W)	$P_{av} = \frac{W}{\pi(r_1^2 - r_2^2)}$	$W = 2\pi C (r_1 - r_2)$
Torque transmitted (T)	$T = n\mu WR$	
No. of pair of friction surfaces (n)	n=2 for Single plate clutch $n = n_1 + n_2 - 1$ for Multi-plate clutch	
Mean radius of friction surface (R)	$R = \frac{2}{3}\left[\frac{r_1^3 - r_2^3}{r_1^2 - r_2^2}\right]$	$R = \frac{r_1 + r_2}{2}$

Problem 4.2: Determine the maximum, minimum and average pressure in a plate clutch when the axial force is 5 kN. The inside radius of the contact surface is 50 mm and the outside radius is 100 mm. Assume uniform wear.

Given

W = 5 kN = 5000 N

r_2 = 50 mm

r_1 = 100 mm

Find

p_{max}, p_{min}, p_{av}

Solution

Total axial spring force (axial load) required

$W = 2\pi C (r_1 - r_2)$

$5000 = 2\pi C (100 - 50)$

C = 15.9 N/mm

We know that,

$C = p_{max} \times r_2$ (or) $p_{min} \times r_1$

$C = p_{max} \times r_2$

$15.9 = p_{max} \times 50$

$p_{max} = 0.318 \text{ N/mm}^2$

$C = p_{min} \times r_1$

$15.9 = p_{min} \times 100$

$p_{min} = 0.159 \text{ N/mm}^2$

$$P_{av} = \frac{W}{\pi(r_1^2 - r_2^2)} = \frac{5000}{\pi(100^2 - 50^2)} = 0.212 \text{ N/mm}^2$$

Problem 4.3: An automobile power unit gives a maximum torque of 13.56 Nm. The clutch is of a single plate dry disc type, having effective clutch lining of both sides of the plate disc. The coefficient of friction is 0.3 and the maximum axial pressure is 8.29 × 10^4 Pa, and external radius of the friction surface is 1.25 times the internal radius. Calculate the dimensions of the clutch plate and the total axial pressures that must be exerted by the clutch springs.

Given data

Single plate dry disc clutch

n = 2

T = 13.56 Nm = 13.56 × 10^3 Nmm

μ = 0.3

p_{max} = 8.29×10^4 pa = 8.29×10^4 N/m² = 0.0829 N/mm²

r_1 = $1.25r_2$

Find

r_1, r_2, W

Solution

Mean radius

$$R = \frac{r_1 + r_2}{2} = \frac{1.25r_2 + r_2}{2} = \frac{2.25r_2}{2} = 1.125r_2$$

Total axial spring force (axial load) required

$$W = 2\pi C(r_1 - r_2)$$
$$= 2\pi p_{max} \times r_2 \times (1.25r_2 - r_2)$$
$$= 2\pi \times 0.0829 \times r_2 \times (0.25r_2)$$
$$= 0.13r_2^2$$

Torque to be transmitted

$$T = n\,\mu\,W\,R$$

$$13.56 \times 10^3 = 2 \times 0.3 \times 0.13r_2^2 \times 1.125r_2$$

$$r_2^3 = 1390769.23$$

$$r_2 = 53.66 \text{ mm} \cong 54 \text{ mm}$$

$$\therefore r_1 = 1.25 \times 54 = 67.5 \text{ mm, and}$$

$$W = 0.13 \times 54^2 = 379.08 \text{ N}$$

Problem 4.4: A single plate clutch effective on both sides is required to transmit 25 kW at 3000 rpm. Determine the outer and inner diameter of frictional surface if co-efficient of friction is 0.255, ratio of diameters is 1.25 and maximum pressure not to exceed 0.1 N/mm². Also determine thrust provided by springs. Assume uniform wear theory.

Given data

Single plate clutch

n = 2

P = 25 kW = 25×10^3 Watts

N = 3000 rpm

μ = 0.255

$$\frac{d_1}{d_2} = 1.25 \Rightarrow \frac{r_1}{r_2} = 1.25$$

p_{max} = $0.1 N/mm^2$

Uniform wear theory

Find

d_1, d_2, W

Solution

Mean radius

$$R = \frac{r_1 + r_2}{2} = \frac{1.25r_2 + r_2}{2} = \frac{2.25r_2}{2} = 1.125r_2$$

Total axial spring force (axial load) required

$$W = 2\pi C (r_1 - r_2)$$
$$= 2\pi p_{max} \times r_2 \times (1.25r_2 - r_2)$$
$$= 2\pi \times 0.1 \times r_2 \times (0.25r_2)$$
$$= 0.157r_2^2$$

$$P = \frac{2\pi NT}{60}$$
$$25\times10^3 = \frac{2\pi\times3000\times T}{60}$$
$$T = 79.621 \text{ Nm} = 79621 \text{ Nmm}$$

Torque to be transmitted

$$T = n \mu W R$$
$$79621 = 2 \times 0.255 \times 0.157r_2^2 \times 1.125r_2$$
$$r_2^3 = 883859.95$$
$$r_2 = 95.96 \text{ mm}$$
$$d_2 = 2 \times 95.96 = 191.92 \text{ @ } 192 \text{ mm}$$
$$d_1 = 1.25 \times 192 = 240 \text{ mm, and}$$
$$W = 0.157 \times 96^2 = 1446.71\text{N}$$

Problem 4.5: A multi-disc clutch has three discs on the driving shaft and two on the driven shaft. The inside diameter of the contact surface is 120 mm. The maximum pressure between the surfaces is limited to 0.1 N/mm². Design the clutch for transmitting 25 kW at 1575 rpm. Assume uniform wear condition and coefficient of friction as 0.3.

Given data

Multi-disc clutch

n_1 = 3

n_2 = 2

d_2 = 120 mm $\Rightarrow r_2$ = 60 mm

p_{max} = $0.1 N/mm^2$

P = 25 kW = 25×10^3 Watts

N = 1575 rpm

Uniform wear condition

μ = 0.3

Solution

Question: Design the clutch. *i.e* finding clutch plates dimensions such as n_1, n_2, r_1 and r_2 and also axial force required to engage the clutch.

Since n_1, n_2, and r_2 are given in the problem, it is required to find r_1.

Power transmitted by the clutch

$$P = \frac{2\pi NT}{60} \text{ Watts}$$

$$25 \times 10^3 = \frac{2\pi \times 1575 \times T}{60}$$

$$T = 151.57 \text{ Nm} = 151.57 \times 10^3 \text{ Nmm}$$

Total axial spring force (axial load) required

$$W = 2\pi C (r_1 - r_2) \qquad C = p_{max} \times r_2$$

$$= 2 \times \pi \times 6 (r_1 - 60) \qquad = 0.1 \times 60$$

$$= 37.69 (r_1 - 60) \qquad = 6 \text{ N/mm}$$

Torque to be transmitted

$$T = n \mu W R$$

$$R = \frac{r_1 + r_2}{2} = \frac{r_1 + 60}{2}$$

$$n = n_1 + n_2 - 1 = 3 + 2 - 1 = 4$$

$$151 \times 10^3 = 4 \times 0.3 \times 37.69 (r_1 - 60) \times \frac{r_1 + 60}{2}$$

$$6677.27 = (r_1 - 60)(r_1 + 60)$$

$$6677.27 = (r_1^2 - 60^2)$$

$$r_1 = 101.37 \cong 102 \text{ mm}$$

$$\therefore W = 37.69 (102\text{-}60)$$

$$= 1582.9 \text{ N}$$

Problem 4.6: A single plate clutch having a single driving plate with contact surfaces on each side is required to transmit 110 kW at 1250 rpm. The outer diameter of the surfaces is to be 300 mm. The co-efficient of friction is 0.4.

i. Assuming a uniform pressure of 0.17 N/mm², determine the inner diameter of the friction surfaces.

ii. Assuming the same dimensions and the same total axial-thrust, determine the maximum torque can be transmitted and the maximum intensity of pressure when uniform wear conditions have been reached.

Given

Single plate clutch

$P = 110 \text{ kW} = 110 \times 10^3$ Watts

$N = 1250$ rpm

$d_1 = 300 \text{ mm} \Rightarrow r_1 = 150$ mm

$\mu = 0.4$

Case (i) Uniform pressure

$P_{av} = 0.17$ N/mm²

$d_{2?}$

Torque to be transmitted

$T = n\,\mu\,W\,R$

Power transmitted by the clutch

$$P = \frac{2\pi NT}{60} \text{ Watts}$$

$$110 \times 10^3 = \frac{2\pi \times 1250 \times T}{60}$$

$T = 840.33 \text{ Nm} = 840.33 \times 10^3$ Nmm

For normal pressure, mean radius of the contact surface

$$R = \frac{2}{3}\left[\frac{150^3 - r_2^{\,3}}{150^2 - r_2^{\,2}}\right]$$

Total axial spring force (axial load) required

$$P_{av} = \frac{W}{\pi(r_1^{\,2} - r_2^{\,2})}$$

$$0.17 = \frac{W}{\pi(150^2 - r_2^{\,2})}$$

$W = 0.534\,(150^2 - r_2^2)$

$$840.33\times10^3 = 2\times0.4\times0.534\,(150^2 - r_2^2)\times\frac{2}{3}\left[\frac{150^3 - r_2^{\,3}}{150^2 - r_2^{\,2}}\right]$$

$= 0.285\ [150^3 - r_2^{\,3}]$

$[150^3 - r_2^{\,3}] = 2948526.31$

$r_2^{\,3} = 426473.68$

$r_2 = 75.27 \text{ mm} \cong 76 \text{ mm}$

$\therefore d_2 = 152$ mm, and

$W = 0.534\,(150^2 - 76^2) = 8930.61$ N

Case (ii) Uniform wear theory

Maximum torque that can be transmitted

$T = n\,\mu\,W\,R \qquad R = \dfrac{r_1 + r_2}{2} = \dfrac{150 + 76}{2} = 113 \text{ mm}$

$= 2 \times 0.4 \times 8930.61 \times 113$

$= 807327.14$ Nmm

$= 807.32$ Nm

Maximum intensity of pressure

$W = 2\pi C\,(r_1 - r_2)$

$W = 2\pi p_{max} \times r_2 \times (r_1 - r_2)$

$8930.61 = 2\pi \times p_{max} \times 76 \times (150\text{-}76)$

$P_{max} = 0.252 \text{ N/mm}^2$

Problem 4.7: A motor car engine develops 5.9 kW at 2100 rpm. Find the suitable size of clutch plate having friction linings riveted on both sides to transmit the power, under the following conditions:

(a) Intensity of pressure on the surface not to exceed 6.87×10^4 pa

(b) Slip torque and losses due to wear etc is 35% of engine torque

(c) Co-efficient of friction on contact surface is 0.3

(d) Inside diameter of the friction plate is 0.55 times the outside diameter

Given

$P = 5.9 \text{ kW} = 5.9 \times 10^3$ Watts

$N = 2100$ rpm

$p_{max} = 6.87 \times 10^4 \text{ pa} = 6.87 \times 10^4 \text{ N/m}^2 = 6.87 \times 10^{-2} \text{ N/mm}^2$

Loss = 35% T, therefore $T_{max} = 1.35$ T

$\mu = 0.3$

$d_2 = 0.55d_1 \Rightarrow r_2 = 0.55r_1$

Find

r_1, r_2

Solution

Torque to be transmitted

$$P = \frac{2\pi N T}{60}$$

$$5.9 \times 10^3 = \frac{2\pi \times 2100 \times T}{60}$$

$T = 26.84$ Nm

$T_{max} = 1.35\ T = 1.35 \times 26.84 = 36.23$ Nm $= 36.23 \times 10^3$ Nmm

$T_{max} = n\ \mu\ W\ R$

$$R = \frac{r_1 + r_2}{2} = \frac{r_1 + 0.55r_1}{2} = \frac{1.55r_1}{2} = 0.775r_1$$

$W = 2\pi C\ (r_1 - r_2)$

$= 2 \times \pi \times 0.03778r_1\ (r_1 - 0.55r_1)$

$= 0.2374r_1\ (0.45r_1)$

$= 0.1068r_1^{\ 2}$ N

$C = p_{max} \times r_2$

$= 6.87 \times 10^{-2} \times 0.55r_1$

$= 0.03778r_1$ N/mm

$\therefore\ 36.23 \times 10^3 = 2 \times 0.3 \times 0.1068r_1^{\ 2} \times 0.775r_1$

$r_1^{\ 3} = 729295.56$

$r_1 = 90$ mm

$\therefore r_2 = 0.55 \times 90 = 49.5$ mm

Problem 4.8: A plate clutch has three discs on the driving shaft and two discs on the driven shaft, providing four pair of contact surfaces. The outside diameter of the contact surfaces is 240 mm and inside diameter 120 mm. Assuming uniform pressure and m = 0.3, find the total spring load pressing the plates together to transmit 23 kW power at 1575 revolution per minute.

If there are 6 springs each of stiffness 13 kN/m and each of the contact surfaces has worn away by 1.25 mm, find the maximum power that can be transmitted, assuming uniform wear.

Given data

$n_1 = 3$

$n_2 = 2$

$n = 4$

$d_1 = 240$ mm $\quad \therefore \quad r_1 = 120$ mm

$d_2 = 120$ mm $\quad \therefore \quad r_2 = 60$ mm

Case (i) Assume uniform pressure

$\mu = 0.3$

$P = 23$ kW

$N = 1475$ rpm

W?

Torque to be transmitted

$T = n \mu W R$

Power transmitted by the clutch

$$P = \frac{2\pi NT}{60} \text{ Watts}$$

$$25 \times 10^3 = \frac{2\pi \times 1575 \times T}{60}$$

$T = 151.57$ Nm $= 151.57 \times 10^3$ Nmm

For normal pressure, mean radius of the contact surface

$$R = \frac{2}{3}\left[\frac{r_1^3 - r_2^3}{r_1^2 - r_2^2}\right]$$

$$= \frac{2}{3}\left[\frac{120^3 - 60^3}{120^2 - 60^2}\right]$$

$= 93.33$ mm

$\therefore$ Torque to be transmitted

$151.57 \times 10^3 = 4 \times 0.3 \times W \times 93.33$

$W = 1353.3$ N

Case (ii) Assume uniform wear

No. of springs = 6

Stiffness of each spring = 13 kN/m = 13×10^3 N/m

Wear on each contact surface = 1.25 mm

P?

Reduction in spring force = Total wear × No. of springs × Stiffness per spring

= (Contact surface of the spring × Wear on each contact surface) × No. of springs × Stiffness per spring

$= (8 \times 1.25) \times 6 \times 13 \times 10^3$

= 780 N

New axial load W = 1353.3 – 780 = 573.3 N

$T = n \mu W R$

For uniform wear $R = \frac{120 + 60}{2} = 90$ mm

$T = 4 \times 0.3 \times 573.3 \times 90$

= 61916.4 Nm

= 61.92 Nm

Maximum power that can be transmitted

$$P = \frac{2\pi NT}{60} \text{ Watts}$$

$$= \frac{2\pi \times 1575 \times 61.92}{60}$$

= 10212.6 Watts

= 10.21 kW

DESIGN OF CONE CLUTCH

θ = Semi-angle of cone (face angle) or angle of friction surface with the axis of the clutch

r_1 = Outer radius of friction surface

r_2 = Inner radius of friction surface

R = Mean radius of friction surface

b = Width of friction surface (face width or cone face)

Case 1: Considering uniform pressure

Normal pressure acting on the friction surface at mean radius

$$P_n = \frac{W}{\pi(r_1^2 - r_2^2)}$$

Torque to be transmitted

$T = \mu W R \operatorname{cosec}\theta$

$R = \frac{2}{3}\left[\frac{r_1^3 - r_2^3}{r_1^2 - r_2^2}\right]$

Case 2: Considering uniform wear

Total axial spring force (axial load) required

$W = 2\pi C\,(r_1 - r_2)$

Torque to be transmitted

$T = \mu W R \operatorname{cosec}\theta$

$R = \frac{r_1 + r_2}{2}$

Normal load acting on the friction surface

$W_n = 2\pi P_n R\, b$

Torque developed by the clutch

$T = 2\pi\mu P_{av} R^2 b$

Axial force required for engaging the clutch

$W_e = W_n (\sin\theta + 0.25\,\mu \cos\theta)$

Axial force required to disengage the clutch

$W_d = W_n (\mu\cos\theta - \sin\theta)$

	Case 1: Uniform pressure	Case 2: Uniform wear
Axial force (load) required (W)	$P_n = \frac{W}{\pi(r_1^2 - r_2^2)}$	$W = 2\pi C\,(r_1 - r_2)$
Torque transmitted (T)	$T = \mu W R \operatorname{cosec}\theta$	
Mean radius of friction surface (R)	$R = \frac{2}{3}\left[\frac{r_1^3 - r_2^3}{r_1^2 - r_2^2}\right]$	$R = \frac{r_1 + r_2}{2}$

PLATE AND CONE CLUTCHES

Time required for the flywheel to attain full speed from the stationary position $T = \frac{I}{g}\frac{\omega}{t}$.

Where,

T = Torque transmitted, Nm

I = Mass moment of inertia, Nm^2

ω = Angular velocity, rad/s

g = acceleration due to gravity, m/s^2

Energy supplied during the period of clutch slip = Tωt

$$\text{KE of flywheel} = \frac{I\omega^2}{2g}$$

$$\text{Energy dissipated (lost) due to clutch slip} = T\omega t - \frac{I\omega^2}{2g}$$

Problem 4.9: A cone clutch with a cone semi-angle of 12° is to transmit 11.19 kW at 750 rpm. The width of the face is 1/4th of the mean diameter and the normal pressure between the contact faces is not to exceed 8.27×10^4 pa. Allowing the coefficient of friction of 0.2, determine the main dimensions of the clutch and the axial force required.

Given

Cone clutch

$\theta = 12°$

$P = 11.19 \text{ kW} = 11.19 \times 10^3$ Watts

$N = 750$ rpm

$$b = \frac{1}{4}D = \frac{r_1 + r_2}{4}$$

$P_{max} = 8.27 \times 10^4 \text{ pa} = 8.27 \times 10^4 \text{ N/m}^2$

$\mu = 0.2$

Find

r_1, r_2 and W

Solution

Assuming uniform wear condition,

Torque to be transmitted

$T = \mu W R \operatorname{cosec}\theta$

$$R = \frac{r_1 + r_2}{2}$$

$$P = \frac{2\pi NT}{60}$$

$$11.19 \times 10^3 = \frac{2\pi \times 750 \times T}{60}$$

$$T = 142.47 \text{ Nm}$$

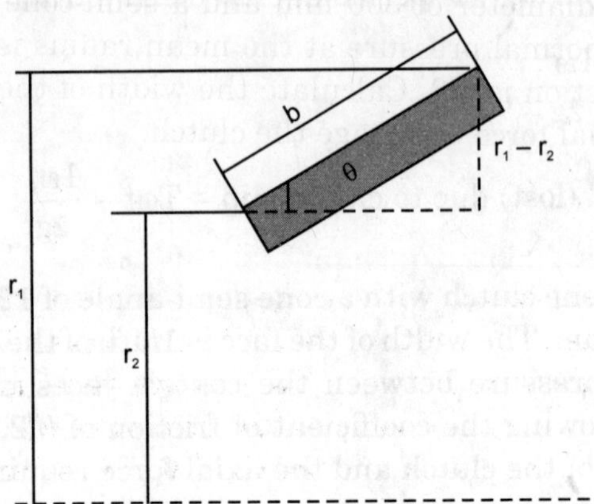

$$\sin\theta = \frac{r_1 - r_2}{b}$$

$$b = \frac{r_1 - r_2}{\sin\theta} \quad \Bigg| \quad b = \frac{r_1 + r_2}{4} \quad \text{(given in the problem)}$$

$$\therefore \quad \frac{r_1 - r_2}{\sin\theta} = \frac{r_1 + r_2}{4}$$

$$\frac{r_1 - r_2}{\sin 12^\circ} = \frac{r_1 + r_2}{4}$$

$$4.8\,(r_1 - r_2) = 0.25\,(r_1 + r_2)$$

$$4.8r_1 - 4.8r_2 = 0.25r_1 + 0.25r_2$$

$$4.55r_1 = 5.05r_2$$

$$r_1 = 1.109r_2$$

Axial force required

$$W = 2\,\pi\,C\,(r_1 - r_2)$$

$$= 2\,\pi\,(P_{max} \times r_2)\,(r_1 - r_2)$$

$$= 2\,\pi\,(8.27 \times 10^4 \times r_2)\,(1.109r_2 - r_2)$$

$$= 56638.5r_2^2$$

$$142.47 = 0.2 \times 56638.5\, r_2^{\,2} \times \frac{1.109 r_2 + r_2}{2} \times \text{cosec}12°$$

$r_2 = 0.135$ m

$r_1 = 1.109 \times 0.135 = 0.149$ m

$W = 56638.5 \times 0.135^2 = 1032.2$ N

Problem 4.10: A torque of 350 Nm is transmitted through a cone clutch having a mean diameter of 300 mm and a semi-cone angle of 15 deg. The maximum normal pressure at the mean radius is 150 kN/m². The coefficient of friction is 0.3. Calculate the width of the contact surface. Also find the axial force to engage the clutch.

Given

$T = 350$ Nm $= 350 \times 10^3$ Nmm

$D = 300$ mm $\Rightarrow R = 150$ mm

$q = 15°$

$P_n = 150$ kNm² $= 150 \times 10^3$ N/m² $= 0.15$ N/mm²

$\mu = 0.3$

Find

b, W_e

Solution

Torque developed by the clutch

$T = 2\pi\mu P_n R^2 b$

$350 \times 10^3 = 2\pi \times 0.3 \times 0.15 \times 150^2 \times b$

$b = 55$ mm

Normal load acting on the friction surface

$W_n = 2\pi P_n R b$

$= 2\pi \times 0.15 \times 150 \times 55$

$= 7775.4$ N

Axial force to engage the clutch

$W_e = W_n (\text{Sin}\theta + 0.25\, \mu\, \text{Cos}\theta)$

$= 7775.4 (\text{Sin}15° + 0.25 \times 0.3 \times \text{Cos}15°)$

$= 2575.7$ N

Problem 4.11: In a cone clutch, the angle of cone is 30°, coefficient of friction is 0.35 and the contact surfaces have an effective mean diameter of 80 mm. If the axial force applied is 196.2 N, find the torque required to produce the slipping of the clutch under uniform wear.

Calculate the time required to attain the full speed and also the energy lost in the slipping of the clutch, if the clutch is employed to connect an electric motor, running uniformly at 1200 rpm with a flywheel which is stationary and has a moment of inertia of 3.4 Nm^2.

Given

Cone clutch - Uniform wear condition

$2\theta = 30°, \ \theta = 15°$

$\mu = 0.35$

$D = r_1 + r_2 = 80 \text{ mm} = 0.08 \text{ m}$

$W = 196.2 \text{ N}$

$N = 1200 \text{ rpm}$

$I = 3.4 \text{ Nm}^2$

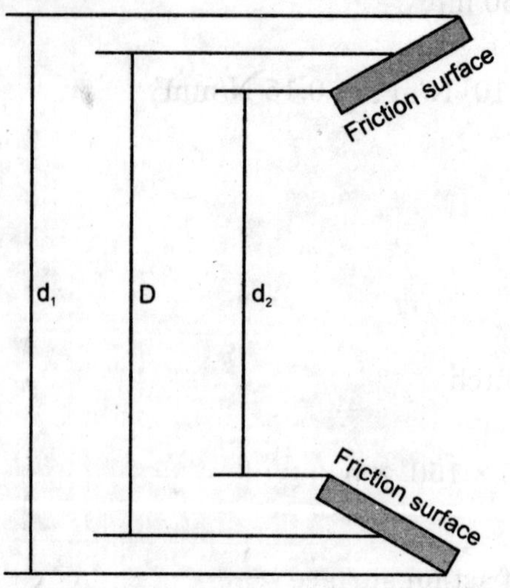

$$D = \frac{d_1 + d_2}{2} = \frac{2r_1 + 2r_2}{2} = \frac{2(r_1 + r_2)}{2} = r_1 + r_2$$

Solution

1. Torque required

$$T = \mu W R \operatorname{cosec}\theta$$

$$= \mu W \left[\frac{r_1 + r_2}{2}\right] \operatorname{cosec}\theta$$

$$= 0.35 \times 196.2 \times \frac{0.08}{2} \times \operatorname{cosec}15°$$

$$= 10.6 \text{ Nm}$$

2. Time lapse to attain the full speed

$$T = \frac{I}{g}\frac{\omega}{t}$$

$$\omega = \frac{2\pi N}{60} = \frac{2\pi \times 1200}{60} = 125.6 \text{ rad/s}$$

$$10.6 = \frac{3.4}{9.81}\frac{125.6}{t}$$

$$t = 4.1 \text{ s}$$

3. Energy lost (dissipated) due to clutch slip

$$E_l = T\omega t - \frac{I\omega^2}{2g}$$

$$= 10.6 \times 125.6 \times 4.1 - \frac{3.4 \times 125.6^2}{2 \times 9.81}$$

$$= 2724.8 \text{ Nm}$$

Gearbox

The gearbox is installed behind the clutch and is usually controlled by the gear shift lever in the driver's compartment. The gearbox changes the engine speed and torque by meshing gears in various combinations.

NEED

In order to achieve a high maximum vehicle speed, together with good acceleration and economy over the whole speed range, a gearing system is employed which permits the engine power to [illegible] corresponding to the best performance. Maximum engine power and economy [illegible] at different engine speeds [illegible] difficult [illegible] variable speed [illegible]

Gearbox

The gearbox is installed behind the clutch and is usually controlled by the gear shift lever in the driver's compartment. The gearbox changes the engine speed and torque by meshing gears in various combinations.

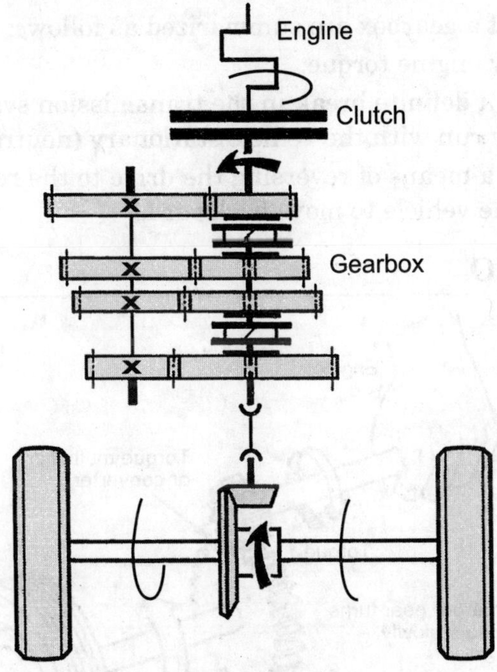

Fig.5.1: *Gear box*

NEED FOR GEARBOX

In order to achieve a high maximum vehicle speed, combined with good acceleration and economy over the whole speed range, a gearing system is required, which permits the engine to operate at the speeds corresponding to its best performance. Maximum engine power, torque and economy all occur at different engine speeds. As a result it becomes difficult to match the gear ratio for best performance, especially when variable operating conditions and driver demands are also to be considered. The engine requirement to suit a given operating condition is as follows.

Operating condition	Engine requirement
Maximum traction	Maximum engine torque
Maximum vehicle speed	Maximum engine power
Maximum acceleration	Maximum engine torque
Maximum economy	Engine at mid-range speed and under light load with a small throttle opening.

FUNCTIONS

The purposes of a gearbox are summarized as follows:

- To multiply engine torque.
- To provide a definite break in the transmission system so that the engine may run with the vehicle stationary (neutral position).
- To provide a means of reversing the drive to the road wheels, thus enabling the vehicle to move backwards.

GEAR RATIO

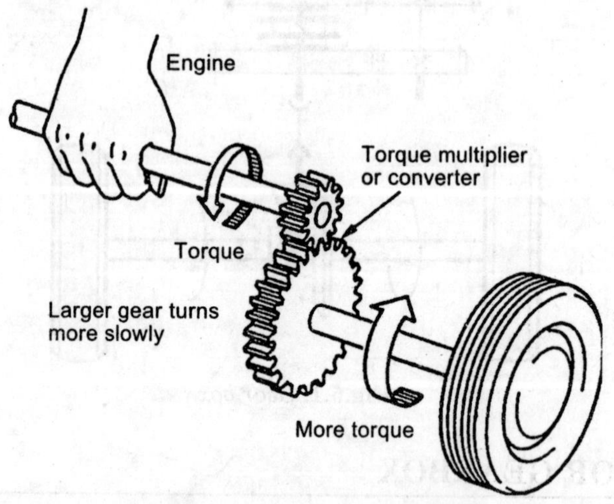

Fig.5.2: *Torque multiplication*

When gear combination shown in Fig.5.2 is used, the wheel rotational speed will be decreased but the torque at the wheel shaft will be increased. The slowing down of the engine speed delivered to the drive wheels is called shifting down and the relation between the speeds is called gear ratio. Gear ratio can be derived from the following equation.

$$\text{Gear ratio} = \frac{\text{Engine speed}}{\text{Gearbox output shaft speed}}$$

Horsepower and torque have the relation shown below. Therefore, in the case the engine is developing the same horsepower, lowering its speed, or in other words raising the gear ratio, will result in increasing the torque (drive power) at the wheels.

$$\text{Horsepower} = K \times T \times N$$

Where, K = proportional constant

T = torque

N = rotational speed

In gear combinations, the gear ratios, in order from the largest ratio, are called low, second, and third. The gear ratio at which the engine speed is the same as the gearbox output shaft speed is called top. When the gear ratio is less than one, that is, when the gearbox output shaft speed is faster than the engine speed, it is called *overdrive*.

TRACTIVE RESISTANCE

When a vehicle is travelling at a constant speed along a road, the resistance to its forward motion is termed "tractive resistance". The tractive resistance, sometimes called total resistance, includes:

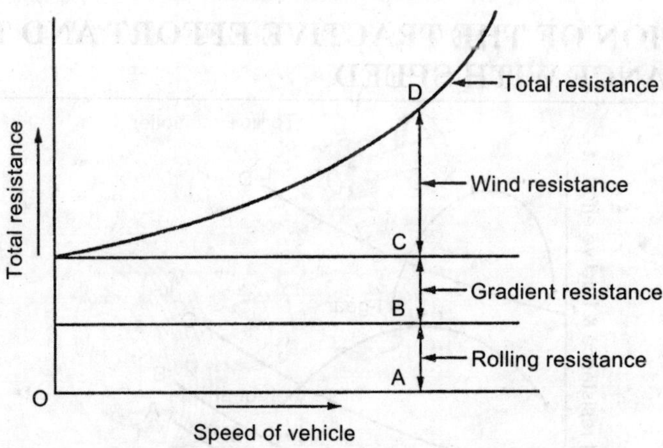

Fig.5.3: *Resistance to the vehicle motion*

Air resistance

Air resistance is due to movement of the vehicle through the air. This depends upon the actual shape and size of the vehicle. A degree of control can be exercised over wind resistance by careful design of the vehicle and by improved streamlining techniques.

Rolling resistance

Rolling resistance which is due to friction between the tyre and road, and largely influenced by the type of road surface. A vehicle travelling off the road on the rough ground will obviously have a greater rolling resistance to overcome than a vehicle travelling on the hard smooth surface of a well-maintained road. The weight of the vehicle, friction between the tyres and the road surface and even the speed, may influence the rolling resistance. Furthermore, transmission friction may also be included in rolling resistance.

Gradient resistance

This is dependent upon the steepness of the hill and the overall weight on the vehicle. This component remains constant at all speeds.

Total resistance

The total resistance to the motion of a vehicle is the sum of the above three resistances as shown in the graph of Fig.5.3, where OA is the speed of the vehicle, AD is the total resistance, the composed of the rolling resistance AB, the gradient resistance BC and the wind resistance CD. If either any of the constant resistance increases or decreases then the curve would simply shift up or down by the amount of increase or decrease.

VARIATION OF THE TRACTIVE EFFORT AND TOTAL RESISTANCE WITH SPEED

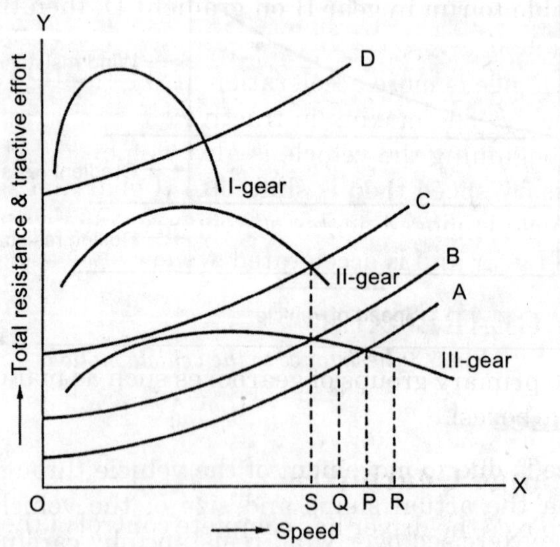

Fig.5.4: *Variation of the tractive effort and total resistance with speed*

The total resistance for the particular road with different gradients may then be represented by curves A, B, C and D shown in Fig.5.4. The higher curve represents steeper gradient. The tractive effort curves for vehicle are I, II and III shown in Fig.5.4. The curves I, II and III represent the tractive effort in first, second and third gears respectively.

The variation of the tractive effort with the variation of the vehicle speed will depend upon the variation of the engine torque with the variation of engine speed. If the gear box ratio is altered, we get another curve of tractive effort.

The curves obtained showing the variation with vehicle speeds, of both the tractive efforts and the total resistance and plotted to the same scales, on the same sheet of the paper. From these curves it is clear that whenever the tractive effort exceeds the total resistance, the vehicle will accelerate to a speed where both tractive effort and total resistance are equal.

As shown, let the vehicle be travelling in III gear on a gradient which gives total resistance curve A, then OP is the stabilizing speed. If the speed in any case is more, say OR. then the excess of resistance will decelerate it to OP, Similarly if the speed at any instant is less, say OQ then the excess of tractive effort will accelerate it to speed OP.

Now in the third gear consider the next higher gradient of curve B, where the stabilizing speed has decreased. Then consider further the curve C which does not cross the curve of gear III. Therefore, it is clear that the vehicle is unable to go at this gradient in the third gear. Then we shift on to II gear and get a stabilizing speed OS. Similarly the vehicle is unable to run in gear II on gradient D, then the vehicle is to be shifted in I gear.

The vehicle needs more acceleration at the time of starting to gain the speed quickly. The maximum tractive effort is available in I gear, therefore, in beginning the vehicle is always moved in I gear. When it obtains necessary speed then is shifted to II gear and so on. Similarly when the vehicle is moved across a steeper gradient then it is always shifted to the I gear and is accelerated at very slow speed.

TYPES OF GEARBOXES

These are two primary groups of gearboxes such as manual change and automatic gearboxes.

Manual change Gearboxes

In these gearboxes the driver has complete control of the gear changing process and can select a gear ratio appropriate to the driving conditions by means of the manual control lever. Generally, these are four to five gear ratio options, apart from the reverse gear.

Automatic Gearboxes

Automatic gearboxes can change gear automatically according to engine load, road speed, driver's demand and in the latest computer controlled units, according to the style of driving. The driver merely selects the direction of movement, and the gear ratio range.

Most current automatic gearboxes use an epicyclic gear system, in which the required gear is obtained by holding or driving a part, or parts of the gear train by means of a friction clutch or brake. Most automatic gear boxes also use a torque converter between the engine and gearbox, in addition to the 3 or 4 speed epicyclic gearbox. The torque converter is generally a third clutch that replaces the conventional friction clutch.

MANUAL GEARBOX

There are three basic types of manual gearbox:

1. Sliding mesh gearbox
2. Constant mesh gearbox
3. Synchromesh gearbox

Of these, the synchromesh is the most common type in use today. It is essentially a combination of the other two types.

SLIDING MESH GEARBOX

Sliding mesh type gearbox is the oldest type of manual transmission used in automobiles. Among the manual gear transmission, this sliding mesh type is the simplest in construction.

However, in this type the mechanical efficiency is very low and the noise level is quite high. Moreover the driver required considerable skill in changing the gears.

CONSTRUCTION

The Fig.5.5 shows the construction of a sliding mesh type transmission having three forward and one reverse speeds. There are two gears (B and C-D) attached on the main shaft and four gears (I, II, III and R) on the counter shaft. The two gears on the main shaft can slide and can mesh with the gears on counter shaft. Therefore, it is called sliding mesh gearbox. In this gearbox, spur gears are used because of the sliding action. A separate gear is mounted on the idler shaft. The gears

B and C-D are mounted on the splined main shaft. These gears can be slided by a shifting yoke.

POWER FLOW OF 4-SPEED SLIDING MESH GEARBOX

Gears in neutral

When the engine is running and the clutch is engaged, the counter shaft is driven by the clutch gear. The low speed and high speed gears are fitted on the transmission main shaft which does not rotate.

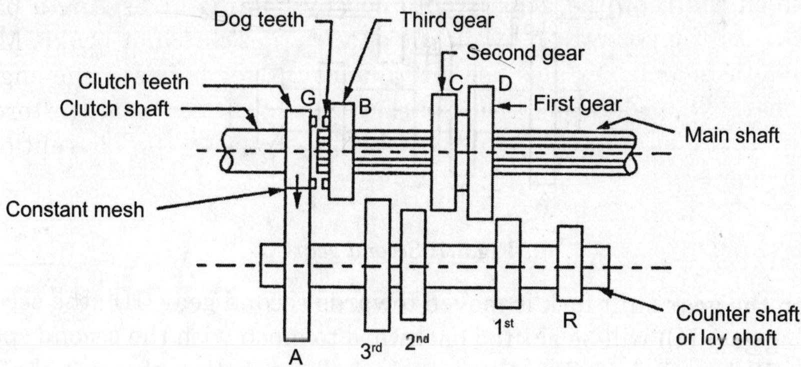

Fig.5.5: *Gears in neutral*

At the same time, they are not engaged with any driving gears. Therefore, there is no motion transmitted from engine to the wheels. Hence, the vehicle is stationary.

First (Low speed) gear

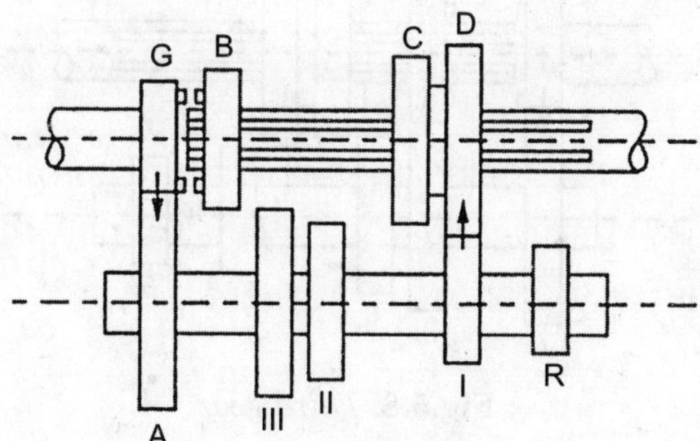

Fig.5.6: *First gear*

When the gear shift fork moves towards first gear (I) by operating the gear shift lever, the sliding gear (D) on the output shaft will be shifted forward to mesh with low speed gear (I) on the countershaft. It results the rotations of input shaft being transmitted in the order G-A-I-D to turn the output shaft. This gear combination is the one that produces the lowest speed from the input shaft and low transmission. In this case the torque output from the gearbox will be as great as that applied at the input.

Second gear

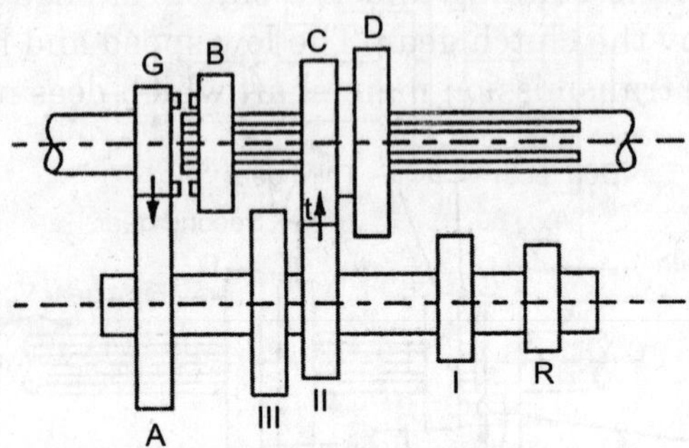

Fig.5.7: *Second gear*

When the gear shift fork is moved towards second gear (II) , the second sliding gear (C) will be shifted backward to mesh with the second speed gear (II) but (I) and (D) are unmeshed. The rotation of input shaft is transmitted in the order G-A-II-C to turn the output shaft. This is the transmission in the second speed.

Third gear

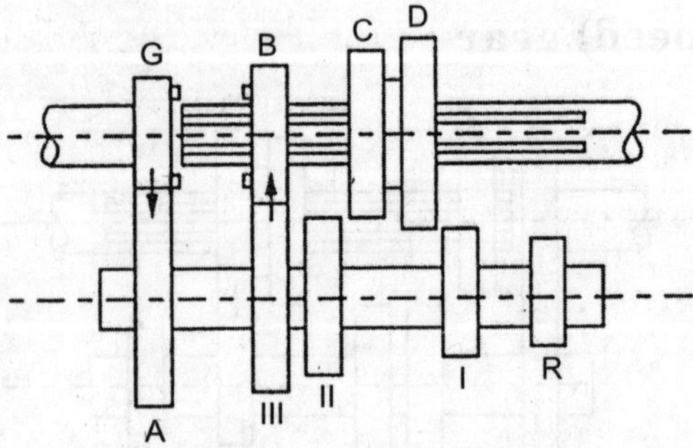

Fig.5.8: *Third gear*

When the gear shift fork is moved towards third gear (III), the second sliding gear (B) will be shifted backward to mesh with the third speed gear (III) but (II) and (C) are unmeshed. The rotation of input shaft is transmitted in the order G-A-III-B to turn the output shaft. This is the transmission in the second speed.

Top gear

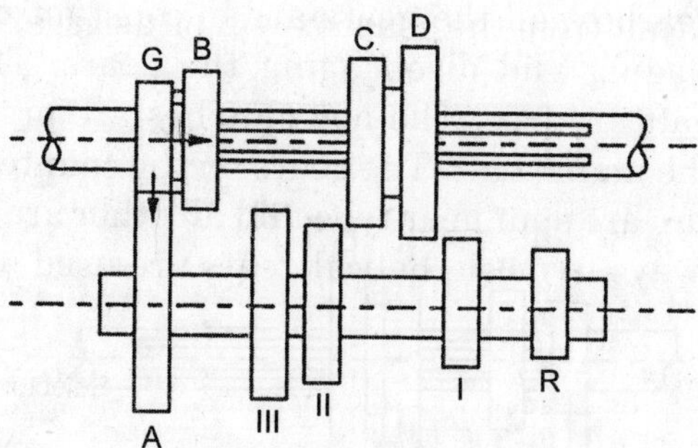

Fig.5.9: *Top gear*

When the gear shift fork is moved towards gear A, the clutch will be meshed but gears (III) and (B) are unmeshed. Due to this, both the input and output shafts are coupled and rotated together with the same speed. This is the transmission in the third or top speed that gives a ratio of 1:1.

Reverse gear

When the gear shift fork is moved toward direction R, the sliding gear (D) will be shifted backward to mesh with the reverse idler gear. Then the rotation of input shaft is transmitted in the order G-A-R-Idler-D to turn the output shaft in reverse direction. This is the transmission in reverse speed.

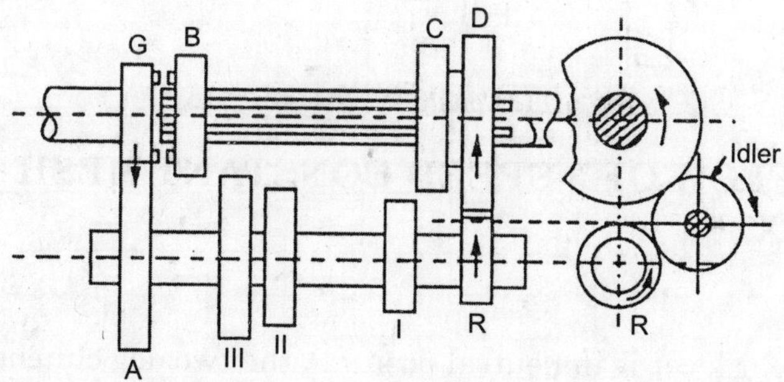

Fig.5.10: *Reverse gear*

This is the same ratio as for the first gear and irrespective of the size of the idler gear it will be seen that the ratio always remains the same. For this reason it is called an idler. It changes the direction, but does not alter the ratio.

CONSTANT MESH GEARBOX

In this type of gearbox, all the gears are in constant mesh having dog clutches for engaging and disengaging the gears. The gears on the splined main shaft are free. The dog clutches are provided which are free to slide on the main shaft. The gears on the counter shaft are fixed. Only reverse gears are spur gear type and all other are helical gears. As the gears are always in mesh, helical gears are used which are quieter running.

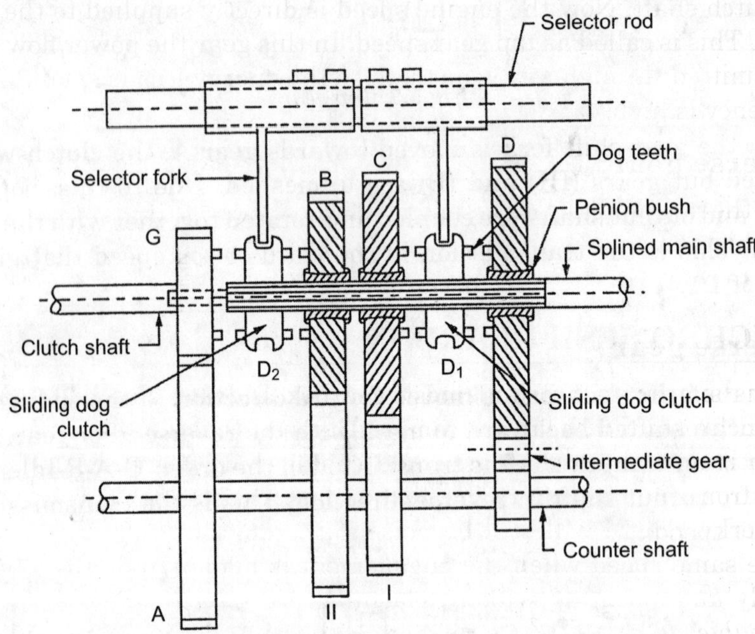

Fig.5.11: *Constant mesh gearbox*

POWER FLOW OF 3-SPEED CONSTANT MESH GEARBOX

Neutral

When the gear lever is in neutral position, the two dog clutches D_1 and D_2 are not in contact with any of the gearwheels and cannot transmit torque to the main shaft.

First gear

The dog clutch (D1) is shifted to left to make engage on (C). Now the power is transmitted through the gear G-A-I-C and dog clutch D1, transmits to the main shaft. Hence, the first gear speed is obtained.

Second gear

Now, the dog clutch (D_1) is disengaged. The dog clutch (D_2) is shifted to right to lock with the gear (D). Therefore, the power is transmitted from clutch shaft through G-A-II-B and dog clutch (D_2) to the main shaft. So, the main shaft rotates with the second gear speed.

Third (or) top gear

In this layout, third gear is a direct drive; namely a gear that gives a ratio of 1:1. The dog clutch (D_2) is moved left to engage with the gear (G) on clutch shaft. Now the engine speed is directly supplied to the main shaft. This is called as top gear speed. In this gear the power flow is not transmitted through any gear teeth so the energy loss is small, i.e. the efficiency is high.

Reverse gear

First the dog clutch D_2 is disengaged. Then the dog clutch D_1 is shifted to right to engage with the gear (D). The idler gear causes the main shaft to rotate in the opposite direction.

SYNCHROMESH GEARBOX

This is similar to constant mesh gearbox and dog clutch is replaced by synchronizer unit. In the simplest type of constant mesh gear box, which is now absolute, gears could be engaged by snatching the gear lever from one position to the next as fast as possible. This was noisy and jerky. There will still be noise if the dog teeth are not rotating at the same speed when the engagement is made. To do the job more quietly and smoothly without clashing, the gears were synchronized by *double de-clutching* i.e. engaging the clutch with the gearbox in neutral and accelerating the engine to speed up the input component, then disengaging the clutch again and engaging the appropriate gear. Drivers today are relieved from the need for double de-clutching by a synchronizing device built into the sliding collars in the gearbox. This synchromesh device is usually fitted to all forward gears.

Synchronizers

Synchronizers equalize the speed of the shaft and gear before they are engaged. They use friction to synchronize the speed of the gear and the shaft before the connection is made.

Neutral position

When the gearbox is in neutral, the synchronizers are in their neutral position and are not rotating with the main shaft.

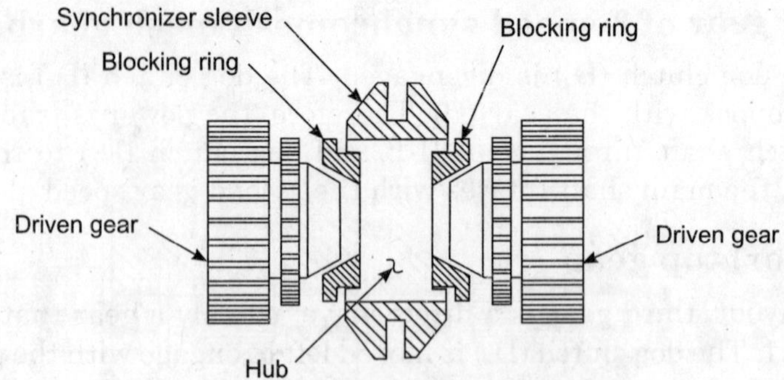

(a) Neutral position

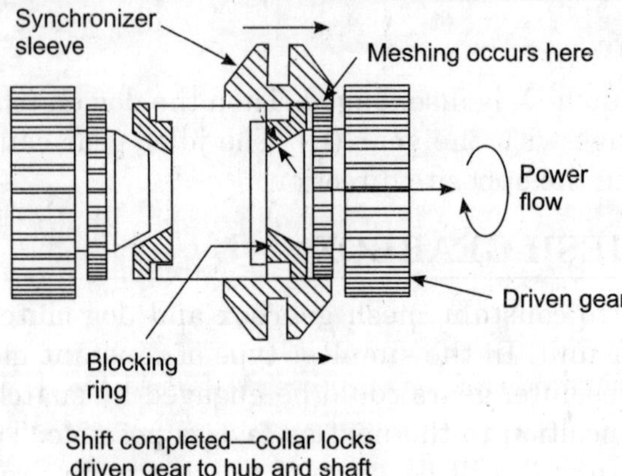

(b) Gear position

Fig.5.12: *Synchronizer operation*

Gear position

When a gear is selected, the shifting fork forces the sleeve toward the selected gear. As the sleeve moves, inserts also moves because they are locked in the sleeve. The movement of the inserts pushes the blocking ring into contact with the shoulder of the gear. When this contact is made, there is a metal to metal contact and begins to rotate.

As the components reach the same speed, the synchronizer sleeve can now slide over the external dog teeth on the blocking ring and then over the dog teeth on the gear's shoulder. This completes the engagement of the synchronizer and the gear is locked to the main shaft. Power now flows as follows:

Input gear – Counter gear – Gear selected – Synchronizer – Main shaft

Power flow of 3-speed synchromesh mesh gearbox

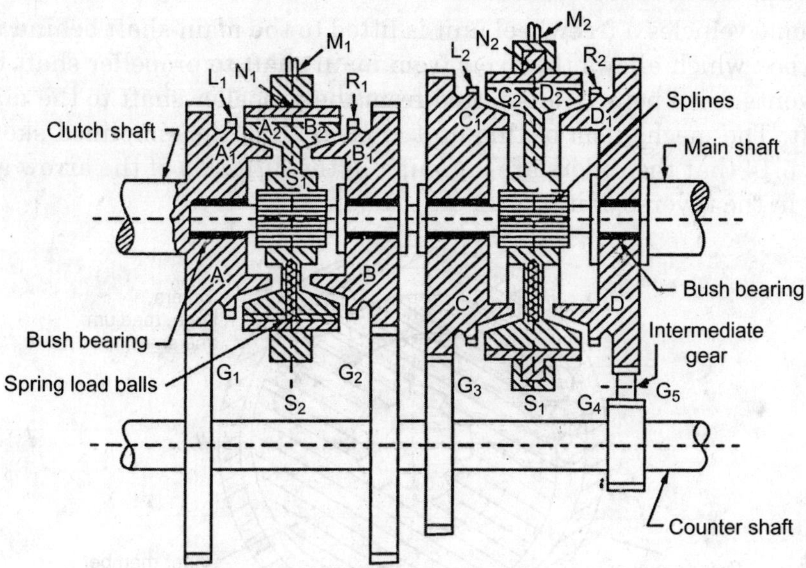

Fig.5.13: *Synchromesh gearbox*

First gear

The synchronizer (S_1) is shifted to left to make engage on (C). Now the power is transmitted through the gear A-G_1-G_3-C and synchronizer S_1, transmits to the main shaft. Hence, the first gear speed is obtained.

Second gear

Now, the synchronizer (S_1) is disengaged. The synchronizer (S_2) is shifted to right to lock with the gear (B). Therefore, the power is transmitted from clutch shaft through A-G_1-G_2-B and synchronizer (S_2) to the main shaft. So, the main shaft rotates with the second gear speed.

Top gear

The synchronizer (S_2) is moved left to engage with the gear (A) on clutch shaft. Now the engine speed is directly supplied to the main shaft. This is called as top gear speed. The top gear provides a 1:1 ratio between the input shaft and the main shaft by locking the two shafts together.

Reverse gear

First the synchronizer S_2 is disengaged. Then the synchronizer S_1 is shifted to right to engage with the gear (D). The idler gear causes the main shaft to rotate in the opposite direction.

FREEWHEEL UNIT

In some vehicles a freewheel unit is fitted to the main-shaft behind the gearbox which allows the drive from main shaft to propeller shaft but prevents drive being transmitted from the propeller shaft to the main shaft. The mechanism of this unit is clear from the simplified sketch Fig. 5.14 that the rollers are jamming in the direction of the arrow and free in the reverse of it.

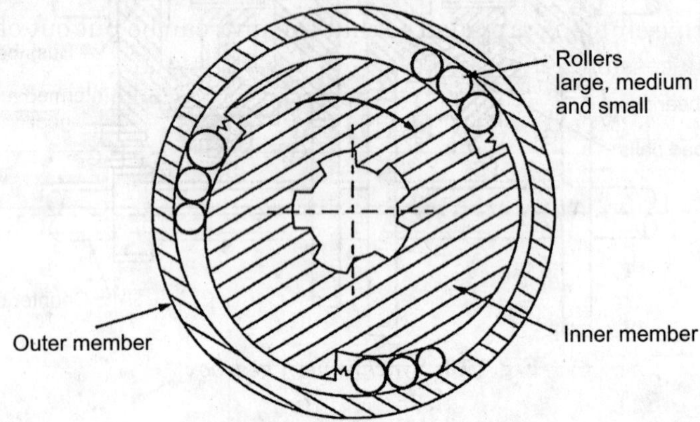

Fig.5.14: *Freewheel unit*

The gearbox shaft is connected to the inner member while the propeller shaft is connected to the outer member. The inner member contains three steps having three rollers of different size as shown in the figure. When the inner member is rotating in the direction shown by the arrow thereby jamming the rollers between the two, the outer member will also be rotating along in the same direction. But when the vehicle is going downhill then the outer member acts as driving member and the inner member as driven member, this connection is automatically disconnect by this unit and the gear box and the engine are isolated from the wheels, which results in fuel economy. The engine effect gives increased fuel economy but, as the engine can exert no retarding force, driving methods must be modified and greater use of the brakes is possible. When there is reverse drive then this unit is needed to be locked. For such cases the locking system is provided. There is a dog clutch provided which is automatically engaged by a connection from the reverse gear selector and can be locked by a manual control for traffic conditions.

Advantages of freewheel unit

1. The engine can be allowed to idle simply by closing the throttle,

without disengaging the gears, and by coasting down hills an improved fuel consumption is obtained.

2. Since the device is placed behind the gear box, then on closing the throttle, the engine and the gear box shafts will slow down and a change of gear can be effected without serious clashing occurring.
3. Less wear on the transmission because the engine and the gear box are disconnected from the propeller shaft when the vehicle is going downhill.
4. A positive clutch is provided so that the unit can be put out of action when necessary.

GEAR SELECTOR MECHANISM

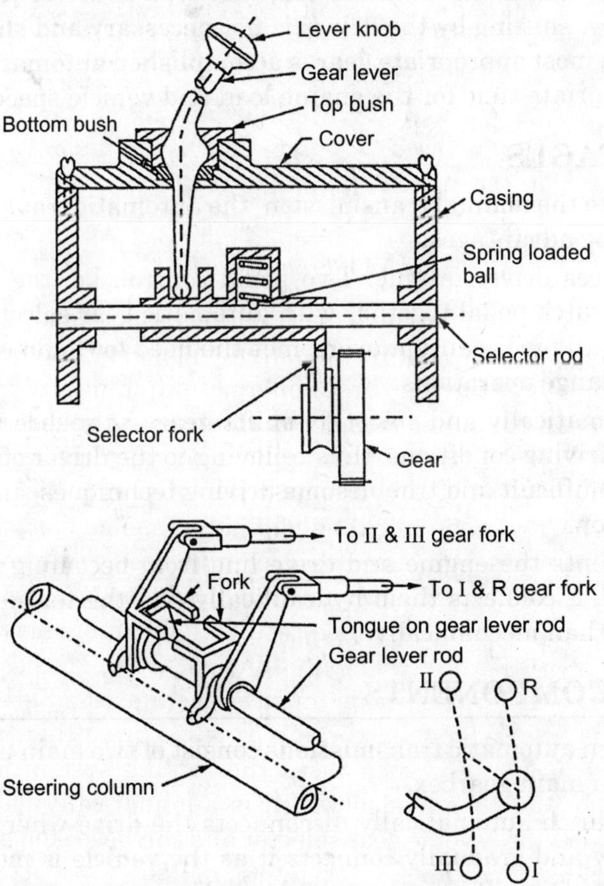

Fig.5.15: *Gear selector mechanism*

To allow the driver to select the most suitable gear ratio, manual gearboxes are fitted with a gear lever (gearbox or steering column mounted) and selector mechanism. The main components of such a selector mechanism are shown in Fig. 5,15. When the driver uses the gear lever to select a particular gear, the movement is transmitted along the appropriate selector rail to the gearwheel. Selector-locking or gear-locating devices hold a gear engaged when it has been selected. To ensure that only one gear can be selected at a time, an interlocking mechanism is incorporated into the gearbox selectors.

AUTOMATIC TRANSMISSION

When a driver is driving a vehicle with a manual transmission, it is necessary for the driver to be constantly aware of the engine load and vehicle speed and to be shifting gears accordingly.

With an automatic transmission, this type of driver judgement is unnecessary, shifting by the driver is not necessary and shifting up or down to the most appropriate gear is accomplished automatically at the most appropriate time for the engine load and vehicle speed.

ADVANTAGES

Compared to the manual transmission, the automatic transmission has the following advantages:

- It reduces driver fatigue. Two pedal control, i.e. the elimination of the clutch pedal together with automatic gear selection reduces driver fatigue because it overcomes the need for tedious clutch and gear change operations.
- It automatically and smoothly shifts gears at speeds appropriate to the driving conditions, thus relieving to the driver of the need to master difficult and troublesome driving techniques such as clutch operation.
- It prevents the engine and drive line from becoming overloaded, because it connects them hydraulically (via the torque converter) rather than mechanically.

MAJOR COMPONENTS

Most modern automatic transmissions consist of two main units: a fluid clutch and a main gearbox.

Fluid clutch automatically disconnects the drive when the engine speed is low and gradually connects it as the vehicle is moved from a stationary position. It is either a fluid coupling or torque converter. The gearbox automatically provides a series of stepped ratios to enable the vehicle to overcome road conditions needing a large driving effort. Also it allows the vehicle to operate over a wide road speed range.

TORQUE CONVERTER

Torque converters, in its simple form, can double the output torque from the engine. In addition the converter is able to reduce the torque amplification gradually as the output speed of the converter increases. The torque converter is generally a third clutch that replaces the conventional friction clutch.

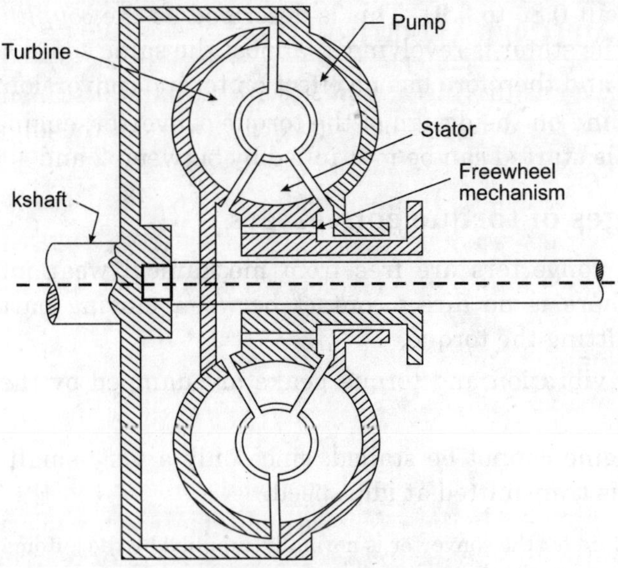

Fig.5.16: *Torque converter*

Torque multiplication

Engine torque + Reaction torque = Output turbine torque

Torque converter operation

When the engine is running, the oil inside the blades of the pump is flung out (throw forcefully) by centrifugal force and deflected by blade curvature into the turbine blades. Oil returned from turbine blades attempts to rotate the stator in the opposite direction to the pump and turbine rotation, but this movement is prevented by the stator free wheel. The oil is therefore deflected by the stator blades such that it now enters the pump at a suitable speed and direction and thus complete the oil circuit.

As the turbine speeds up, the difference between its speed and that of the pump decreases, so that the oil flow strikes the stator blades at a different angle. This in turn reduces the additional forces on the turbine blades, so that the torque is reduced.

The stator then begins to turn slowly in the oil flow, when the pump impeller and turbine speeds are almost identical, the oil flow striking the stator blades has changed its angle to such that torque at the turbine is almost the same as the torque input from the engine. In other words, no further torque conversion takes place at this speed and the unit acts as a fluid coupling.

At this stage in vehicle operation, the ratio of impeller to turbine speed is about 0.86 to 0.91. This is described as the *coupling point.* At this point, the stator is revolving at almost the same speed as the other two wheels and therefore has no effect on torque conversion.

Depending on the design of the torque converter, engine torque as the vehicle is started can be multiplied by between 2 and 4.5 times.

Advantages of torque converters

1. Torque converters are free from mechanical wear in operation, since there is no direct contact between moving parts or those transmitting the torque.
2. Engine vibration and torque peaks are damped by the hydraulic fluid.
3. The engine cannot be stalled, since only a very small amount of torque is transmitted at idle speed.

Note: The fluid for the converter is normally supplied by the automatic gearbox and it is generally a low viscosity mineral oil which contains additives to improve lubrication and resist frothing. Cavitation noise caused by air in the converter is minimized by pressurizing the fluid to about 1.38 bar.

AUTOMATIC GEARBOX

Power flow

P – Park
R – Reverse
N – Neutral
D – Drive
2 – Second gear
1 or L – First (or) Low gear

R is reverse and backward movement of the vehicle is achieved in this position. P is parking position. In neutral (N) no drive is applied. This gear position is used when the vehicle is being towed. D is the normal automatic mode which covers the full range of forward gears (1-2-3 or 3-2-1 according to the engine load and vehicle speed).

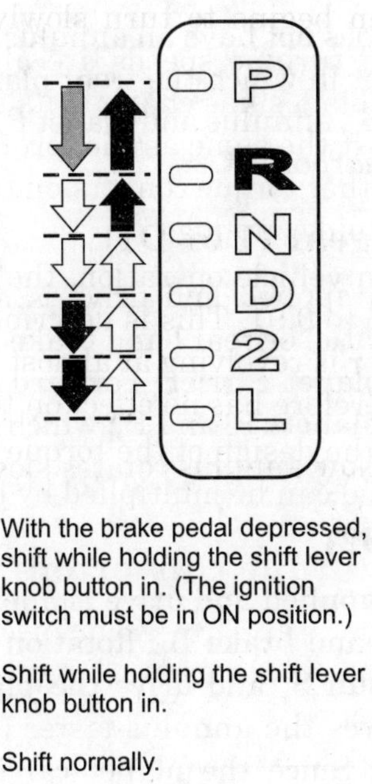

PLANETARY GEAR SET

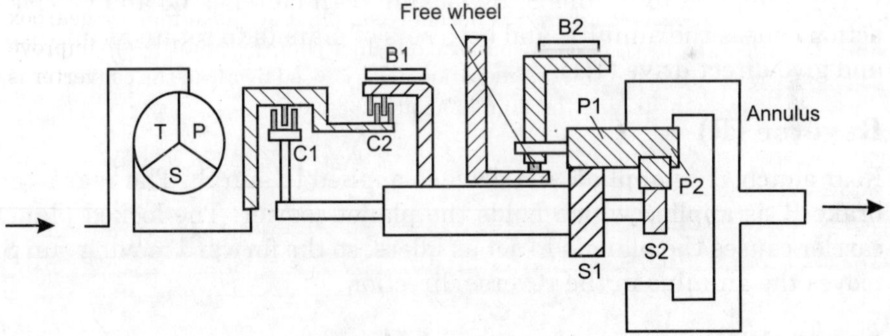

Fig.5.17: *Ravigneaux gearbox*

The majority of automatic gearboxes use planetary gearing to give three forward ratios and one reverse. Borg-Warner model 35 uses two epicyclic gear sets. Power flow through the train is controlled by:

- Two multiplate wet clutches, front (C_1) and rear (C_2)
- Two brake bands, front (B_1) and rear (B_2)
- A free wheel (one way clutch), fitted between the casing and planet carrier. Free wheel will unlock if planet carrier rotates in the forward direction.

The front set does not have an annulus. Instead it uses a compound planet arrangement in which the front planet P_1 mesh with three gear members: the sun S_2, annulus and planet P_2. This combination is known as a *Ravigneaux gearbox* set.

First (or) Low gear (1 or D1)

Front clutch C_1 is applied and drive is applied to sun S_2. The free wheel (fully automatic) or rear band brake B_2 (first gear selected by the driver) holds the planet carrier. Forward motion of sun S_2 transmits the drive through planets P_1 and P_2, which act as idlers, to the annulus and output shaft. Now annulus rotates slower and gives first gear.

Second gear (D2)

Front clutch C_1 is applied and drive is applied to sun S_2. The sun S_1 is held by the front band brake B_1. Rotation of P_1 by P_2 causes it to roll around the fixed sun S_1 and drive the planet carrier in the forward direction. This drives the annulus faster (compared to first gear) and gives second gear. Since the planet carrier is moving in a forward direction, the free wheel offers no resistance.

Direct drive (D3)

Drive is applied to S_1 and S_2 by locking both clutches C_1 and C_2. This action causes the annulus and both sets of planets to rotate as one unit and give direct drive (1:1).

Reverse (R)

Rear clutch C_2 is applied and drive is applied to sun S_1. The rear band brake B_2 is applied which holds the planet carrier. The locked planet carrier causes the planets to act as idlers, so the forward moving sun S_1 moves the annulus in the reverse direction.

Neutral (N)

All clutches and brake bands are released, so no drive is transmitted.

Park (P)

A parking pawl, mounted to the casing, is moved into a gear connected to the gearbox output shaft. This provides a powerful lock that positively prevents movement of the transmission.

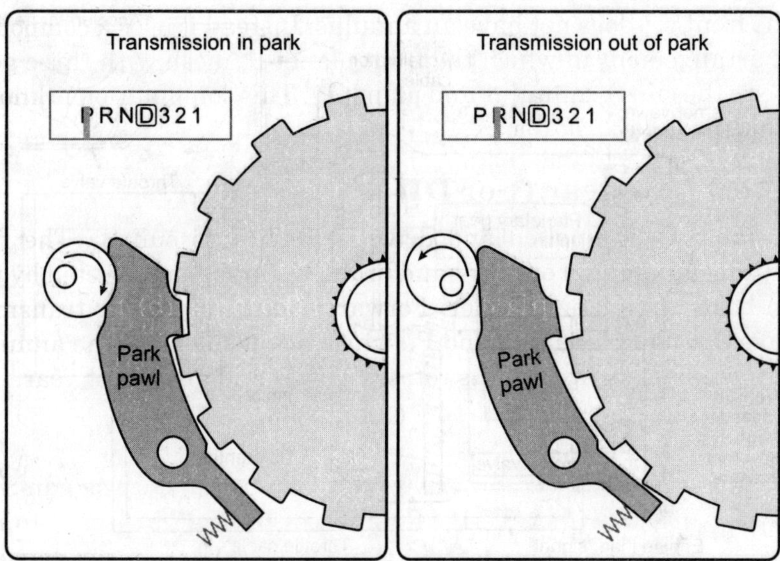

Fig.5.18: *Parking mechanism*

HYDRAULIC CONTROL SYSTEM

The hydraulic control system consists of the oil pan, which acts as the fluid reservoir; the oil pump, which generates the hydraulic pressure; various valves having various functions; and fluid passages and tubes, which deliver the transmission fluid to the clutches, brakes and other components of the hydraulic control system. Most valves in the hydraulic control system are housed in the valve body assembly under the planetary gears.

MANUAL SHIFT VALVE

This valve is located in the hydraulic control unit and actuated by the driver's selector lever. It determines which gear ratio is to be selected. The position of the selector lever can be used to vary the working pressure in the shift valves. Operation of the lever moves the valve and uncovers the fluid port that supplies hydraulic pressure to the clutches and brake bands to obtain the desired gear position.

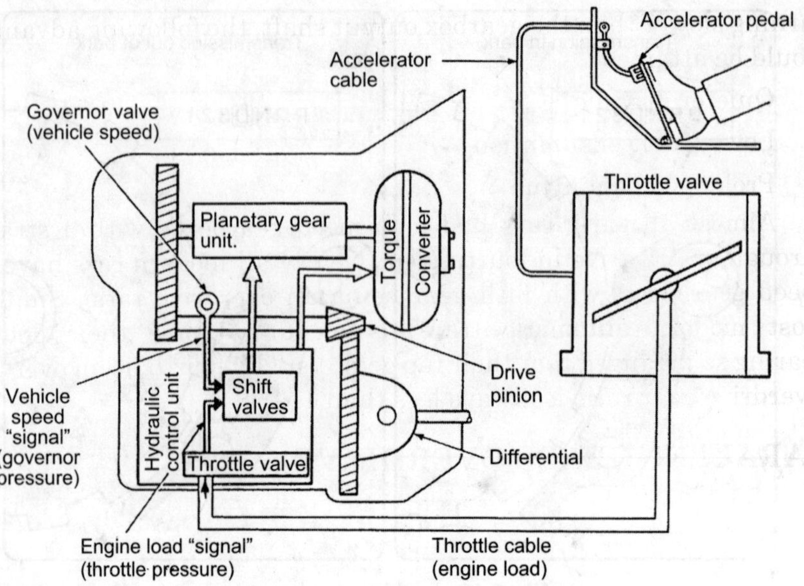

Fig.5.19: *Hydraulic control system*

Speed sensing

The centrifugal governor is mounted on the transmission output shaft and supplies a governor pressure dependent on road speed, which increases with the speed of rotation. This pressure is applied to one side of the spool shift valves in the valve body. Above a certain road speed, governor pressure rises and accordingly the shift valves supply pressure to the clutches and brake bands which cause the next higher gear to be selected. If road speed drops, governor pressure also falls and the drive shifts to the next lower gear.

Load sensing

The function of the regulating valve is to control, in a separate line, the fluid pressure that increases as the accelerator pedal is depressed. This throttle pressure applied to the shift valves and as a result alters the position of the shift valves with respect to governor pressure.

OVERDRIVE

The overdrive is a speed increasing device that allows the gearbox output shaft to rotate faster than the engine.

When driving at comparatively high speed on roads where acceleration is not specially required, such as on level road with small running resistance, the engine will have considerable surplus power. If at this time the engine speed is lowered so that the engine will be turning slower than the gearbox output shaft, the following advantages would be attained.

1. Quieter driving
2. Lower fuel consumption
3. Prolonged engine life

Almost all early cars had four-speed gearboxes with a straight-through or 1:1 drive in fourth gear. Nearly all modern cars have five-speed gearboxes, with fifth gear being an overdrive ratio. Similarly, most modern automatic transmission units have their epicyclic gearing so arranged that their top gear, usually fourth, is an overdrive. Overdrive gear ratio is always less than 1.

PARALLEL GEARS OVERDRIVE

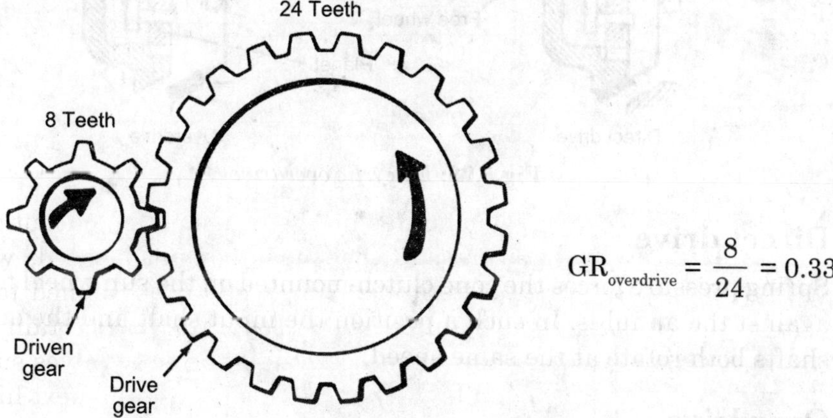

Fig.5.20: *Overdrive gear ratio*

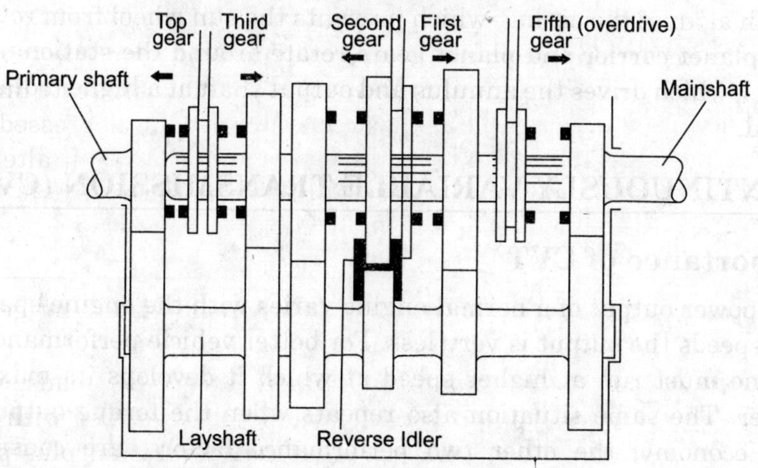

Fig.5.21: *Overdrive in a manual gearbox*

EPICYCLIC OVERDRIVE

There are, however, some vehicles still use a four-speed manual gearbox, but with a separate overdrive unit fitted behind or in an extension to the gearbox. Such separate units normally use an epicyclic gear train to provide the overdrive ratio.

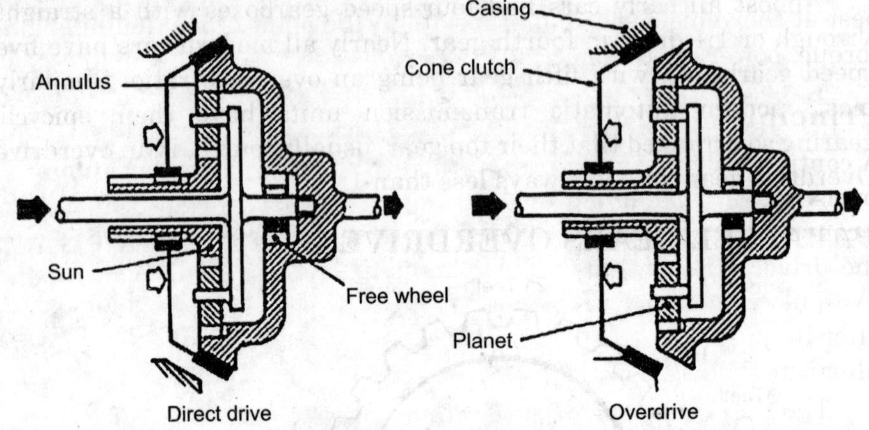

Fig.5.22: *Epicyclic overdrive*

Direct drive

Spring pressure forces the cone clutch mounted on the sun wheel firmly against the annulus. In such a position the input shaft and the output shafts both rotate at the same speed.

Overdrive

When the overdrive is selected, hydraulic pressure forces the cone clutch against the casing, which prevents the sun wheel from rotating. The planet carrier and planet gears rotate around the stationary sun wheel, which drives the annulus and output shaft at a higher rotational speed.

CONTINUOUSLY VARIABLE TRANSMISSION (CVT)

Importance of CVT

The power output of a normal engine varies with the engine speed. At low speeds the output is very less. For better vehicle performance, the engine must run at higher speed at which it develops its maximum power. The same situation also repeats when the torque output and fuel economy, the other two performance factors, are considered. Maximum torque occurs at a different speed from that for maximum economy and also neither of these speeds coincides with the point of maximum power.

The attainment of the constant engine speed required to achieve any one of the three performance factors is not possible with a conventional gearbox, because the engine speed is required to be continually changed to match the vehicle speed. Therefore, the engine only performs its best at the vehicle speed appropriate to the point of maximum engine torque, power or economy.

Principle

A continuously variable transmission (CVT) is a particular automatic transmission capable of providing a smoothly varying gear ratio. Unlike a conventional automatic gearbox the CVT has no fixed gears. It varies the drive ratio continuously by changing the operating diameters of two pulleys that are linked by a steel V-belt. The trans-mission can alter its ratio imperceptibly, without any interruption of drive. Figure illustrates the basic principle of the CVT system.

The CVT offers the driving comforts associated with an automatic gearbox as well as the efficiency of a manual transmission, without any loss of vehicle performance. Unlike a conventional automatic, no torque converter is used and hence no hydraulic slippage losses occur in this system. The CVT is capable of responding instantly to throttle pressure, providing smooth and rapid acceleration. Unfortunately, due to the difficulties of manufacturing a reliable and durable system only a few car makers incorporate such a transmission.

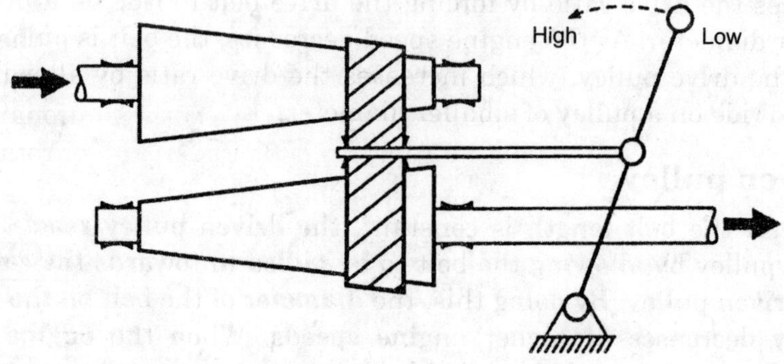

Fig.5.23: *Principle of CVT*

OPERATION

Drive pulley works on centrifugal force principle while the driven pulley use compression spring to change their effective diameter.

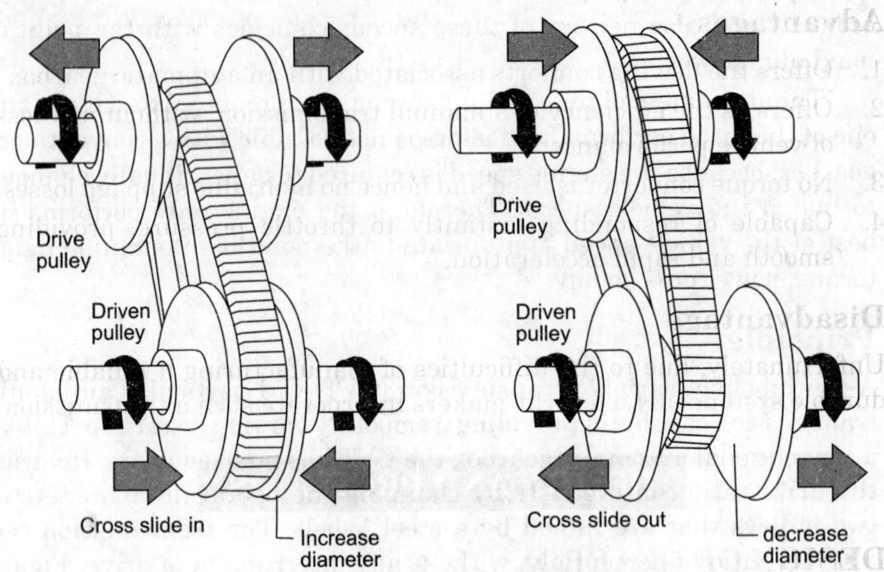

Fig.5.24: *Operation of CVT*

Drive pulley

The drive pulley consists of a fixed and a movable face. The movable face has the capability to slide toward the drive face. There is also a ramp plate incorporated in the drive pulley, which pushes weight rollers against the drive face. As engine speed increases, centrifugal force pushes the movable face toward the fixed face, which in turn pushes the drive belt upward toward the top of the drive pulley. This reduces the drive ratio by forcing the drive belt to ride on a pulley of larger diameter. As the engine speed decreases, the belt is pulled back into the drive pulley, which increases the drive ratio by allowing the belt to ride on a pulley of smaller diameter.

Driven pulley

Because the belt length is constant, the driven pulley reacts to the drive pulley by allowing the belt to be pulled in towards the centre of the driven pulley. By doing this, the diameter of the belt on the driven pulley decreases at higher engine speeds. When the engine speed decreases and the belt is pulled back into the drive pulley, the spring in the driven pulley moves the movable driven pulley face towards its original position. This increases the effective diameter of the driven pulley and pushes the belt back to the circumference of the driven pulley.

Advantages

1. Offers the driving comforts associated with an automatic gearbox
2. Offers as the efficiency of a manual transmission, without any loss of vehicle performance.
3. No torque converter is used and hence no hydraulic slippage losses.
4. Capable of responding instantly to throttle pressure, providing smooth and rapid acceleration.

Disadvantage

Unfortunately, due to the difficulties of manufacturing a reliable and durable system only a few car makers incorporate such a transmission.

DESIGN AND PROBLEMS

DESIGN OF GEARBOX

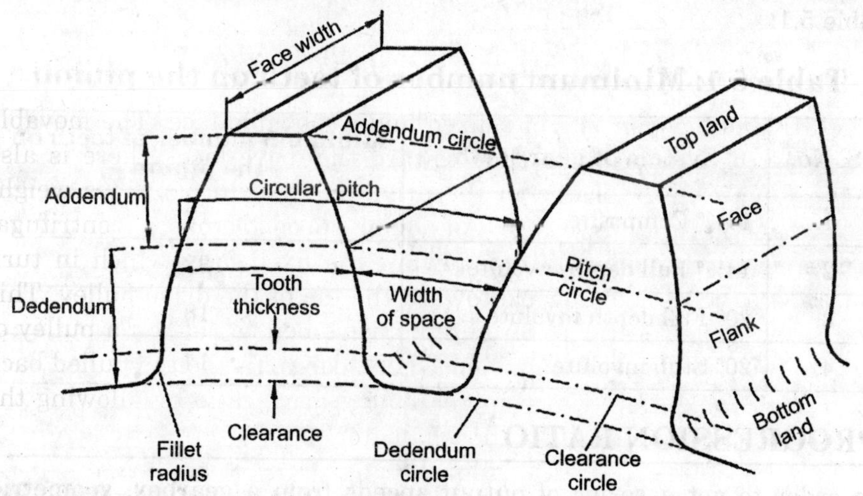

Circular pitch

It is the distance measured on the circumference of the pitch circle from a point of one tooth to the corresponding point on the next tooth. It is usually denoted by p_c.

Mathematically,

Circular pitch, $p_c = \pi D/T$

Where

D = Diameter of the pitch circle, and

T = Number of teeth on the wheel.

Note: The two gears will mesh together correctly, if the two wheels have the same circular pitch.

Pitch circle diameter

It is the diameter of the pitch circle. The size of the gear is usually specified by the pitch circle diameter. It is also called as pitch diameter.

Module

It is the ratio of the pitch circle diameter in millimetres to the number of teeth. It is usually denoted by m. Mathematically,

Module, m = D / T

Minimum number of teeth without interference and undercutting

The minimum number of teeth on the pinion which will mesh with any gear (also rack) without interference are given in the following table 5.1:

Table 5.1: Minimum number of teeth on the pinion

S. No.	System of gear teeth	Minimum number of teeth on the pinion
1.	14½° Composite	12
2.	14½° Full depth involute	32
3.	20° Full depth involute	18
4.	20° Stub involute	14

PROGRESSION RATIO

In order to get a series of output speeds from a gearbox, geometric progression is used. It is also known as step ratio.

$$\text{Step ratio } (\phi) = \frac{\text{Maximum speed of output shaft}}{\text{Minimum speed of output shaft}}$$

By using geometric progression the speed is reduced uniformly in different stages.

$$G2 = \sqrt{G1 \times G3}$$

Similarly,

$$G3 = \sqrt{G2 \times G4}$$

CENTRE DISTANCE BETWEEN THE SHAFTS

$$a = \frac{D_p}{2} + \frac{D_g}{2} = \frac{D_A}{2} + \frac{D_B}{2} = R_A + R_B$$

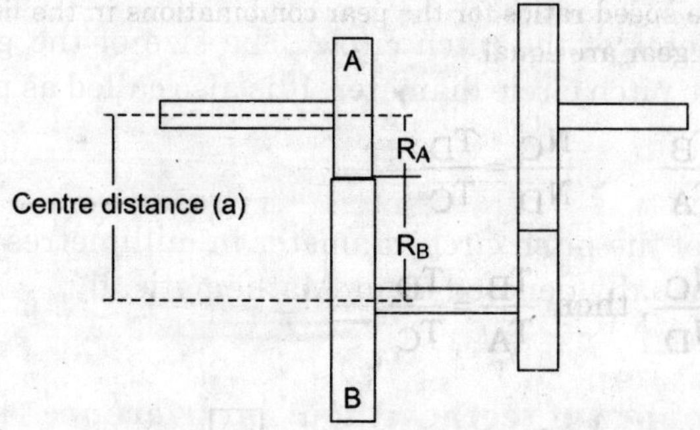

Note: Since the pitch is same for all gears and the centre distance is the same for all pair of mating gears, the total number of teeth must be same for each pair.

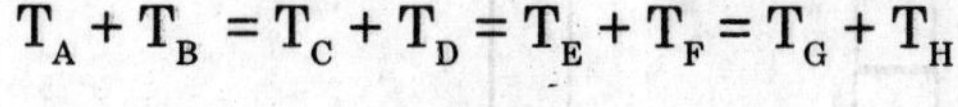

$$T_A + T_B = T_C + T_D = T_E + T_F = T_G + T_H$$

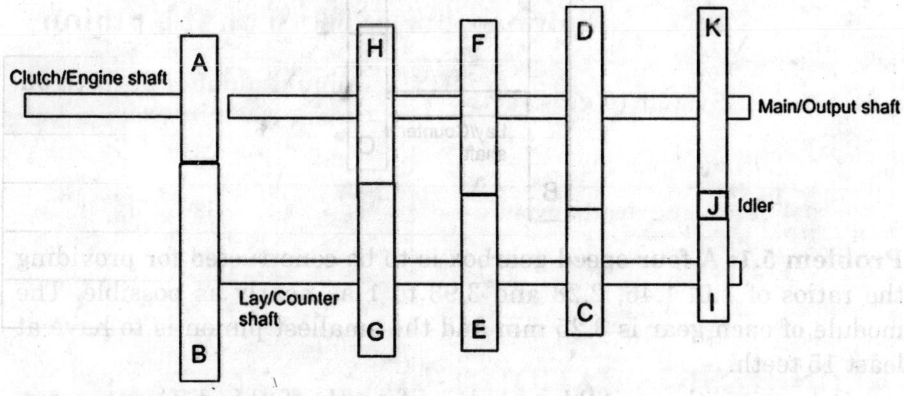

REVERSE GEAR RATIO

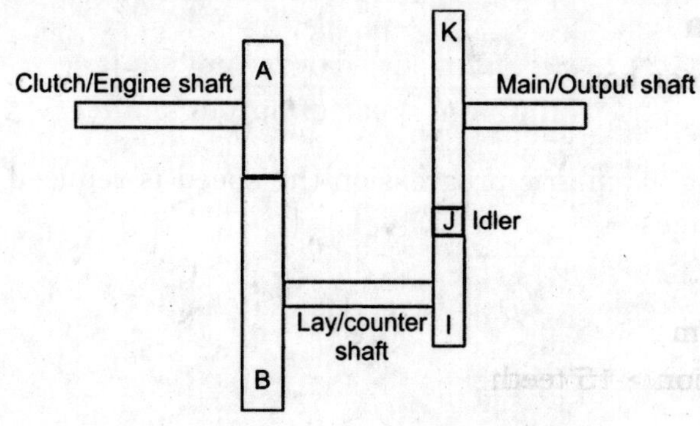

$$GR = \frac{T_B\, T_J\, T_K}{T_A\, T_I\, T_J} = \frac{T_B\, T_K}{T_A\, T_I}$$

Note: If the speed ratio between engine shaft and lay shaft is not given in the problem, assume speed ratios for the gear combinations in the first gear (lower gear) and clutch gear are equal.

$$\frac{N_A}{N_B} = \frac{T_B}{T_A} \qquad \frac{N_C}{N_D} = \frac{T_D}{T_C}$$

If $\frac{N_A}{N_B} = \frac{N_C}{N_D}$, then $\frac{T_B}{T_A} = \frac{T_D}{T_C}$

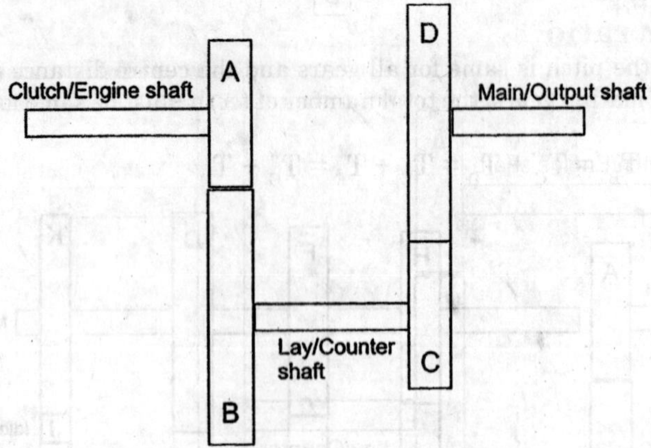

Problem 5.1: A four speed gearbox is to be constructed for providing the ratios of 1.0, 1.46, 2.28 and 3.93 to 1 as nearly as possible. The module of each gear is 3.25 mm and the smallest pinion is to have at least 15 teeth.

Determine the suitable number of teeth of the different gears. What is the distance between the main shaft and lay shaft?

Given data

Four speed gear box

G4 = 1:1

G3 = 1.46:1

G2 = 2.28:1

G1 = 3.93:1

m = 3.25 mm

Smallest pinion > 15 teeth

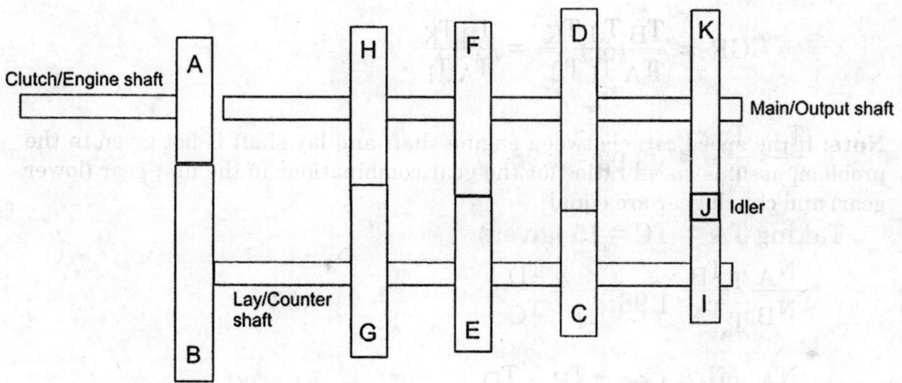

SOLUTION

First gear ratio

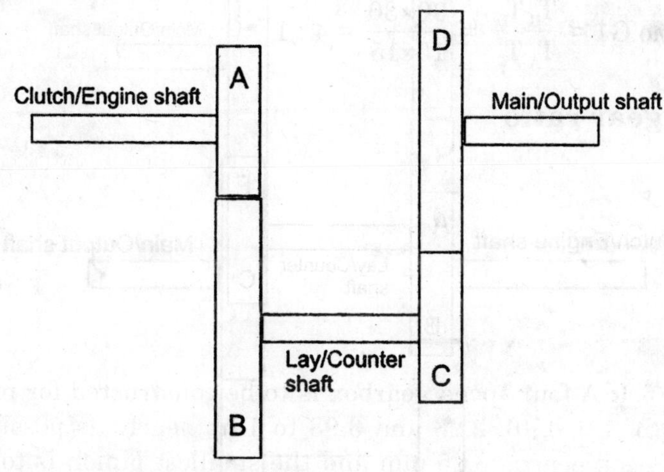

$$G1 = \frac{T_B \, T_D}{T_A \, T_C} = 3.93$$

Assuming $\frac{T_B}{T_A} = \frac{T_D}{T_C}$

$$\therefore \quad G1 = \frac{T_B \, T_B}{T_A \, T_A} \text{ (or) } \frac{T_D \, T_D}{T_C \, T_C} = 3.93$$

$$= \left(\frac{T_B}{T_A}\right)^2 \text{ (or) } \left(\frac{T_D}{T_C}\right)^2 = 3.93$$

$$= \frac{T_B}{T_A} \text{ (or) } \frac{T_D}{T_C} = \sqrt{3.93}$$

$$\frac{T_B}{T_A} = \frac{T_D}{T_C} = \sqrt{3.93} = 1.98$$

Taking TA = TC = 15 (given)

$$\frac{T_B}{T_A} = 1.98 \qquad \frac{T_D}{T_C} = 1.98$$

$$T_B = 1.98 \text{ TA} \qquad T_D = 1.98 \text{ TC}$$

$$= 1.98 \times 15 \qquad = 1.98 \times 15$$

$$= 27.9 \sim 30 \qquad = 27.9 \sim 30$$

$$\text{Actual ratio G1} = \frac{T_B\, T_D}{T_A\, T_C} = \frac{30\times 30}{15\times 15} = 4 : 1$$

Second gear ratio

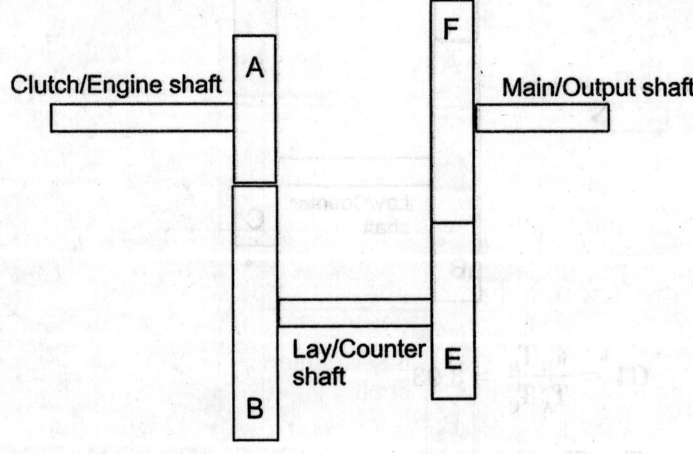

$$G2 = \frac{T_B\, T_F}{T_A\, T_E} = 2.28$$

$$\frac{30\, T_F}{15\, T_E} = 2.28$$

$$2\, \frac{T_F}{T_E} = 2.28$$

$$\frac{T_F}{T_E} = 1.14$$

$$T_F = 1.14\ T_E$$

We know that,

$$T_A + T_B = T_C + T_D = T_E + T_F = T_G + T_H$$

$$T_A + T_B = T_E + T_F$$

$$15 + 30 = T_E + 1.14\ T_E$$

$$45 = (1 + 1.14)\ T_E$$

$$T_E = 21$$

$$T_F = 1.14 \times 21 = 23.94 \sim 24$$

$$\text{Actual ratio G2} = \frac{T_B\ T_F}{T_A\ T_E} = \frac{30 \times 24}{15 \times 21} = 2.285 : 1$$

Third gear ratio

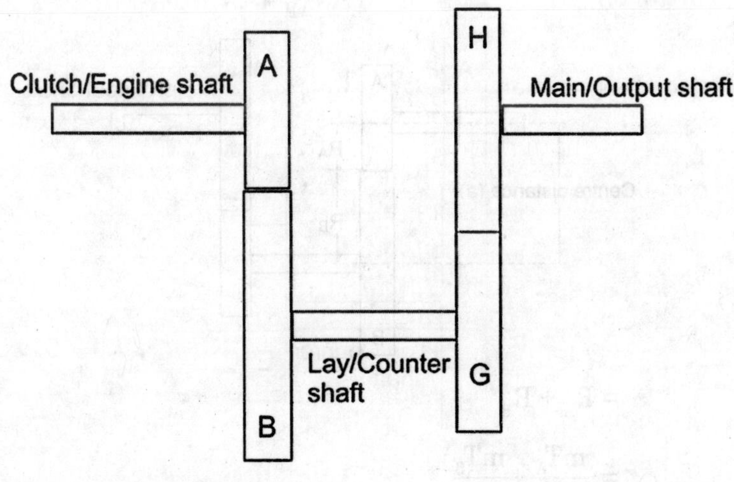

$$G3 = \frac{T_B\ T_H}{T_A\ T_G} = 1.46$$

We know that,

$$T_A + T_B = T_G + T_H$$

$$\frac{30 T_H}{15 T_G} = 1.46$$

$$\frac{T_H}{T_G} = 0.73$$

$$T_H = 0.73\ T_G$$

$$T_A + T_B = T_G + T_H$$

$$15 + 30 = T_G + 0.73T_G$$

$$45 = (1 + 0.73)\ T_G$$

$$T_G = 26$$

$$T_H = 0.73 \times 26 = 18.98 \sim 19$$

$$\text{Actual ratio G3} = \frac{T_B\ T_H}{T_A\ T_G} = \frac{30\times19}{15\times26} = 1.461 : 1$$

Top gear ratio

$$G4 = 1.1$$

Center distance between the shafts

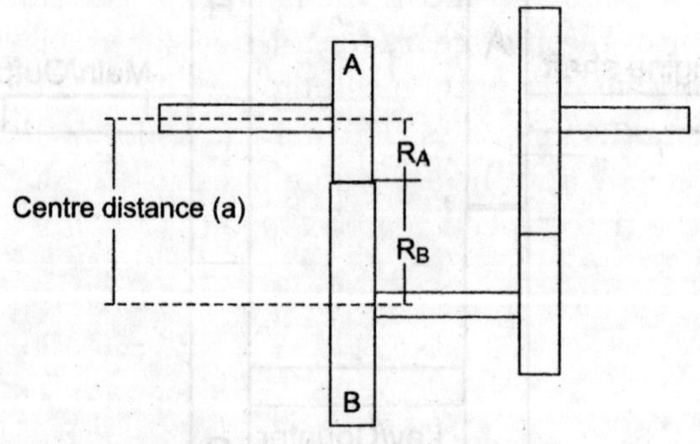

$$a = R_A + R_B$$

$$= \frac{mT_A}{2} + \frac{mT_B}{2}$$

$$= \frac{m(T_A + T_B)}{2}$$

$$= \frac{3.25(45)}{2}$$

$$= 73.125 \text{ mm}$$

LEWIS EQUATION

Face width of the gear can be found by modified Lewis formula,

$$F = \frac{cfb}{1000}\frac{m}{1000}$$

$$T_e = F\frac{D}{2000}$$

Where,

T_e = Engine torque, Nm
F = Tangential tooth load in gear, N
D = Pitch diameter of gear, mm
c = Form factor
f = Allowable stress, N/m^2
b = Face width of gear, mm
m = Module of gear, mm

Problem 5.2: The maximum gearbox ratio of an engine 75 mm bore and 100 mm stroke is 4. The pitch diameter of the constantly meshing gear is 75% of the piston stroke. If the module is 4.25 mm, calculate the size and number of teeth of gears for a three speed gearbox. Calculate the face width of the constantly meshing gear using the modified Lewis formula. The engine torque is 910 kgf-cm, value of constant in the Lewis formula is 0.07 and the allowable stress is 900 kgf/cm^2. Draw the neat sketch of the three speed gearbox layout.

Given data

d = 75 mm
l = 100 mm
G_1 = 4
D_A = 75%l = 0.75 × 100 = 75 mm
m = 4.25 mm
T_e = 910 kgf-cm = 9100 N-cm = 91 Nm
c = 0.07
f = 900 kgf/cm^2 = 9000 N/cm^2 = 9000×10^4 N/m^2
3-speed sliding mesh gearbox

Solution

(i) No. of teeth

We know that,

G1 = 4 and G3 = 1

$$G2 = \sqrt{G1G3} = \sqrt{4\times1} = 2$$

First gear ratio

$$G_1 = \frac{T_B\,T_D}{T_A\,T_C} = 4$$

Assuming speed ratios are equal, $\frac{T_B}{T_A} = \frac{T_D}{T_C} = \sqrt{4} = 2$

From table 5.1, for 20° full depth involute profile, $T_A = T_C = 18$ to avoid interference

$$T_B = T_D = 36$$

Second gear ratio

$$G_2 = \frac{T_B\,T_E}{T_A\,T_F} = 2$$

$$2\,\frac{T_E}{T_F} = 2$$

$$\frac{T_E}{T_F} = 1$$

$$T_E = T_F$$

Assuming pitch, module and centre distance are same

$$T_A + T_B = T_C + T_D = T_E + T_F = 18 + 36$$

$$T_A + T_B = T_C + T_D = T_E + T_F = 54$$

$$T_E + T_F = 54$$

$$T_F + T_F = 54$$

$$T_E = T_F = 27$$

(ii) Size of the gears

We know that,

$$m = \frac{D}{T}$$

Pitch circle diameter of gear A (D_A) = 75 mm (given)

Pitch circle diameter of gear B (D_B) = m × T_B = 4.25×36 = 153 mm

Pitch circle diameter of gear C (D_C) = m × T_C = 4.25×18 = 76.5 mm
Pitch circle diameter of gear D (D_D) = m × T_D = 4.25×36 = 153 mm
Pitch circle diameter of gear E (D_E) = m × T_E = 4.25×27 = 115 mm
Pitch circle diameter of gear F (D_F) = m × T_F = 4.25×27 = 115 mm

(iii) Face width of the constantly meshing gear

Te, Nm; F, N; D, mm; f, N/m²; b, mm; m, mm

$$T_e = F \frac{D}{2000}$$

$$91 = F \frac{75}{2000}$$

$$F = 2426.66 \text{ N}$$

$$F = \frac{cfb}{1000}\frac{m}{1000}$$

$$2426.66 = \frac{0.07 \times 9000 \times 10^4 \times b}{1000}\frac{4.25}{1000}$$

$$b = 90.63 \text{ mm}$$

Problem 5.3: If the number of teeth on a spur gear is 20 and has a module of 5 mm, determine the reference diameter, tip circle diameter and root circle diameter.

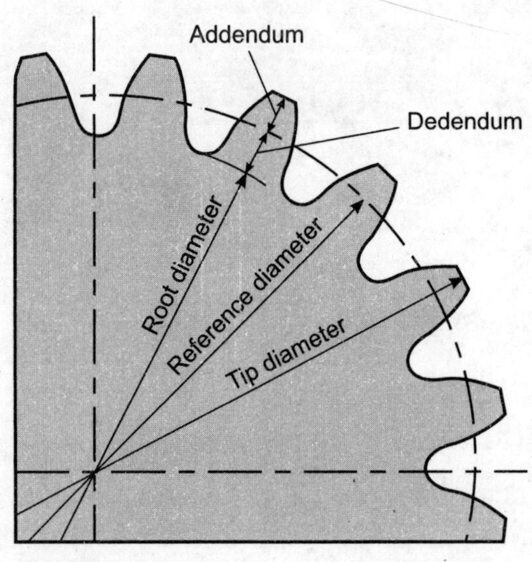

Reference diameter (Pitch circle diameter) $d = mT$

$= 5 \times 20$

$= 100$ mm

Tip circle diameter $d_a = d + 2m$

$= 100 + 2(5)$

$= 110$ mm

Root circle diameter $d_f = d - 2.5m$

$= 100 - 2.5(5)$

$= 87.5$ mm

Propellershaft Assembly

The propellershaft assembly on front engine rear drive vehicles and 4WD vehicles transmits power from the gearbox to the final drive and differential assembly. It consists of three main parts as given under:

1. Propellershaft
2. Universal joints
3. Sliding joint

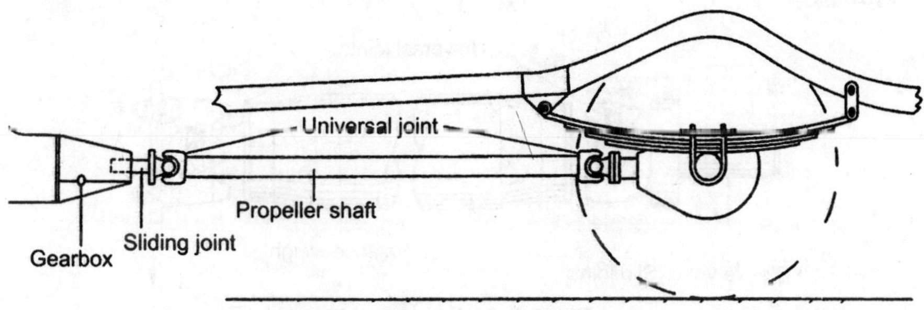

Fig.6.1: *Propellershaft assembly*

PROPELLERSHAFT

Propellershaft, sometimes called a Cardan shaft, is usually a light weight hollow carbon steel tube which is strong enough to resist twisting and bending. Only one propellershaft is normally used. A two-piece and a three-piece type propellershaft are sometimes used incorporating center bearing(s) which is designed to reduce vibration and noise.

Requirements of a propellershaft

For achieving efficient functions, the following requirements are expected in a propellershaft.

1. High torsional strength.
2. Toughened and hardened.
3. Efficiently jointed.
4. Dynamically balanced.

5. Least stress induction.
6. Reduced thrust loads.
7. Higher fatigue life.

Hollow propellershaft

A tubular-section propellershaft is normally used because it has

(a) Low weight,

(b) Large resistance to misalignment, especially sag,

(c) Good torsional strength, and

(d) Low resistance (low inertia) to changes in angular speed, which arise when a hooks' type coupling is used to drive the shaft.

Balance weight

A balance weight is installed outside the tube to balance out rotational weight.

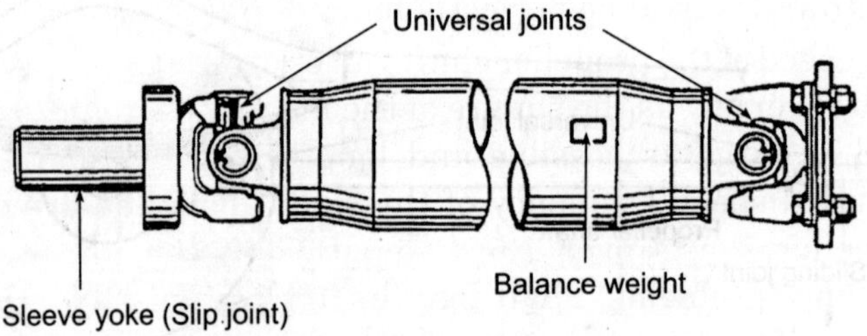

Fig.6.2: *Balance weight*

Whirling speed of propellershaft

Even after a perfect static alignment, shaft sags (i.e. forms a bow) at the centre due to its own weight. When this sagging becomes excessive, rotation of the shaft causes the bow to increase due to the centrifugal effect. This deformation, or whip of the shaft, sets up a vibration that becomes severe as it approaches the whirling speed. The speed at which shaft run with additional deflection from axis of rotation became infinite, known as critical or whirling speed. Hence, the operating speed of shaft is less (about 40% theoretically, about 10% experimentally) than critical speed to avoid failure and violent vibrations.

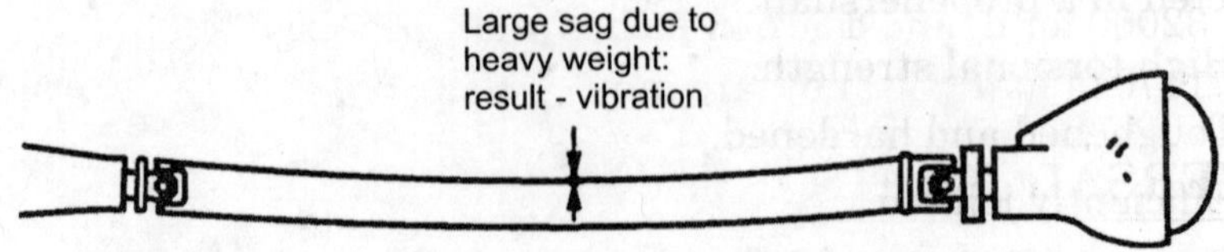

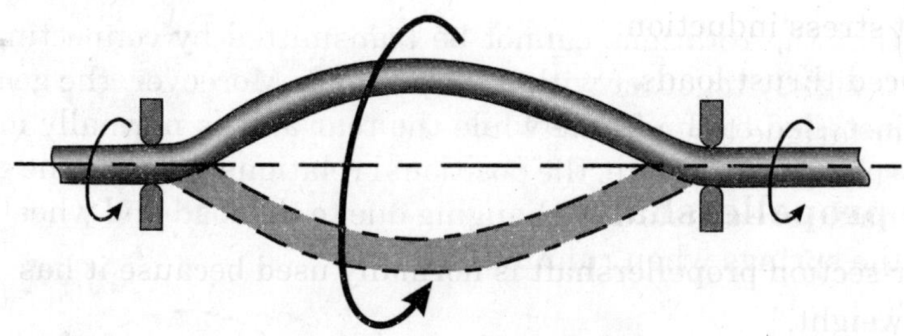

Fig.6.3: *Whirling speed*

Since propeller shafts of road vehicles are sufficiently long and operate in general at high speed, whirling may occur at certain critical speed. This produces bending stresses in the material that are higher than the shearing stresses caused by transmitted torque. While the critical speed increases with decrease in the mass of the shaft, the moment of inertia of the section increases. The tendency for the propeller shaft to whirl should be reduced and to do so. it should be made tubular and should be perfectly balanced.

Critical speed of the propeller shaft varies directly as the diameter of the tube and inversely as the square of the length. Therefore, diameters are selected as large as possible and lengths as short as possible to keep the critical speed frequency of the shaft above the driving speed range. Propeller shafts over 1.5 m length between universal joints cause unbalance problems. Shaft lengths are minimised by using long transmission extension housing and centre universal joint with two-piece propeller shafts. When used, the centre universal joint is supported by a centre support bearing which is insulated from the vehicle chassis. The propeller shaft tubing is usually rolled from flat sheet, straightened within 0.25 mm, run-out and balanced within 0.00018 kg-m. This keeps the centre mass very nearly on the longitudinal axis centre to minimise whirl. The critical speed is given by,

$$N_c = C\frac{\sqrt{d_o^{\,2}+d_i^{\,2}}}{\ell^2}$$

Where,

d_o = Outside diameter of the tubular shaft

d_i = Inside diameter of the tubular shaft

C = 32000 for d_0 and d_i are in inches and ℓ in feet

= 117044 for d_0, d_i and ℓ are in m

UNIVERSAL JOINTS

The gearbox and the rear axle are generally not at the same center line so that the rotations cannot be transmitted by connecting these two parts rigidly together with a single shaft. Moreover, the gearbox is rigidly installed to the frame while the rear axle is normally mounted on rear springs. Therefore, the positional relations between the gearbox and rear axle are constantly changing due to the load and wheel shocks flexing the springs when running.

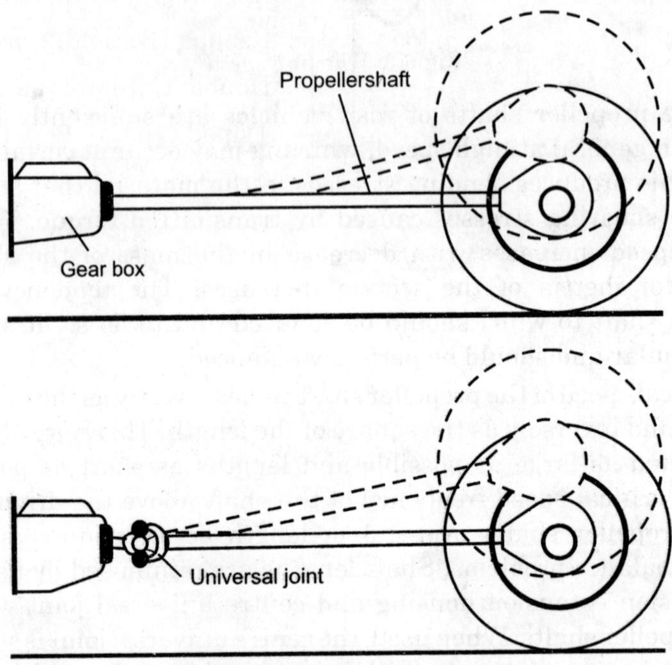

Fig.6.4: *Need of universal joints*

Requirements of Universal Joints

Universal joints must meet the following requirements:

1. It must transmit power without changing angular velocity even when the angle of the propellershaft relative to the gearbox and differential is large.
2. It must transmit power smoothly and without noise.
3. It must be simple in construction, and trouble free.

UNIVERSAL JOINTS TYPES

There are three types of universal joints: Hooke's type, flexible type and constant velocity type.

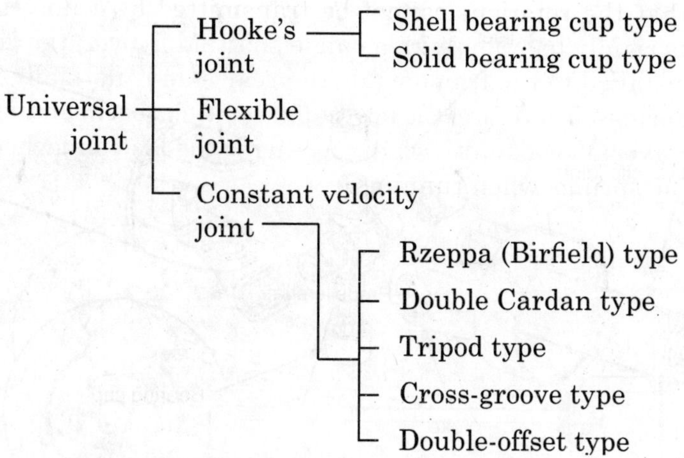

Hooke's joint

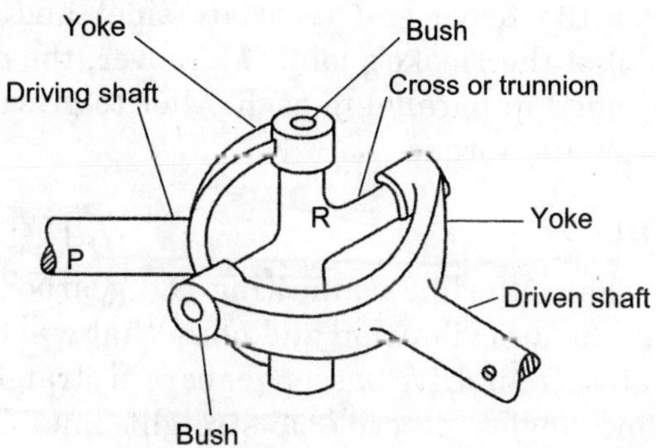

Fig.6.5: *Hooke's universal joint*

Hooke's joints are commonly used because of their simple construction and functional accuracy. One of the two yokes is welded to the propellershaft, and the other yoke forms an integral part of a joint flange or a slip joint, while a spider is installed between them via bearings. The spider is forged from a special strong type of steel, and has four surface hardened journals to give high strength and wear resistance.

Needle roller bearings are installed in each bearing cup, which is press fit into the yoke mounting hole, in order to minimise operating resistance between the journal and the yoke. In order to prevent the bearing cup from flying off when the propellershaft is turning at high speed, either a snap ring or a lock plate is used to fasten the bearing cup in the solid bearing cup type. In the shell bearing cup type, the bearing cup is crimped. Therefore, the shell bearing cup type cannot be disassembled.

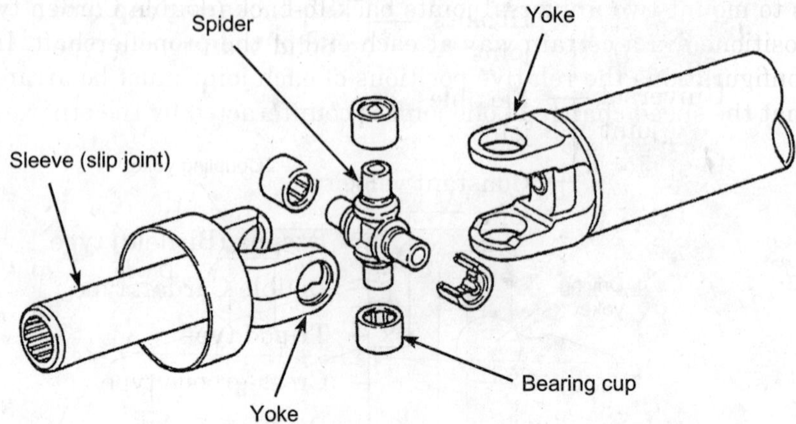

Fig.6.6: *Parts of Hooke's joint*

Variations in the angular velocity are cancelled out by universal joints located at the drive end (gearbox side) and the driven end (differential side) of the Hooke's joint. Moreover, the drive and driven shafts are positioned in parallel to each other to prevent variations in the rotating speed and torque.

Flexible joint

The straighter the centerline connecting the gearbox, propellershaft, and differential, the less vibration and noise that will occur. Therefore, in some of the latest front drive passenger cars, a straight propellershaft (one whose joint angles describe a straight line) is used. Such a propellershaft also has flexible joints to ensure less vibration and noise.

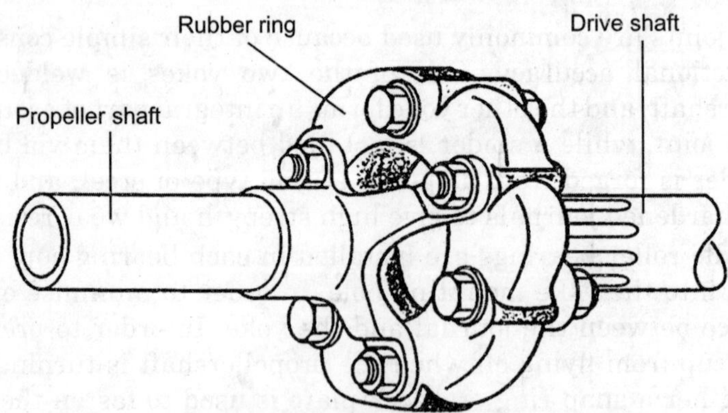

Fig.6.7: *Flexible joint*

Constant velocity universal joints

One method of achieving a constant speed output from the propellershaft is to mount two universal joints back-to-back (double Carden type) or positioned in a certain way at each end of the propellershaft. In both configurations the relative positions of each joint must be arranged so that the speed change of one joint is counteracted by the other.

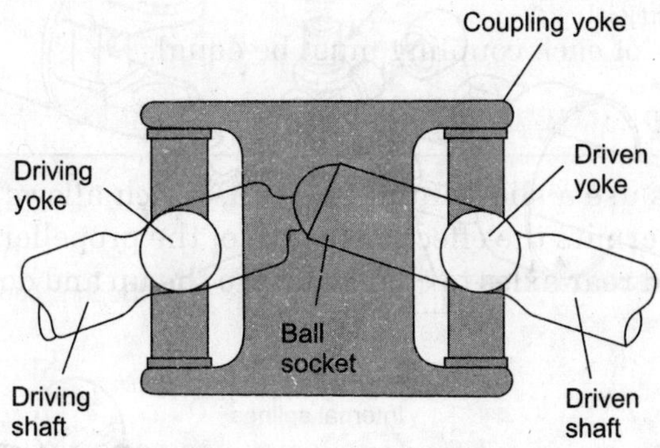

Fig.6.8: *Double Carden joint*

A constant velocity universal joint (double Carden type) used in front engine rear wheel drive is shown in Fig.6.8. It consists of two universal joints linked by a ball and socket. The ball and socket splits the angle of the drive and driven shafts between the two universal joints of the constant-velocity unit. Because the two universal joints operate at the same angle, the speed-up and slow-down action is canceled out. The speed-up resulting at any instant from the action of one universal joint is canceled out by the slow-down of the other. Therefore no speed change occurs between the two shafts connected to a constant-velocity universal joint.

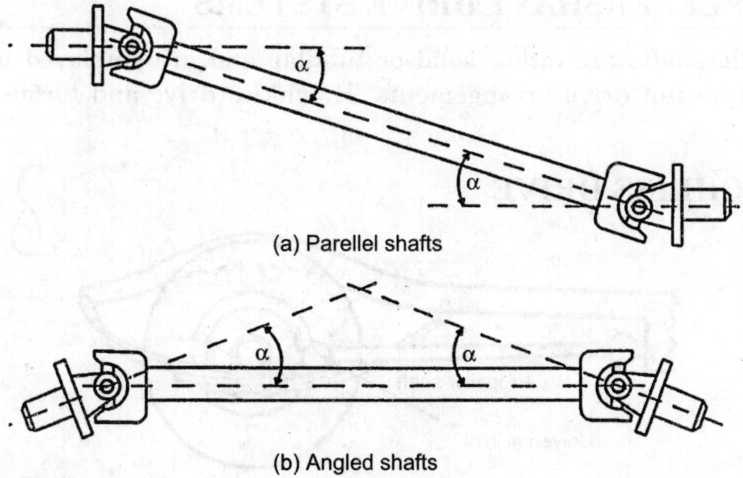

Fig.6.9: *Driveline layouts*

The phasing of universal joints, as applied to two separate driveline layouts, is shown in Fig.6.9. This diagram shows that to restore a constant speed, two conditions must be satisfied:

- yokes at each end of the propellershaft must be positioned in the same plane.
- drive angle of each coupling must be equal.

SLIP JOINT

Propellershafts use a slip joint at one end, which allows it to lengthen or shorten. It permits the effective length of the propellershaft between the gearbox and rear axles to change due to the up and down movement of the wheels.

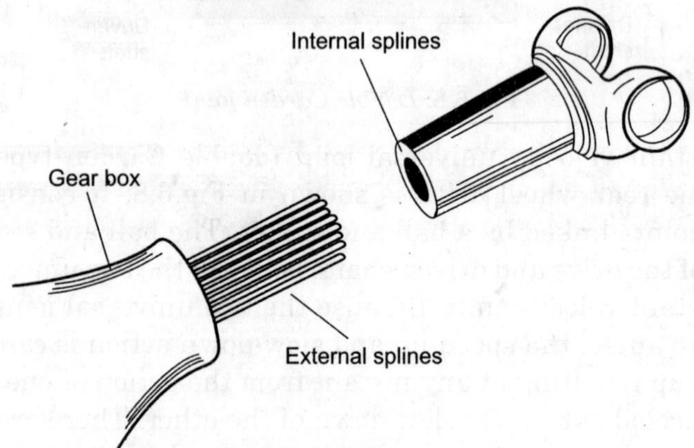

Fig.6.10: *Slip Joint and its function*

PROPELLERSHAFT DRIVE SYSTEMS

Propellershafts are either solid or tubular and are employed in two main types of drive arrangements: Hotchkiss drive and torque tube drive.

HOTCHKISS DRIVE

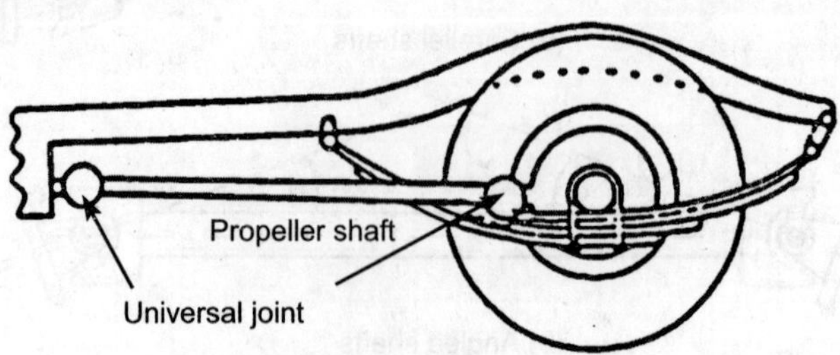

Fig.6.11: *Hotchkiss drive*

This system employs a tubular or hollow propellershaft and is the simplest type of drive. The propellershaft connecting the gearbox to the rear axle is exposed and provided with universal joint at each end. This type is also known as open shaft drive. In this case the suspension springs besides taking weight of the body, also take the torque reaction, driving thrust and the side thrust.

Application

This system is more commonly used in commercial vehicles. Such as buses, trucks, etc.

TORQUE TUBE DRIVE

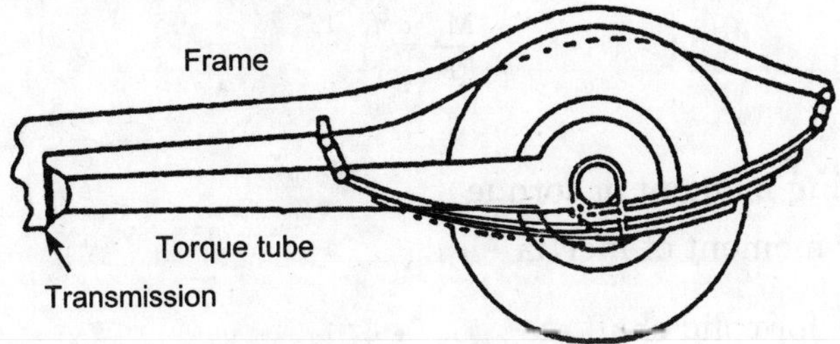

Fig.6.12: *Torque tube drive*

In this type, the spring takes only the side thrust besides supporting the body weight. The torque reaction, braking torque and the driving thrust are taken by another member which is called the torque tube. Here a solid shaft is used and is enclosed in a tube or casing referred to as a "torque tube". The rear end is bolted to the rear axle casing and the front end is supported by a ball and socket joint, the socket being secured to the gearbox casing. An arrangement of this type allows the rear end of the torque tube to move up and down keeping the distance between gearbox and rear axle constant. One universal joint is fitted to the front of the propellershaft connecting it to the gearbox main shaft.

Application

Although the torque tube drive system was used extensively on early vehicles, it is rarely employed in modern vehicle design.

DESIGN AND PROBLEMS

DESIGN OF PROPELLER SHAFT

Design of propeller shaft is basically determination of the diameter to give sufficient strength under various operating conditions.

Shafts whose cross-section is uniform throughout the length may be designed on the basis of two conditions.

1. Strength conditions
2. Stiffness or rigidity conditions

Propeller shaft design on the basis of strength

Strength means whether the shaft is strong enough to transmit the required power, support the required load and so on.

When the propeller shaft is subjected to twisting moment or torque only (pure torsion), the diameter of the shaft may be obtained by using the torsion equation.

$$\frac{M_t}{J} = \frac{\tau}{r}$$

Where,

M_t = twisting moment or torque

J = Polar moment of inertia

$= \frac{\pi}{32} d^4$ for solid shaft

$= \frac{\pi}{32}(d_0^4 - d_i^4)$ for hollow shaft

τ = Torsional shear stress

r = Radius of the shaft

Torque to be transmitted (M_t)

Engine torque

$$P = \frac{2\pi N T_e}{60} \text{ Watts}$$

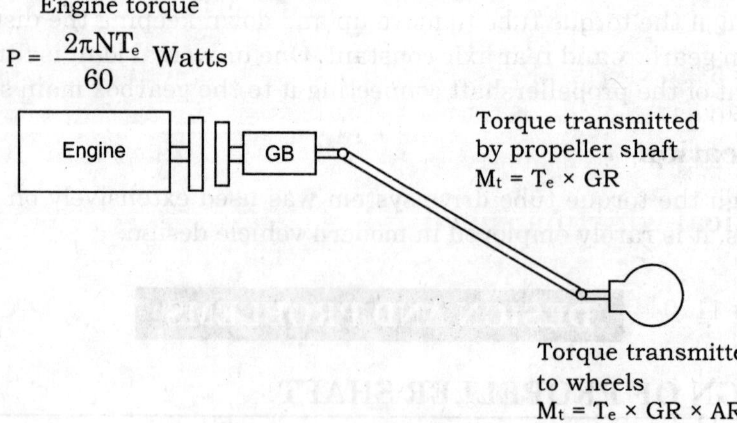

Problem 6.1: An automobile engine develops 28 kW at 1500 rpm and its bottom gear ratio is 3.06. If a propeller shaft of 40 mm outside diameter is to be used, determine the inside diameter of mild steel tube to be used, assuming a safe shear stress of 55×10^3 kpa for the MS.

If the length of the propeller shaft is 1.54 m, find the critical speed of the shaft.

Given data

$P = 28\text{ kW} = 28\times10^3$ Watts

$N = 1500$ rpm

$GR = 3.06$

$d_o = 40\text{ mm} = 0.04\text{ m}$

$\tau = 55\times10^3\text{ kpa} = 55\times10^6\text{ pa} = 55\times10^6\text{ N/m}^2$

Solution

The diameter of the shaft may be obtained by using the torsion equation.

Engine torque

$$P = \frac{2\pi N T_e}{60}\text{ Watts}$$

$$28\times10^3 = \frac{2\pi\times1500\times T_e}{60}$$

$$T_e = 178.25\text{ Nm}$$

Torque transmitted by the propeller shaft

$$Mt = T_e \times GR$$

$$= 178.25\times3.06$$

$$= 545.45\text{ Nm}$$

$$\frac{M_t}{J} = \frac{\tau}{r}$$

$$\frac{M_t}{\frac{\pi}{32}(d_o^4 - d_i^4)} = \frac{\tau}{\frac{d_o}{2}}$$

$$\frac{545.45}{\frac{\pi}{32}(0.04^4 - d_i^4)} = \frac{55\times10^6}{\frac{0.04}{2}}$$

$$2.0203\times10^{-6} = 0.04^4 - d_i^4$$

$$d_i^4 = 5.397\times10^{-7}$$

$$d_i = 0.0271\text{ m}$$

$$= 27.1\text{ mm @ 28 mm}$$

Critical whirling speed

$$N_c = 117044\,\frac{\sqrt{d_o^2 + d_i^2)}}{\ell^2}$$

$$= 117044\,\frac{\sqrt{0.04^2 + 0.028^2}}{1.54^2}$$

$$= 2409.68 \cong 2450\text{ rpm}$$

Problem 6.2: An automobile engine develops a maximum torque of 162 Nm. The low gear ratio of transmission is 2.75, while the back axle ratio is 4.25. The effective wheel radius is 0.325 m and the co-efficient of friction between the tyre and the road surface is 0.6. If the permissible shear stress is 32373×10⁴ pa, determine the maximum shaft diameter, assuming that the load is nearly torsional. What is the maximum load permissible on each wheel?

Given data

$T_e = 162$ Nm

GR = 2.75

AR =4.25

R = 0.325 m

$\mu = 0.6$

$\tau = 32373\times10^4$ pa $= 32373\times10^4$ N/m²

Solution

(a) Shaft diameter

For a shaft subjected to torsional load,

$$\frac{M_t}{J} = \frac{\tau}{r}$$

$$\frac{M_t}{\frac{\pi}{32}d^4} = \frac{\tau}{\frac{d}{2}}$$

$$\frac{1893.37}{\frac{\pi}{32}d^4} = \frac{32373\times10^4}{\frac{d}{2}}$$

Torque transmitted to wheels

$$M_t = T_e \times GR \times AR$$

$$= 162 \times 2.75 \times 4.25$$

$$= 1893.37 \text{ Nm}$$

$d^3 = 2.9786\times10\text{-}5$

$d = 0.0309$ m

$= 30.9$ mm $\cong 31$ mm

(b) Maximum load permissible on each wheel

We know that

$$\text{Co-efficient of friction } \mu = \frac{F}{W}$$

$$M_t = F \times R$$

$$1893.37 = F \times 0.325$$

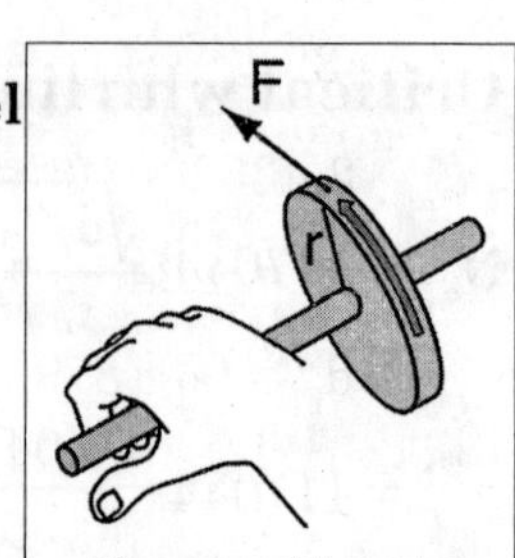

Torque T = Fr

$$F = 5825.75 \text{ N}$$

$$\therefore \quad 0.6 = \frac{5825.75}{W}$$

$$W = 9709.58 \text{ N}$$

Problem 6.3: An automobile engine develops 33.75 kW at 1500 rpm and its bottom gear ratio is 3.06:1. If a propeller shaft of 40 mm outside diameter is to be used, determine the inside diameter of mild steel tube to be used, assuming a safe shear stress of 55×10^3 kpa for the mild steel. If the length of the propeller shaft is 1.54 m, find the critical speed of the shaft. Also ascertain whether the critical speed is sufficient to avoid whirling.

Given data

P = 33.75 kW = 33.75×10^3 Watts

N = 1500 rpm

GR = 3.06

d_o = 40 mm = 0.04 m

t = 55×10^3 kpa = 55×10^6 pa = 55×10^6 N/m^2

Solution

(a) Inside diameter of a propeller shaft

The diameter of the shaft may be obtained by using the torsion equation.

$$\frac{M_t}{J} = \frac{\tau}{r}$$

$$\frac{M_t}{\frac{\pi}{32}(d_0^4 - d_i^4)} = \frac{\tau}{\frac{d_0}{2}}$$

$$\frac{657.46}{\frac{\pi}{32}(0.04^4 - d_i^4)} = \frac{55 \times 10^6}{\frac{0.04}{2}}$$

$$2.4352 \times 10^{-6} = 0.04^4 - d_i^4$$

$$d_i^4 = 1.248 \times 10^{-7}$$

$$d_i = 0.01879 \text{ m}$$

$$= 18.79 \text{ mm} \cong 19 \text{ mm}$$

Engine torque

$$P = \frac{2\pi N T_e}{60} \text{ Watts}$$

$$33.75 \times 10^3 = \frac{2\pi \times 1500 \times T_e}{60}$$

$$T_e = 214.85 \text{ Nm}$$

Torque transmitted by the propeller shaft

$$Mt = T_e \times GR$$

$$= 214.85 \times 3.06$$

$$= 657.46 \text{ Nm}$$

(b) Critical speed of the propeller shaft

$$N_c = 117044 \frac{\sqrt{d_o^{\,2} + d_i^{\,2}}}{\ell^2}$$

$$= 117044 \frac{\sqrt{0.04^2 + 0.019^2}}{1.54^2}$$

$$= 2185.47 \cong 2186 \text{ rpm}$$

Propeller shaft design on the basis of rigidity

The equation established for torsional rigidity is

$$\frac{M_t}{J} = \frac{G\theta}{\ell}$$

Where,

M_t = twisting moment or torque

J = Polar moment of inertia

$$= \frac{\pi}{32} d^4 \text{ for solid shaft}$$

$$= \frac{\pi}{32} (d_0^{\,4} - d_i^{\,4}) \text{ for hollow shaft}$$

G = Modulus of rigidity of the shaft material

θ = Torsional deflection or angle of twist in radians

l = Length of the propeller shaft

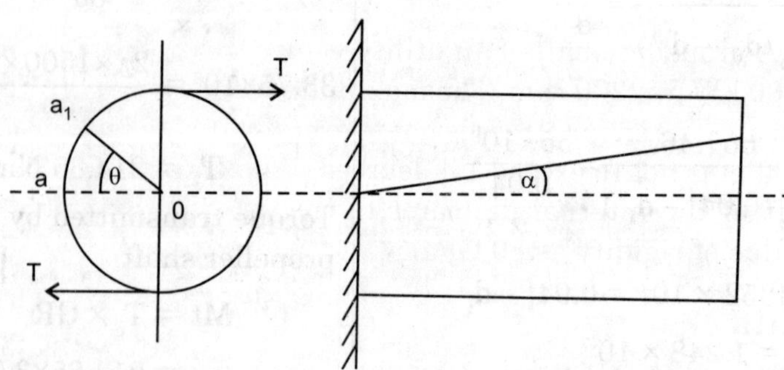

Problem 6.4: A solid propeller shaft is to transmit 112 kW at 500 rpm with provision for a 20% overload. The angle of twist must not exceed 1 degree in a length of 15 diameters. For practical reasons, the diameter

must not be more than 85 mm. Design the shaft. Take modulus of rigidity as 8×10^4 N/mm^2.

Given data

$P = 112$ kW $= 112\times10^3$ Watts

$N = 500$ rpm

$M_t = 1.2\ T_e$

$\theta = 1° = \pi/180$ radians

$l = 15d$

$d < 85$ mm

$G = 8\times10^4$ N/mm^2

Engine torque

$$P = \frac{2\pi N T_e}{60} \text{ Watts}$$

$$112\times10^3 = \frac{2\pi\times500\times T_e}{60}$$

$T_e = 2139$ Nm

$= 2139\times10^3$ Nmm

Maximum torque

$Mt = 1.2$

$T_e = 1.2\times2139\times10^3$

$= 2567\times10^3$ Nmm

Solution

The equation established for torsional rigidity is

$$\frac{M_t}{J} = \frac{G\theta}{\ell}$$

$$\frac{2567\times10^3}{\frac{\pi}{32}d^4} = \frac{8\times10^4\times\frac{\pi}{180}}{15d}$$

$d^3 = 280898.79$

$d = 65.49$ mm

$\cong 66$ mm < 85 mm

$\therefore$ Design is safe

Problem: A propeller shaft of an utility vehicle rear wheel driven is to transmit 100 kW at 2200 rpm. The maximum torque being 20% greater than the mean. The shear stress is not to exceed 60 Mpa and the twist in a length of 3 m is not to exceed 1.4 degrees. Find the external diameter of the shaft, if the internal diameter to the external diameter is 3/8. Take modulus of rigidity as 80 GPa.

Given data

$P = 100$ kW $= 100\times10^3$ Watts

$N - 2200$ rpm

$M_t = 1.2T_e$

$\tau = 60$ Mpa $= 60\times10^6$ pa $= 60\times10^6$ N/m²

$\theta = 1.4° = 1.4 \times \frac{\pi}{180} = 0.0244$ rad

$l = 3$ m

$\frac{d_i}{d_o} = \frac{3}{8} = 0.375$

$d_i = 0.375d_0$

$G = 80$ Gpa $= 80\times10^9$ pa $= 80\times10^9$ N/m²

Solution

τ and θ are given. Hence the problem can be solved on the basis of both strength and rigidity.

(a) On the basis of strength

$$\frac{M_t}{J} = \frac{\tau}{r}$$

$$\frac{M_t}{\frac{\pi}{32}(d_0^4 - d_i^4)} = \frac{\tau}{\frac{d_0}{2}}$$

$$\frac{520.87}{\frac{\pi}{32}(d_0^4 - (0.375d_0)^4)} = \frac{60\times10^6}{\frac{d_0}{2}}$$

$d_0^3 = 4.5115\times10^{-5}$

$d_0 = 0.03559$ m

$= 35.59$ mm

$\cong 36$ mm

$d_i = \frac{3}{8} \times 36$

$d_i = 13.5 \cong 14$ mm

Engine torque

$$P = \frac{2\pi N T_e}{60} \text{ Watts}$$

$$100\times10^3 = \frac{2\pi \times 2200 \times T_e}{60}$$

$T_e = 434.05$ Nm

Maximum torque

Mt $= 1.2$

$T_e = 1.2\times434.05$

$= 520.87$ Nm

(b) On the basis of rigidity

The equation established for torsional rigidity is

$$\frac{M_t}{J} = \frac{G\theta}{\ell}$$

$$\frac{520.87}{\frac{\pi}{32}(d_0^4 - (0.375d_0)^4)} = \frac{80 \times 10^9 \times 0.0244}{3}$$

$d_0^4 = 8.3185 \times 10^{-6}$

$d_0 = 0.0537$ m

$= 53.7$ mm @ 54 mm

$d_i = \frac{3}{8} \times 54$

$= 20.25 \cong 21$ mm

Select the higher value

$d_o = 54$ mm

$d_i = 21$ mm

Final Drive and Differential Assembly

FINAL DRIVE

The final gear of the differential assembly consists of the drive pinion and ring gear (crown wheel). The purpose of final drive is to:

- provide a constant speed reduction, irrespective of what gear is engaged in the gearbox (permanent gear reduction) to reduce the speed of the axle shafts to the value required by the road wheels while increasing the torque applied to them.
- turn the drive through 90°, where an in-line engine is employed.

FINAL DRIVE RATIO (AXLE RATIO)

The ring gear is three to four times larger than the pinion gear. When the pinion turns the ring gear, it reduces the speed of the axle shafts while increasing the torque applied to them. For most light vehicles, a final drive gear ratio of approximately 4:1 is common, while for heavier vehicles the reduction may be as high as 20:1. The actual gear reduction is calculated by dividing the number of teeth on the crown wheel (ring gear) by the number of teeth on the pinion gear. That is:

$$\text{Gear ratio} = \frac{\text{Number of teeth on crown wheel}}{\text{Number of teeth on pinion}}$$

TYPES OF FINAL DRIVE

The position and type of final drive varies according to the vehicle's layout. With a transverse engine the final drive is driven directly from the gearbox output shaft and is known as reduction gearing. When longitudinal engine layout is employed, a crown wheel and pinion is used. Some heavy vehicles are equipped with worm and wheel drive.

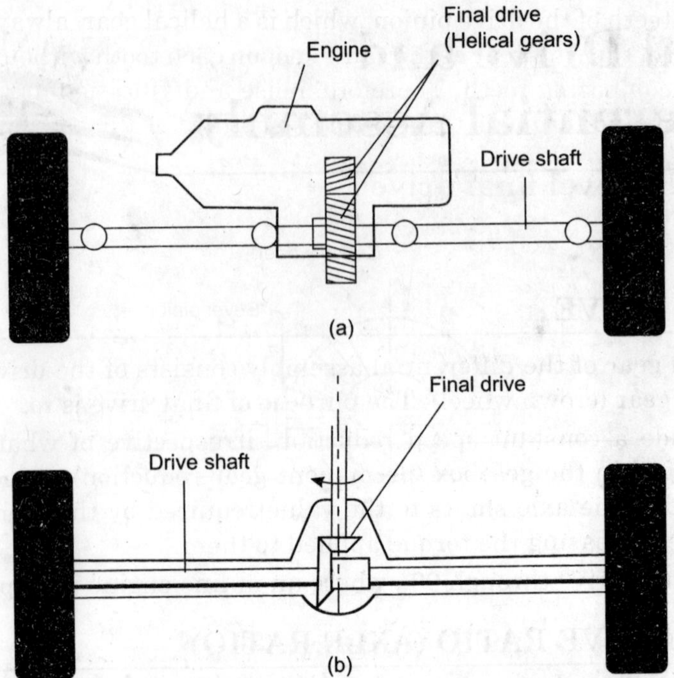

Fig.7.1: *Final drive (a) for front-wheel drive and (b) for rear-wheel drive*

It is generally a pair of helical gears on front-wheel-drive vehicles while bevel gears on rear-wheel-drive vehicles.

Helical gear final drive

The popularity of transverse, front-wheel drive vehicles means that a pair of helical gears is now one of the most common types of final drive. It employs two helical-cut gears mounted in substantial bearings.

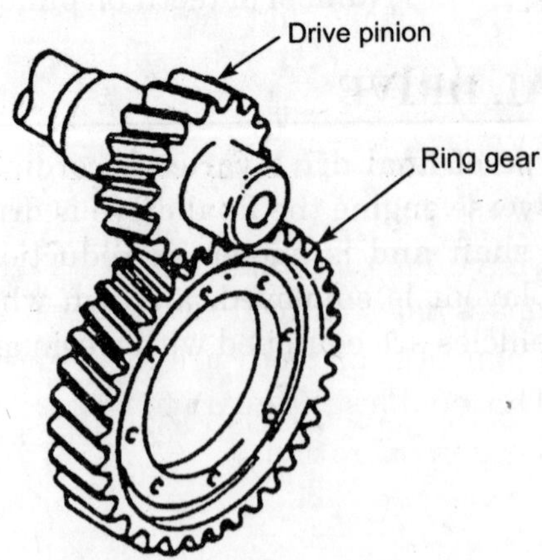

Fig.7.2: *Helical gear final drive*

The teeth of the drive pinion, which is a helical gear, always contact the teeth of the ring gear at the same spot on each tooth without slipping along the opposing teeth. Therefore, noise and vibration are kept to a minimum, and torque is transmitted smoothly.

Straight bevel final drive

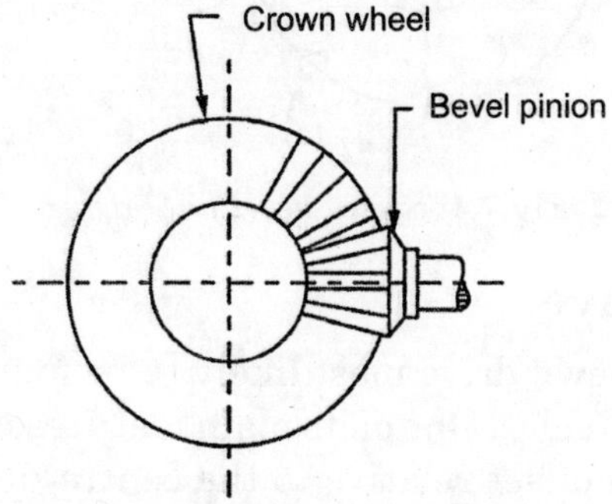

Fig.7.3: *Straight bevel final drive*

In this design, straight-cut bevel teeth are machined onto both the cone-shaped pinion and the crown wheel.

The gear teeth are specially hardened and tempered to withstand the torsional stresses and shock loadings that may be imposed when transmitting the engine's torque. There are several disadvantages with this type of drive:

- the axes of the pinion and the crown wheel must coincide, which requires a high floor level or deep propeller shaft tunnel.
- with straight-cut gears the drive is noisy.
- only one pair of gear teeth are in contact at any one time, which limits the amount of torque, that the final drive can transmit.

Because of these limitations, straight-cut teeth are no longer used on final drives.

Spiral bevel final drive

Spiral bevel drives have the teeth machined at an inclined angle. The use of spiral bevel gears has technical advantages:

- more than one pair of gear teeth are meshing with each other at any one time. This enables the gears to take a higher loading.
- the drive is quieter in operation.

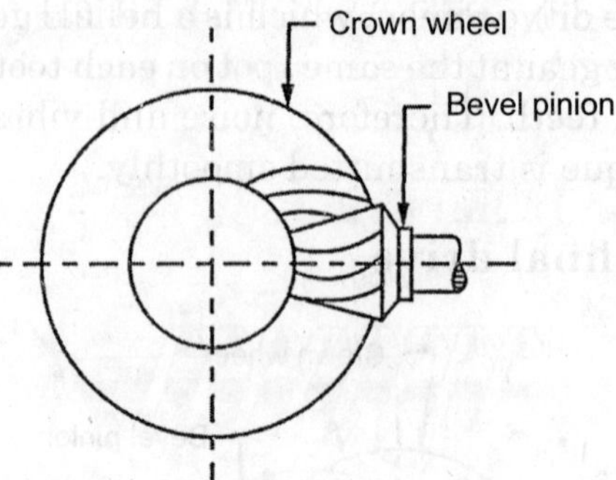

Fig.7.4: *Spiral bevel final drive*

Hypoid final drive

This is the type of bevel drive most likely to be found on current front-engine rear-drive vehicles. One of its main features is that the centerline axis of the pinion is offset relative to the centre-line axis of the crown wheel. An offset of one-fifth of the crown wheel diameter is common.

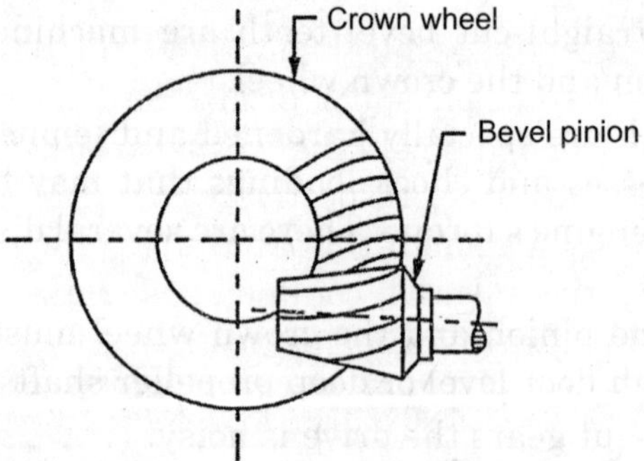

Fig.7.5: *Hypoid final drive*

The pinion offset may be used to improve body design. By lowering the propeller shaft level it enables the vehicle body floor to be dropped. This allows the vehicle to be designed with,

- a body having normal ground clearance which helps to improve stability of the vehicle.
- gears are meshed at a larger meshing ratio and the gear operation is very quiet. It also allows more teeth to be in contact to carry the load.

The pinion is one-third larger and stronger than the spiral bevel type and failure due to breakage is rare.

Since these gears are constantly sliding and rolling against each other, the hypoid bevel gear must be lubricated with hypoid gear oil (extreme pressure oil), which has a high oil film strength. Additives such as chlorine and sulphate prevent the gear teeth from 'scuffing' as well as providing adequate lubrication at all likely temperatures.

Worm and wheel final drive

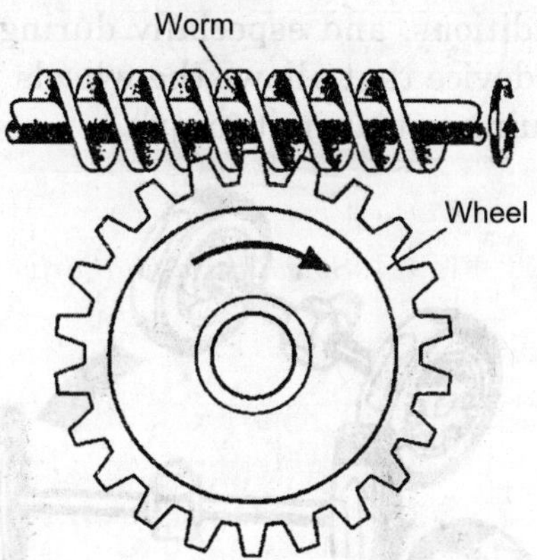

Fig.7.6: *Worm and wheel final drive*

This drive consists of a spiral-shaped worm and a toothed wheel. The wheel is made from phosphor bronze and the worm is normally made from case-hardened steel. The worm can be placed either above or below the wheel to give, respectively, a higher ground clearance or a lower floor line. Although this final drive provides a larger gear reduction in a relatively small, confined space, the materials used in its construction are expensive. This restricts its use to heavy goods vehicles and buses.

DIFFERENTIAL

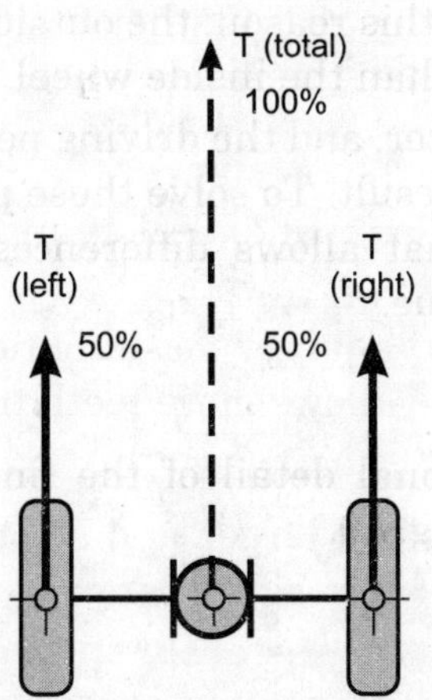

Fig.7.7: *Principle of differential*

The differential is a device that enables two drive shafts to be driven at different speeds, but with equal torque.

Need for differential

The right and left wheels do not always rotate at the same speed because of road conditions, and especially during cornering. There is a need for a special device that allows the wheels to rotate at different speeds while continuing to transmit torque.

Fig.7.8: *Need for differential unit*

In comparing the path of the inside wheel with the path of the outside wheel of a vehicle when it turns along a curve, as illustrated in Fig.7.8, the distance of travel of the outside wheel is longer than that of the inside wheel. For this reason, the outside wheel is forced to move faster and rotate more than the inside wheel.

Therefore, wear faster, and the driving performance of the vehicles would be affected as a result. To solve these problems, there is needed a differential device that allows differences in rpm to occur while transmitting equal torque.

Construction

The general constructional detail of the final drive and differential assembly is shown in Fig.7.9.

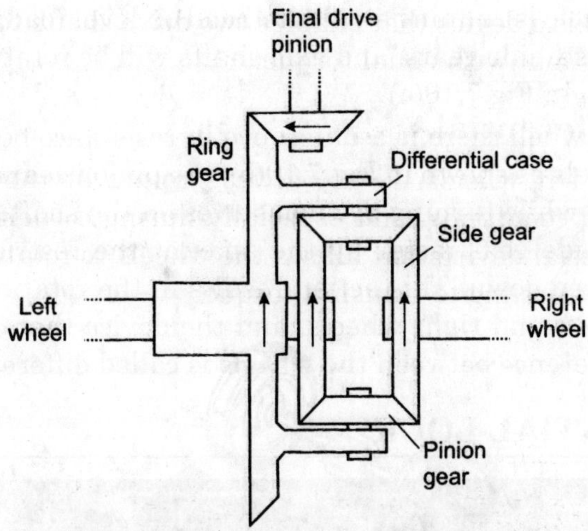

Fig.7.9: *Differential assembly*

The ring gear (or crown wheel) of the final drive is attached to a differential case which contains four bevel-type gear pinions all facing inwards, meshing with each other in the form of a box. Two of the bevel pinions opposite each other are splined to the half shafts and are referred to as the sun gears. The other opposed pair of pinions are free to rotate upon a pinion shaft and are known as pinion gears. The pinion shaft which acts as a pivot for the pinion gears is mounted in the differential housing which is driven by the ring gear. There is no direct connection between the ring gear and the half shafts.

Operation

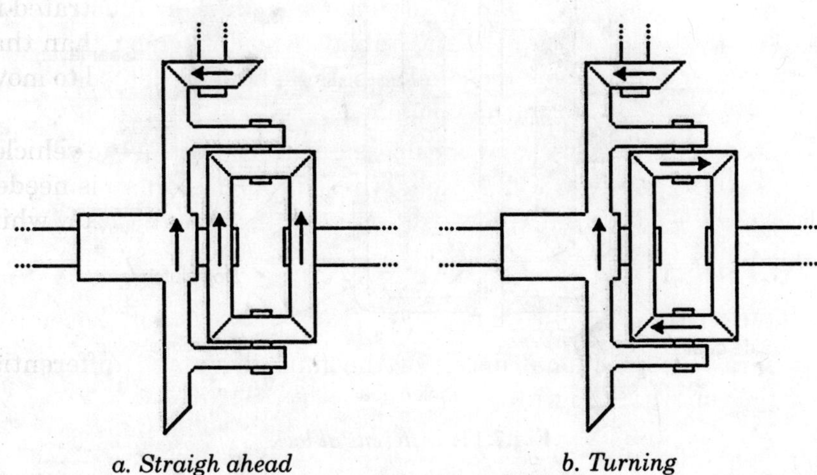

a. Straigh ahead *b. Turning*

Fig.7.10: *Differental operation*

When the vehicle is running straight ahead on level road, the ring gear, differential case, side gears, and axle shafts will be rotating as a single body as shown in Fig.7.10(a).

However, when there is a difference in resistance between the left and right wheels as shown in Fig.7.10(b), the pinion gears revolve along the side gears while turning by themselves on the shaft, and drive the unrestricted side gear faster by the amount the restricted side gear had been slowed down. This action results in the rotational difference between the left and right wheels, and the device that creates such a rotational difference between the wheels is called differential gears.

DIFFERENTIAL LOCK

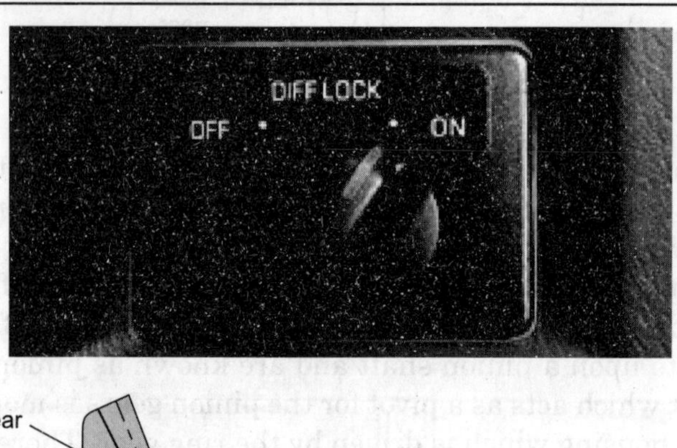

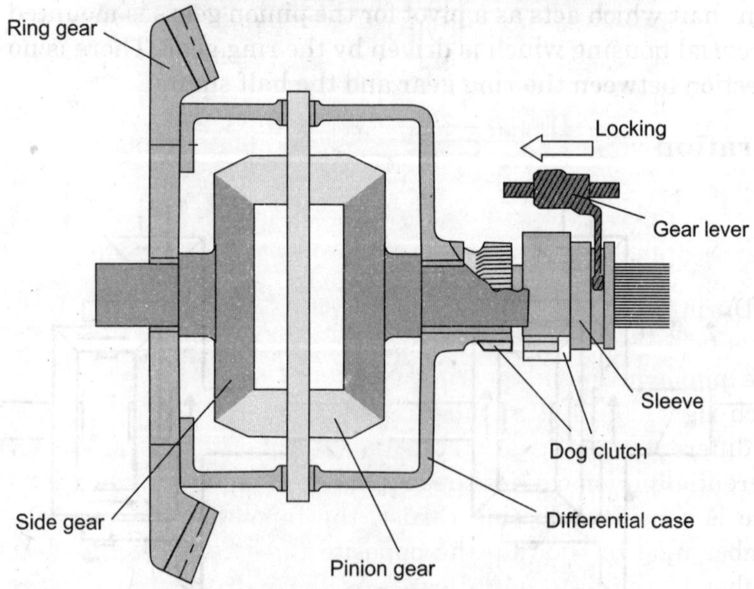

Fig.7.11: *Differential lock*

There is also a disadvantage in providing the differential action. This disadvantage occurs when the wheel at one side becomes stuck in mud or sand. The free wheel will have smaller resistance so that it will turn without allowing the stuck wheel to turn. This prevents the vehicle from freeing itself when mired.

This difficulty can be solved by stopping the self-rotation of the pinion gears so as to allow the same amount of torque to be transmitted to the side gears. A device performing such action is called differential lock.

LIMITED SLIP DIFFERENTIAL

If one of the drive wheels of a vehicle with an ordinary differential slips into a ditch or gets stuck in the mud, that wheel will begin to spin at high speed. The other wheel will stop and the vehicle will not be able to get out of the ditch or mud. The limited slip differential is designed to help alleviate these problems by adding a slip-limiting function to the ordinary differential.

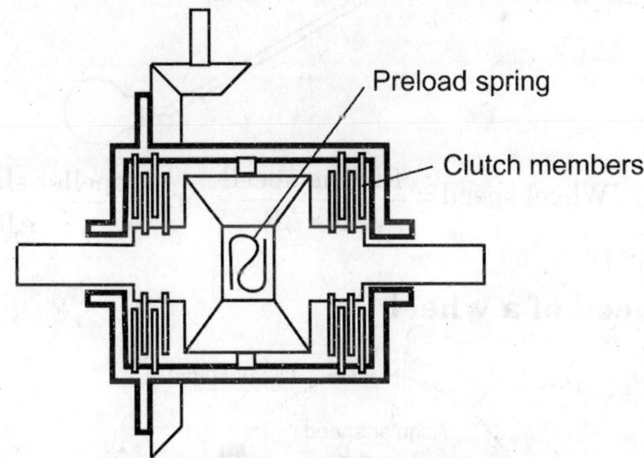

Fig.7.12: *Limited slip differential*

During normal straight road driving, power flow is same as the conventional differential. The turning force is transmitted from the drive pinion through the ring gear, and the differential case to the clutch right, and left members, and through the differential pinions, and differential side gears to the rear axle shafts. Note that the turning differential pinion carries the clutch left member around with it. Since there is considerable side thrust, the taper lands on the clutch left member tend to slide up the opposite taper lands of the clutch right member in the differential case. As these drive up, these are forced outwardly, and this force is transmitted to the differential side gears. The side gears thus lock the differential side gear shafts. Then, if one wheel should encounter a patch of ice or mud that causes it to lose traction temporary, it will not spin since it it cannot turn faster than the other wheel.

When rounding a curve, the differential acts in a conventional manner to permit the outer wheel to rotate a little faster than the inner wheel. This action is permitted by slipping of the differential side gears.

DESIGN AND PROBLEMS

DESIGN OF FINAL DRIVE

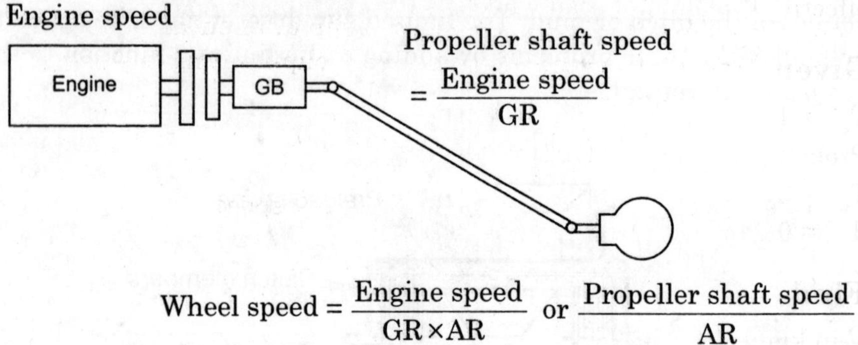

$$\text{Wheel speed} = \frac{\text{Engine speed}}{\text{GR} \times \text{AR}} \text{ or } \frac{\text{Propeller shaft speed}}{\text{AR}}$$

Linear speed of a wheel

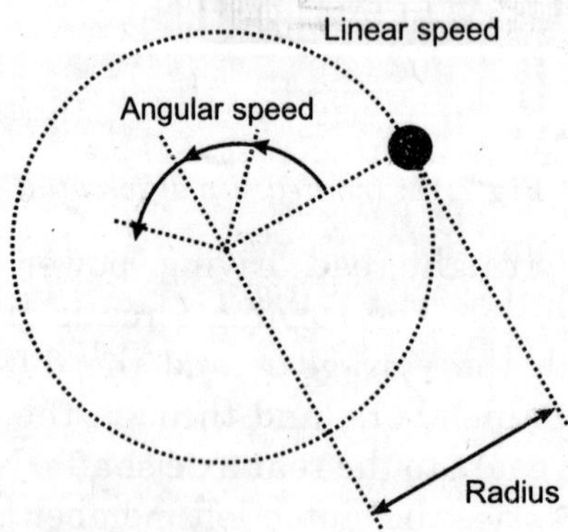

Linear speed of a wheel

$$V = \text{Wheel speed} \times \text{Circumference of wheel}$$

$$= \text{Wheel speed} \times \pi d$$

Axle ratio

(a) For bevel gears final drive

$$AR = \frac{\text{No. of teeth on crown wheel}}{\text{No. of teeth on bevel pinion}} = \frac{T_c}{T_b}$$

(b) For worm and worm wheel final drive

$$AR = \frac{\text{No. of teeth on worm wheel}}{\text{No. of start on worm}} = \frac{T_w}{T_s}$$

Problem 7.1: A four start worm is driven by a propeller shaft at 2500 rpm and meshes with a crown wheel on the rear axle shaft. The crown wheel has 21 teeth. Calculate the speed of the vehicle in kmph if the effective diameter of road wheels is 0.7 m.

Given

$T_s = 4$

Propeller shaft speed = 2500 rpm

$T_w = 21$

$d = 0.7$ m

Find

V in kmph

Solution

$$AR = \frac{\text{No. of teeth on worm wheel}}{\text{No. of start on worm}} = \frac{T_w}{T_s} = \frac{21}{4} = 5.25:1$$

$$\text{Wheel speed} = \frac{\text{Propellershaft speed}}{AR} = \frac{2500}{5.25} = 476.19 = 477 \text{ rpm}$$

$$\begin{aligned}\text{Linear speed} &= \text{Wheel speed} \times \text{Circumference of wheel}\\ &= \text{Wheel speed} \times \pi d\\ &= 477 \times (\pi \times 0.7)\\ &= 1048.97 \text{ m/min}\\ &= 1048.97 \times \frac{60}{1000}\\ V &= 62.93 \text{ Kmph}\end{aligned}$$

Problem 7.2: A vehicle speed has a third gear ratio of 1.5 to 1 and a rear axle ratio of 4.5 to 1. Calculate (a) the overall gear ratio and (b) the number of revolutions made by the crown wheel per minute if the engine speed is 2700 rpm.

Given

GR = 1.5

AR = 4.5

Engine speed = 2700 rpm

Find

Overall gear ratio, wheel speed

Solution

(a) Overall gear ratio $= GR \times AR = 1.5 \times 4.5 = 6.75$

(b) Wheel speed $= \dfrac{\text{Engine speed}}{GR \times AR} = \dfrac{\text{Engine speed}}{\text{Overall gear ratio}}$

$= \dfrac{2700}{6.75}$

$= 400$ rpm

Axle

FRONT AXLE

The front axle supports the front of the vehicle and is also used for steering. It also absorbs shocks due to road irregularities. This usually takes the form of an I-section beam which is machined from a heat-treated steel forging. The beam is provided with spring seats for mounting the front road springs. Both ends of the beam are upswept and of oval section to resist the action of the braking torque which attempts to turn the axle with the road wheel. To keep the chassis height low the center portion of the front axle is given a downward sweep as shown in the Fig.8.1. A stub axle or swivel pin assembly is secured to each end of the beam by means of a king pin.

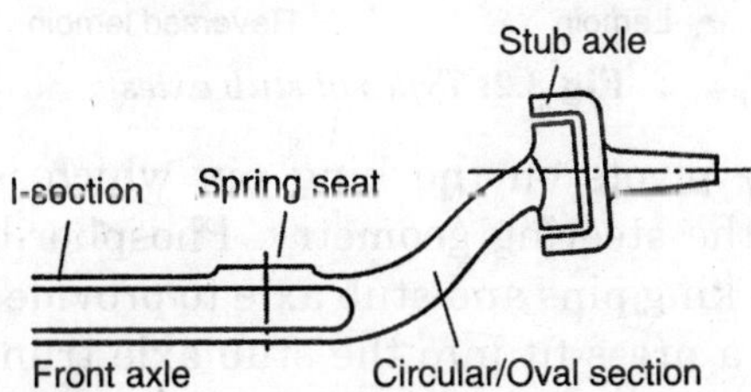

Fig.8.1: *Front axle of commercial vehicles*

Simple tubular type non-driving axles are also sometimes used on commercial vehicles. In this type the centre section of the axle is manufactured from tempered seamless steel tube. The stub axle eyes and spring seats are electrically welded into position on the tube and become integral parts of the axle assembly.

There are two types of front axles: Dead front axle, and Live front axle

Dead front axle

Dead axles are those axles, which do not rotate. These axles have sufficient rigidity and strength to take the weight. The ends of front axle are suitably designed to accommodate stub axles.

Live front axle

Live axles are used to transmit power from gearbox to front wheels.

STUB AXLE

Vehicles are equipped with stub axles mounted on kingpins or steering knuckles. The stub axle of the steered wheel is pivoted around the steering axis. The wheelbase remains more or less the same. As a result, the vehicle remains stable even at the maximum steering lock angles.

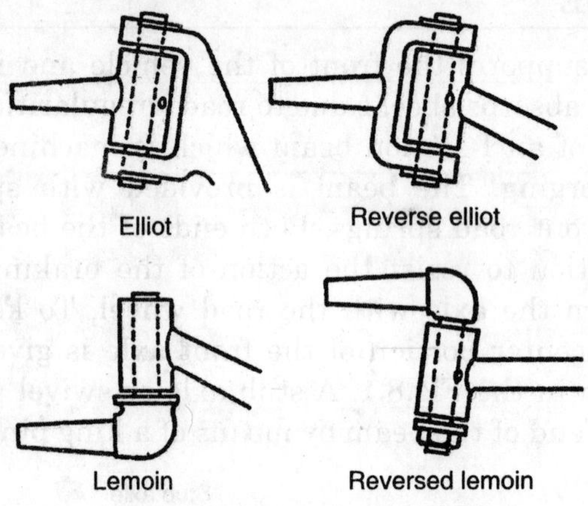

Fig.8.2: *Types of stub axles*

The stub axle pivots on the king pin which is mounted at the correct angle for the steering geometry. Phosphor-bronze bushes are fitted between the king pins and stub axle to provide a bearing surface. These bushes are a press fit into the stub axle trunnion eyes and are reamed to be sliding fit to the actual king pin, which must pass through the three eyes to provide a firm but free swivel action. The vertical load is taken on a hardened-steel thrust washer or ball-thrust bearing which is located on the underside of the axle eye.

Each stub axle is machined to accommodate the steering arms. These are connected by a track rod to ensure that each road wheel is turned through the correct angle. On right-hand drive vehicles the offside stub axle carries an intermediate steering arm which is connected to the drag link. Each end of the track rod and drag link must be fitted with a ball joint to allow the wheels to turn.

To prevent the road wheels from being turned through too great an angle, steering stops are mounted on each stub axle. In practice these restrict the steering movement to prevent the wheels fouling the chassis on full lock.

Stub axles that are used on motor vehicles can be grouped into several types as shown above in the Fig.8.2. The most popular is reversed Elliot type.

DESIGN AND PROBLEMS

LOAD DISTRIBUTION

No	Type	Load	
		Front axle	Rear axle
1.	Front engine-Rear drive	55%	45%
2.	Front engine-Front drive	60%	40%
3.	Rear engine-Rear drive	35%	65%

TYPES OF AXLE BEAM

Straight axle beam

Fig.8.3: *Straight axle beam*

Dropping the center part of axle beam

This lowers centre of gravity. Here central part of axle beam dropped immediately after spring seat.

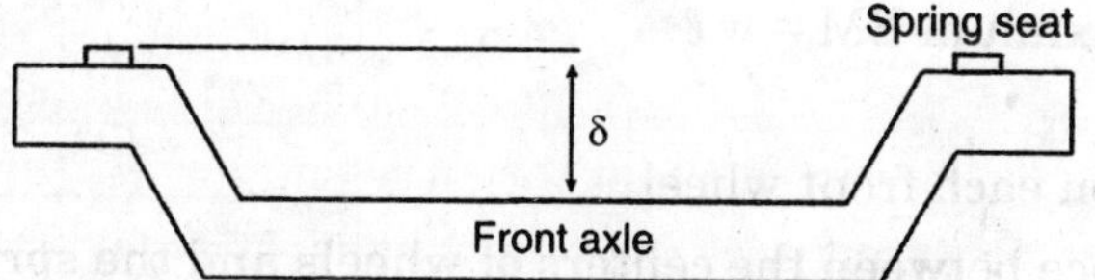

Fig.8.4: *Dropping the center part of axle beam*

Dropping the spring seat and center part of the axle beam

This lowers CG further and hence vehicle safety is improved.

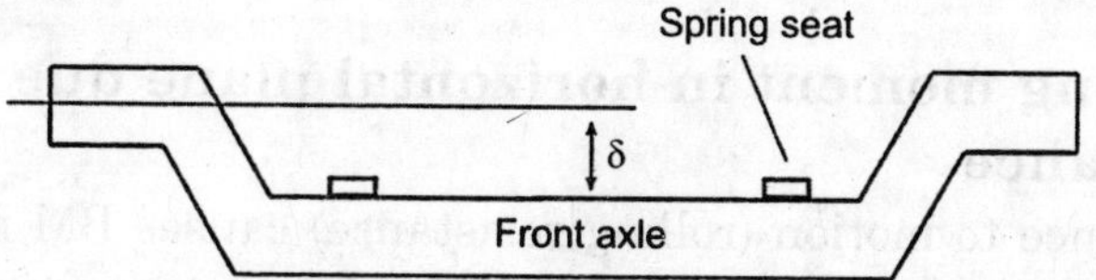

Fig.8.5: *Dropping the spring seat and center part of the axle beam*

LOADS AND STRESSES ON AXLE BEAM

1. Bending in vertical plane due to static loading

Front axles are subjected to bending and shear stresses. In the static condition, the axle may be considered as a beam supported vertically at the ends and at the center of the wheel and loaded vertically downward at the center of the spring pads. The vertical bending moment (BM) thus caused is zero at the point of support and rises linearly to a maximum at the point of loading and then remains constant.

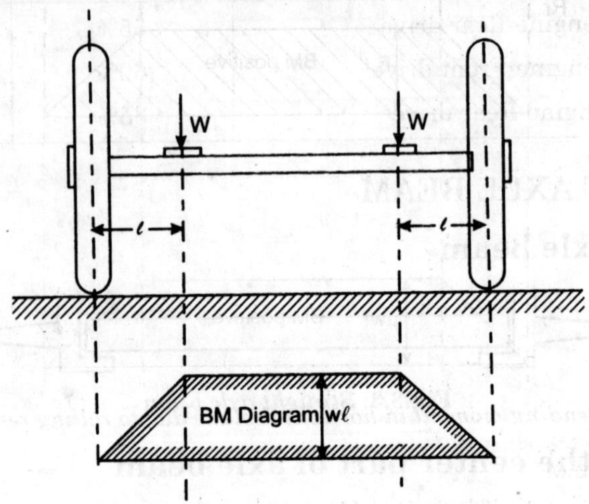

Fig.8.6: *Load carried and BM*

Thus the maximum BM = w ℓ

Where,

w = Load on each front wheel

ℓ = Distance between the centers of wheels and the spring pad

2. Bending in vertical plane due to vertical components of dynamic loads namely road shocks

Under dynamic conditions, the vertical BM is increased due to road roughness. But its estimation is difficult and hence can be accounted through a factor of safety.

3. Bending moment in horizontal plane due to rolling resistance

The resistance to motion (rolling resistance) causes BM in horizontal plane. It also causes a torque in the case of drop type front axle. Thus the portions projected after the spring pads are subjected to combined bending and torsion.

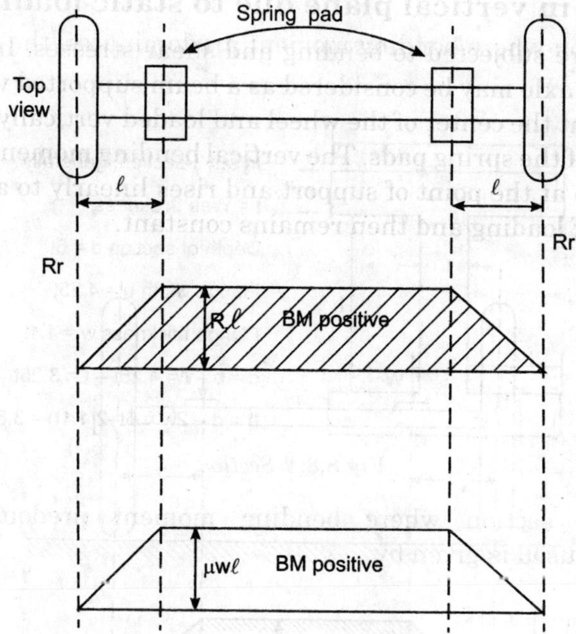

Fig.8.7: *Bending moment in horizontal plane due to rolling resistance*

The magnitude of:

BM $= R_r \ell$

Torque $= R_r \delta$

Where,

δ = The drop from the spring axis to the center of the section

4. Bending moment in horizontal plane due to brake application

BM $= \mu W \ell$

5. The shear stress in the axle due to braking torque

Braking torque $= \mu\ W R_w$

The braking torque is lower for the section lying between the spring pads and is equal to μ W (R_w–d). In this portion the bending moment predominates whereas at the steering head, torsion predominates. Thus I-section is used for portion where bending moment predominates and is gradually changed to circular, oval or rectangular section at the steering head.

DESIGN OF FRONT AXLE

I Section

For the section where bending moment predominates, the relationship used is given by

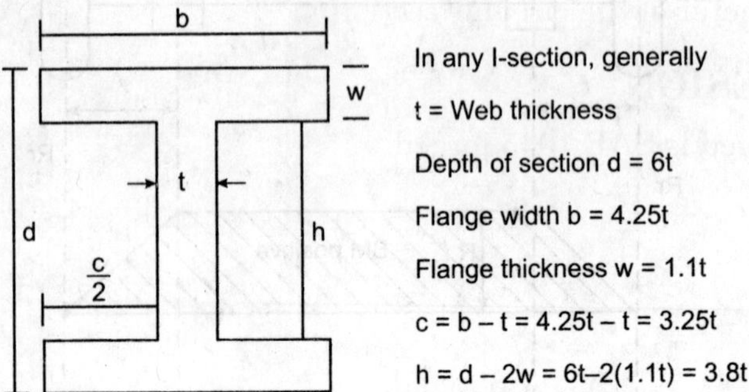

Fig.8.8: *I-Section*

For the section where bending moment predominates, the relationship used is given by

$$\frac{M}{I} = \frac{\sigma_b}{y}$$

Where,

M = Maximum bending moment

I = Moment of inertia of the I-section above neutral axis

$$= \frac{bd^3 - ch^3}{12}$$

y = Maximum distance of the fibre from the neutral axis = d/2

σ_b = Allowable bending stress

Circular/Oval section

For the section where torsion predominates the relation used is given by

$$\frac{M_t}{J} = \frac{\tau}{y}$$

Where,

M_t = Twisting moment or torque

J = Polar moment of inertia

$$= \frac{\pi}{32} d^4 \text{ for circular section}$$

$= \frac{\pi}{32} d^3 b$ for oval section with minor axis 'b'

τ_s = Torsional shear stress

y = Distance from the neutral axis to the outer most fibre of the axle = d/2

d = Diameter of the circular section

AXLE DESIGN

Material used is SAE 1045 forged.

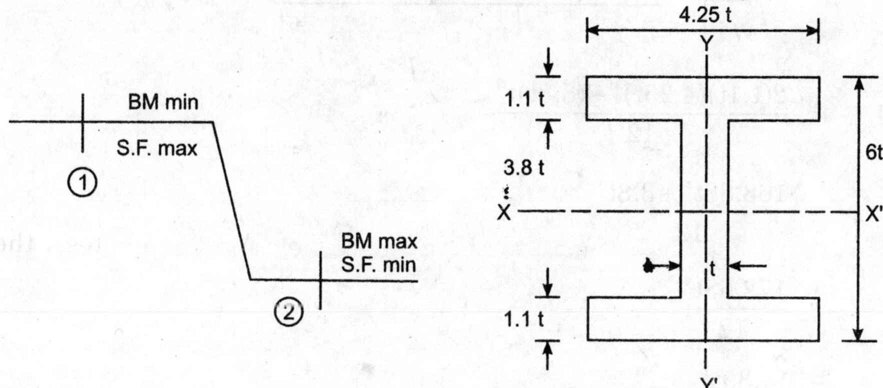

Fig.8.9: *Front axle design*

Section between spring seats

Maximum BM $= W\ell$ (1)

$= \sigma_b z$

Considering the Fig.8.9,

Moment of inertia

$$I_{xx'} = \frac{(4.25t)(6t)^3}{12} - \frac{(3.25t)(3.8t)^3}{12}$$

$$= \frac{918t^4}{12} - \frac{178.33t^4}{12}$$

$$= \frac{t^4}{12}\left[918 - 178.33\right]$$

$$= 61.64t^4$$

Section modulus

$$Z_{xx'} = \frac{I_{xx'}}{y}$$

$$= \frac{61.64t^4}{3t}$$

$$= 20.547t^3 \quad (2)$$

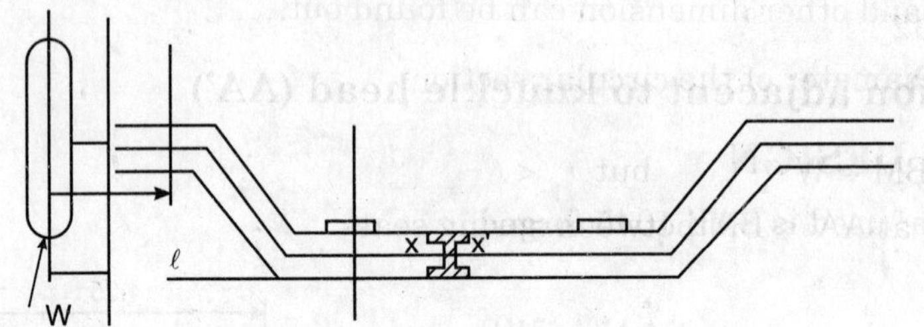

$$I_{yy'} = \frac{2(1.1t)(4.25t)^3 + (3.8t)t^3}{12}$$

$$= \frac{168.88t^4 + 3.8t^4}{12}$$

$$= \frac{172.68t^4}{12}$$

$$= 14.39t^4$$

$$Z_{yy'} = \frac{I_{yy'}}{y}$$

$$= \frac{14.39t^4}{2.125t}$$

$$= 6.77t^3 \quad (3)$$

Here, $Z_{xx'} > Z_{yy'}$, hence we should consider only $Z_{xx'}$

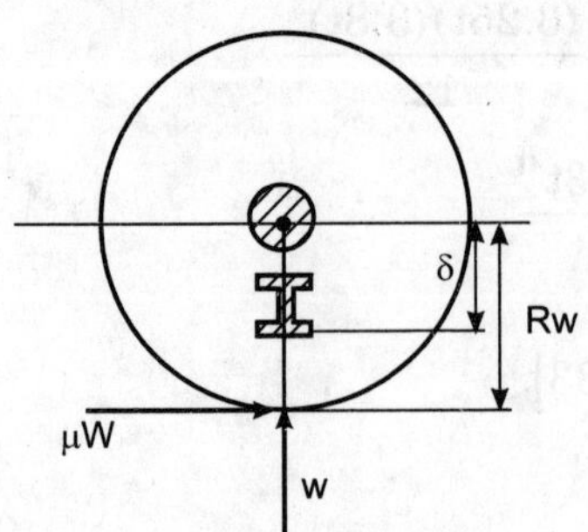

Torsional moment at a section located at spring seat

$$= \mu W(R_w - \delta) \quad (4)$$

Comparing (1) and (4)

BM > TM

Hence we should consider only BM

Now, BM = $W\ell = \sigma_b z$

$\therefore$ z, t and other dimension can be found out.

Section adjacent to knuckle head (AA')

Here BM = $W\ell_s$ but $\ell_s < \ell$

$\therefore$ BM at AA' < BM between spring seats

Now,

Torsional moment at AA' = μWR_w (5)

Here, $\mu WR_w > w\ell_s$

$\therefore$ TM > BM

Therefore, the section adjacent to knuckle head is designed by considering torsional moment since it is greater than bending moment.

Generally for BM the section will be 'I' and for TM the section will be circular.

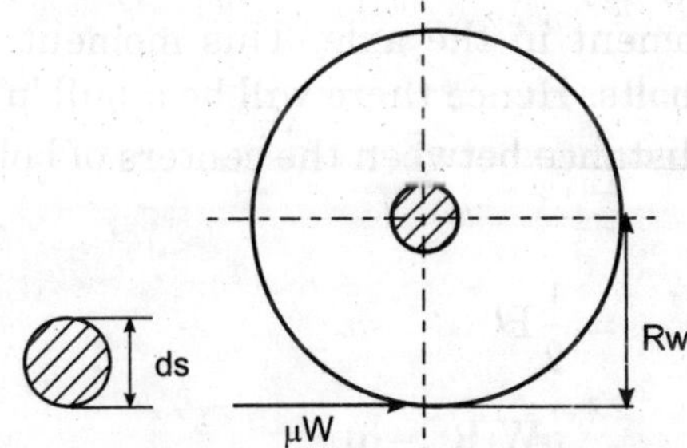

For the section where torsion predominates the relation used is given by

$$\frac{M_t}{J} = \frac{\tau_s}{y}$$

$$\frac{M_t}{\frac{\pi d_s^4}{32}} = \frac{\tau_s}{\frac{d_s}{2}}$$

$$M_t = \frac{\pi\tau d_s^3}{16} \quad (6)$$

Equating (5) and (6)

$$\frac{\pi \tau_s d_s^3}{16} = \mu W R_w$$

$$d_s = \sqrt[3]{\frac{16 \mu W R_w}{\pi \tau_s}} \quad (7)$$

Spring bolt diameter

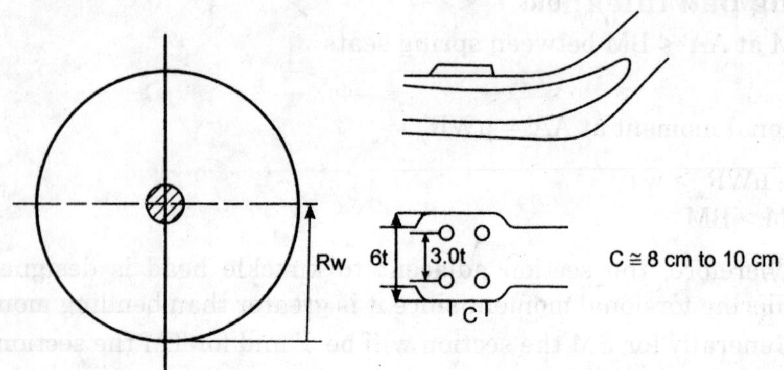

Consider the figure, when brake is applied, the braking torque will give a twisting moment in the axle. This moment is resisted by the spring and spring bolts. Hence there will be a pull 'p' in the bolts.

Let 'C' be the distance between the centers of bolts. Then

$$PC = \mu W(R_w - \delta)$$

$$\therefore \text{ Pull in each bolt } = \frac{1}{2}P$$

$$= \frac{\mu W(R_w - \delta)}{2C} \quad (8)$$

$$\text{Also, Pull } = \frac{\pi d^2}{4}\sigma_t \quad (9)$$

Where, d = Diameter of bolt

σ_t = Permissible tensile stress

= 10000 to 15000 N/cm² for nickel steel

Equating (8) and (9)

$$\frac{\mu W(R_w - \delta)}{2C} = \frac{\pi d^2}{4}\sigma_t$$

$$d^2 = \frac{2\mu W(R_w - \delta)}{\pi C \sigma_t}$$

$$d = \sqrt{\frac{2\mu W(R_w - \delta)}{\pi C \sigma_t}}$$

Hence standard bolt can be chosen.

Spring pad thickness

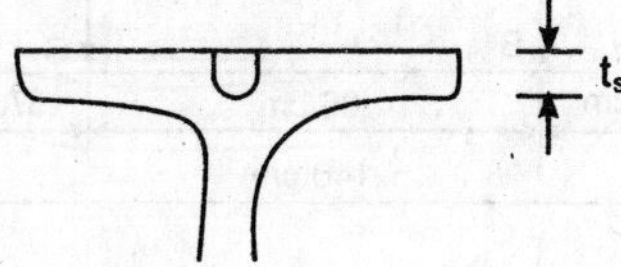

Thickness of flange at spring seat

$$t_s = \sqrt{\frac{W_S}{K}}$$

Where

W_S = Load on spring seat, kg

K = Constant

= 280 for light weight fast moving vehicles

= 420 for heavy weight slow moving vehicles

Problem 8.1: The load distribution between the front and rear axle of a motor vehicle weighing 1350 Kgf is 48% load to the front axle and the remaining to the rear. The front wheel track is 140 cm and the distance between spring pads is 66 cm. Design a suitable I-section for the axle beam assuming a working stress of 915 kgf/cm^2.

Given

Total load = 1350 kgf = 13500 N

Load on front axle = 48% of Total load

= 0.48 × 13500

= 6480 N

σ_b = 915 kgf/cm^2 = 9150 N/cm^2

Solution

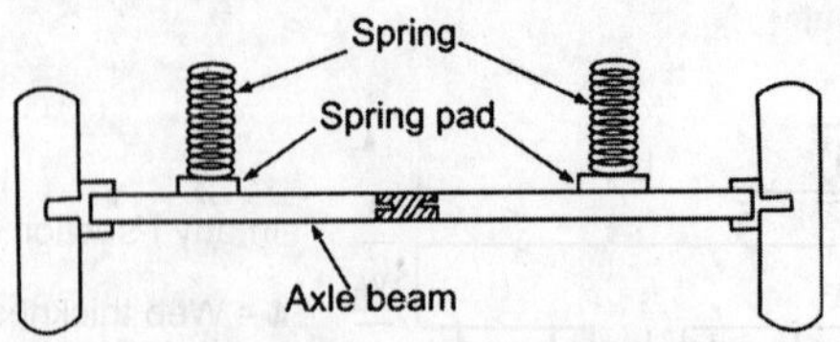

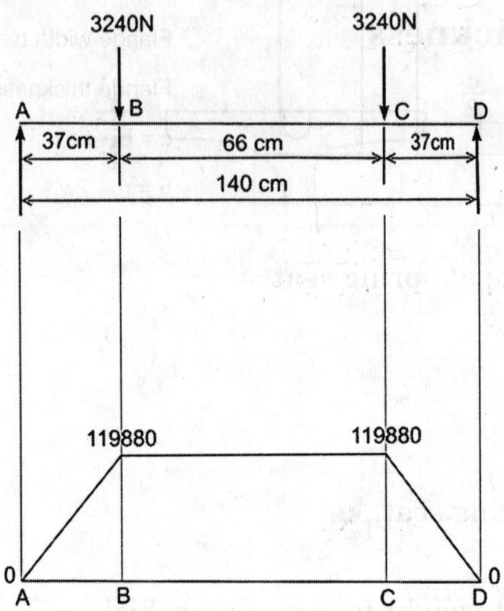

Equating forces

$R_A + R_D = 3240+3240$

$R_A + R_D = 6480$ N

Taking moments along A

$R_D \times 140 = (3240\times 103)+(3240\times 37)$

$R_D = 3240$ N

$\therefore R_A = 3240$ N

Bending moment

BM at A and D = 0

BM at C $= R_D \times 37 = 3240\times 37 = 119880$ Ncm

BM at B $= (R_D\times 103)-(3240\times 66) = (3240\times 103) - (3240\times 66)$

$= 119880$ Ncm

Maximum BM

M = 119880 Ncm

For the section where bending moment predominates, the relationship used is given by

$$\frac{M}{I} = \frac{\sigma_b}{y}$$

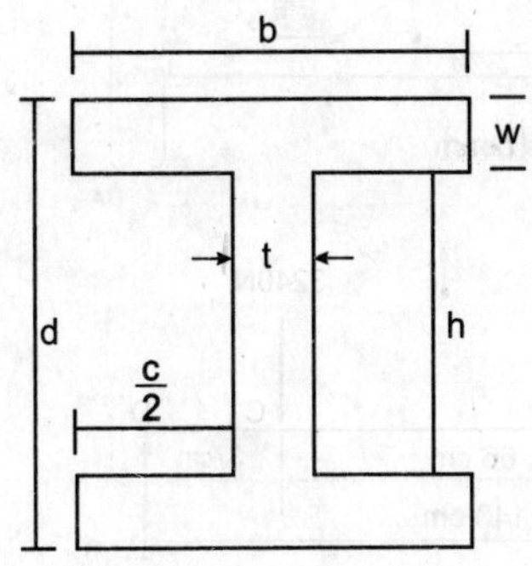

In any I-section, generally

t = Web thickness

Depth of section d = 6t

Flange width b = 4.25t

Flange thickness w = 1.1t

c = b – t

h = d – 2w

$$c = b - t = 4.25t - t = 3.25t$$

$$h = d - 2w = 6t - 2(1.1t) = 3.8t$$

$$I = \frac{bd^3 - ch^3}{12} = \frac{(4.25t)(6t)^3 - (3.25t)(3.8t)^3}{12} = \frac{(4.25t)(216t^3) - (3.25t)(55t^3)}{12}$$

$$= \frac{918t^4 - 179t^4}{12} = \frac{739t^4}{12} = 61.58t^4$$

$$\frac{119880}{61.58t^4} = \frac{9150}{(6t/2)}$$

$$\frac{119880}{61.58t^4} = \frac{9150}{3t}$$

$$t^3 = 0.6382$$

$$t = 0.86 \text{ cm}$$

Dimension of I-Section

$$d = 6t = 6 \times 0.86 = 5.16 \text{ cm}$$

$$b = 4.25t = 4.25 \times 0.86 = 3.65 \text{ cm}$$

$$w = 1.1t = 1.1 \times 0.86 = 0.946 \text{ cm}$$

$$c = 3.25t = 3.25 \times 0.86 = 2.79 \text{ cm}$$

$$h = 3.8t = 3.8 \times 0.86 = 3.26 \text{ cm}$$

Problem 8.2: The load distribution between the front and rear axle of a motor vehicle weighing 1350 Kgf is 48% load to the front axle and the remaining to the rear. The front wheel track is 140 cm and the distance between spring pads is 66 cm. Design a suitable I-section for the axle beam assuming that the width of the flange and its thickness are 0.6

and 0.2 of the overall depth of the section. The thickness of the web is 0.25 of the width of the flange. Assume a working stress of 915 kgf/cm^2.

Given

Total load = 1350 kgf = 13500 N

Load on front axle = 48% of Total load

$= 0.48 \times 13500$

$= 6480$ N

$b = 0.6d$

$w = 0.2d$

$t = 0.25b$

$\sigma_b = 915$ kgf/cm^2 = 9150 N/cm^2

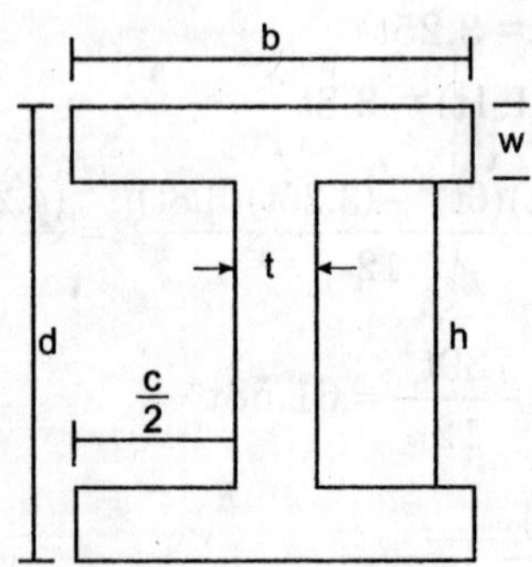

Solution

Equating forces

$R_A + R_D = 3240+3240$

$R_A + R_D = 6480$ N

Taking moments along A

$R_D \times 140 = (3240\times103)+(3240\times37)$

$R_D = 3240$ N

$\therefore R_A = 3240$ N

Bending moment

BM at A and D = 0

BM at C $= R_D\times37 = 3240\times37 = 119880$ Ncm

BM at B $= (R_D\times103) - (3240\times66) = (3240\times103) - (3240\times66)$

$= 119880$ Ncm

Maximum BM

M = 119880 Ncm

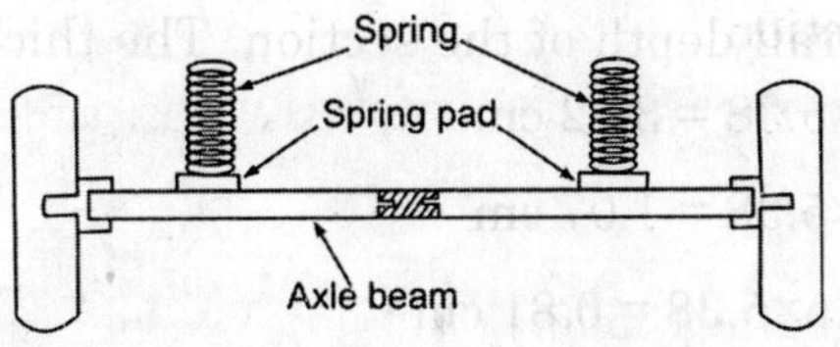

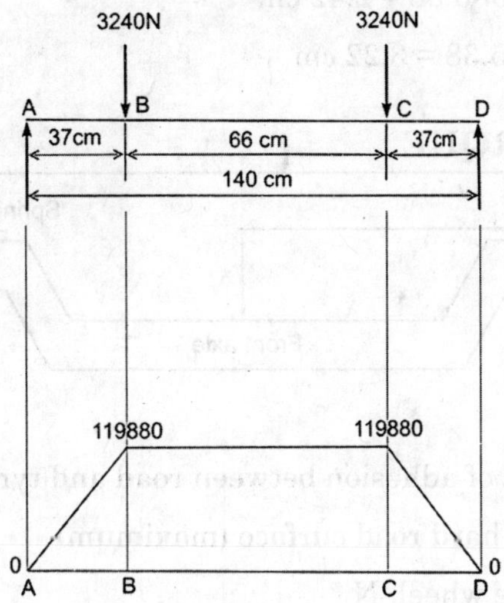

For the section where bending moment predominates, the relationship used is given by

$$\frac{M}{I} = \frac{\sigma_b}{y}$$

b = 0.6d

w = 0.2d

t = 0.25b = 0.25(0.6d) = 0.15d

c = b–t = 0.6d–0.15d = 0.45d

h = d–2w = d–2(0.2d) = 0.6d

$$I = \frac{bd^3 - ch^3}{12} = \frac{(0.6d)(d)^3 - (0.45d)(0.6d)^3}{12}$$

$$= \frac{0.6d^4 - 0.0979d^4}{12} = \frac{0.5028d^4}{12} = 0.0419d^4$$

$$\frac{119880}{0.0419d^4} = \frac{9150}{(d/2)}$$

d^3 = 156.34

d = 5.38 cm

Dimension of I-Section

b = 0.6d = 0.6×5.38 = 3.22 cm

w = 0.2d = 0.2×5.38 = 1.07 cm

t = 0.15d = 0.15×5.38 = 0.81 cm

c = 0.45d = 0.45×5.38 = 2.42 cm

h = 0.6d = 0.6×5.38 = 3.22 cm

BRAKING TORQUE

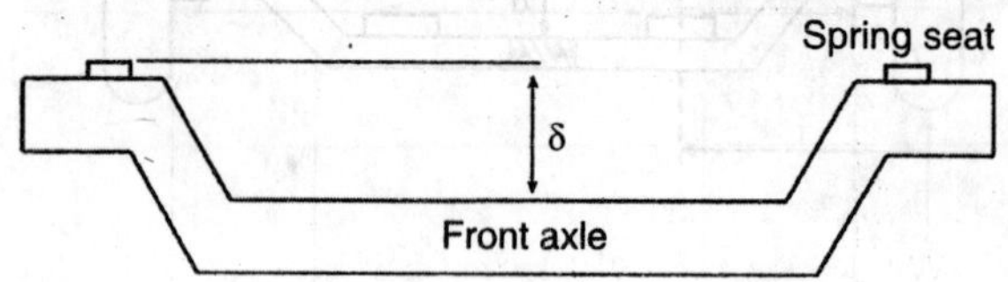

Let,

μ = Co-efficient of adhesion between road and tyre

= 0.6 for dry, hard road surface (maximum)

W = Load on one wheel, N

R_w = Wheel radius

δ = Dropping

Case 1: Torsional moment for the section between spring pad and stub axle.

$$M_t = \mu W R_w$$

Case 2: Torsional moment for the section between spring pads.

$$M_t = \mu W (R_w - \delta)$$

Problem 8.3: The load distribution between the front and the rear axle of a motor vehicle weighing 50kN is that 60% of the load is taken by the front axle. The axle has a drop of 12 cm, the wheel radius of 40 cm. The wheel track is I40 cm. The distance from the center plane of wheel to the nearest spring pad is 20 cm. Take of coefficient of adhesion between tyre and road is 0.6; maximum permissible bending stress is 160 N/mm² and maximum shear stress is 120 N/mm². Design a suitable I-section at the center and circular section at the ends for the front axle assuming that the width of the flange and its thickness are 0.6 and 0.2 of the overall depth of the section respectively and the thickness of the web 0.25 of the width of the flange.

Given

Total load = 50 kN = 50000 N

Load on front axle = 60% of Total load

= 0.6 × 50000

= 30000 N

W = 30000/2 = 15000 N

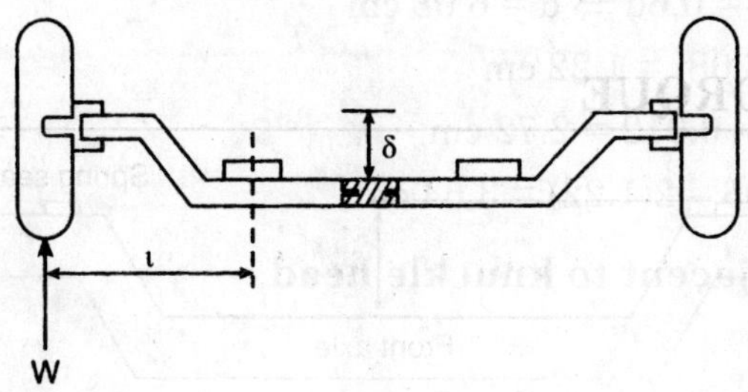

δ = 12 cm

R_w = 40 cm

T = 140 cm

ℓ = 20 cm

μ = 0.6

σ_b = 160 N/mm² = 16000 N/cm²

τ_s = 120 N/mm² = 12000 N/cm²

b = 0.6d

w = 0.2d

t = 0.25b

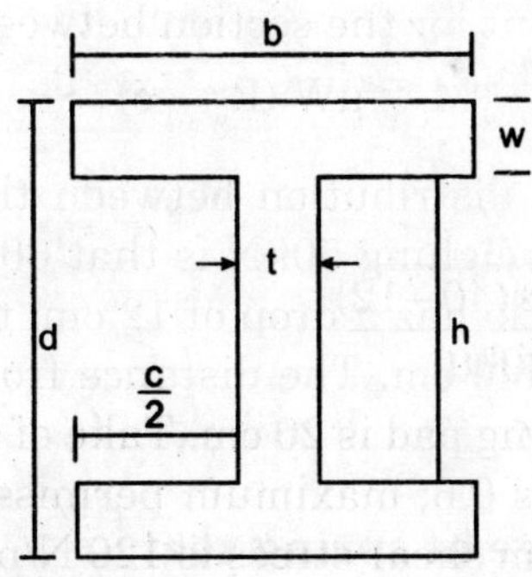

(a) Section between spring pads

BM at spring seat = $W\ell$ = 15000×20 = 300000 Ncm

Generally I-Section is preferred between the spring pads

We know that

$W_\ell = \sigma_b z$

$300000 = 16000 \times 20.547t^3$

$t = 0.912$ cm

Dimension of I-Section

$t = 0.25b \Rightarrow 0.912 = 0.25b \Rightarrow b = 3.64$ cm

$b = 0.6d \Rightarrow 3.64 = 0.6d \Rightarrow d = 6.08$ cm

$w = 0.2d = 0.2(6.08) = 1.22$ cm

$c = b - t = 3.64 - 0.912 = 2.72$ cm

$h = d - 2w = 6.08 - 2(1.22) = 3.64$ cm

(b) Section adjacent to knuckle head

We know,

$$d_s = \sqrt[3]{\frac{16\mu W R_w}{\pi \tau_s}}$$

$$= \sqrt[3]{\frac{16 \times 0.6 \times 15000 \times 40}{\pi \times 12000}}$$

$$d_s = 5.34 \text{ cm}$$

(c) Spring bolt diameter

We know

$$d = \sqrt{\frac{2\mu W (R_w - \delta)}{\pi C \sigma_t}}$$

Assuming

$C = 10$

$\sigma_t = 12000$ N/cm^2

$$= \sqrt{\frac{2 \times 0.6 \times 15000 \times (40 - 12)}{\pi \times 10 \times 12000}}$$

$$= 1.15 \text{ cm}$$

(d) Thickness of flange at spring seat

$W_s = 15000$ N $= 1500$ kg

$$t_s = \sqrt{\frac{W_S}{K}} = \sqrt{\frac{1500}{280}} = 2.31 \text{ cm}$$

Circular section at the ends

For the section where torsion predominates, the relation used is given by,

$$\frac{M_t}{J} = \frac{\tau}{y}$$

We know that, Torsional moment for the section between spring pad and stub axle

$$M_t = \mu W R_w$$
$$= 0.6 \times 15000 \times 40$$
$$= 360000 \text{ Ncm}$$
$$= 3600000 \text{ Nmm}$$

Polar moment of inertia $J = \dfrac{\pi d^4}{32}$

$$\frac{3600000}{\frac{\pi d^4}{32}} = \frac{120}{\frac{d}{2}}$$

$d = 53.46$ mm

Problem 8.4: Design a drop type front axle for a passenger car whose centre has a drop of 20 cm, the wheel radius 30 cm and the wheel load is 360 kg. The front track is 150 cm and the distance between spring pads is 100 cm. Assume other data.

Given data

Drop type front axle

$\delta = 20$ cm

$R_w = 30$ cm

$W = 360 \text{ kg} = 3600$ N

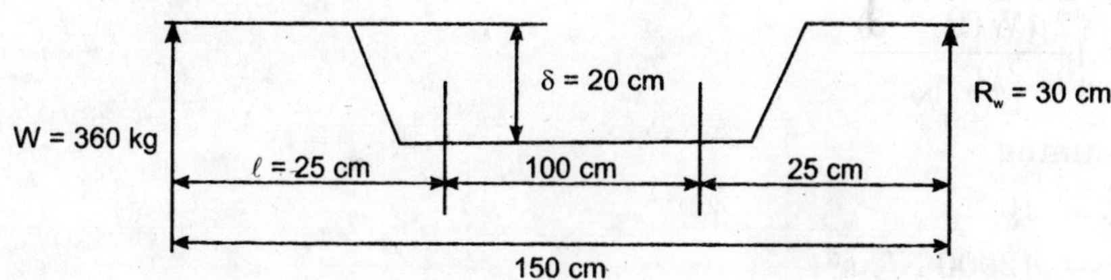

Design

(a) Section between spring pads

BM at spring seat = $W\ell = 3600 \times 25 = 90000$ Ncm

Generally I-Section is preferred between the spring pads

We know that

$W\ell = \sigma_b z$

Assuming $\sigma_b = 9500$ N/cm^2

$90000 = 9500 \times 20.547t^3$

$t = 0.7725$ cm

Dimension of I-Section

$d = 6t = 6 \times 0.7725 = 4.63$ cm

$b = 4.25t = 4.25 \times 0.7725 = 3.28$ cm

$w = 1.1t = 1.1 \times 0.7725 = 0.849$ cm

$c = 3.25t = 3.25 \times 0.7725 = 2.51$ cm

$h = 3.8t = 3.8 \times 0.7725 = 2.93$ cm

(b) Section adjacent to knuckle head

We know,

$$d_s = \sqrt[3]{\frac{16\mu W R_w}{\pi \tau_s}}$$

Assuming

$\mu = 0.6$

$\tau_s = 8300$ N/cm^2

$$= \sqrt[3]{\frac{16 \times 0.6 \times 3600 \times 30}{\pi \times 8300}}$$

$d_s = 3.41$ cm

(c) Spring bolt diameter

We know

$$d = \sqrt{\frac{2\mu W (R_w - \delta)}{\pi C \sigma_t}}$$

Assuming

$C = 10$

$\sigma_t = 12000$ N/cm^2

$$= \sqrt{\frac{2 \times 0.6 \times 3600 \times (30 - 20)}{\pi \times 10 \times 12000}}$$

$= 0.338$ cm

(d) Thickness of flange at spring seat

For W_s in kg, K = 280 for light weight fast moving vehicles

$$t_s = \sqrt{\frac{W_S}{K}} = \sqrt{\frac{360}{280}} = 1.13 \text{ cm}$$

Problem 8.5: Design a drop type of front axle for a passenger car for the following data:

Drop of the axle = 30 cm, wheel diameter = 72 cm, GVM = 45 kN of which 45% is acting at the front, the front track = 275 cm and the distance between spring pads=175 cm. Assume μ = 0.6, f_b = 95000 kpa, f_s = 80000 kpa, f_t = 120000 kpa, C = 10 and K = 280

Given data

Drop type front axle

d = 30 cm

D_w = 72 cm ⇒ R_w = 36 cm

Total load = 45 kN = 45000 N

Load on front axle = 45% of Total load

= 0.45 × 45000

= 20250 N

W = 10125 N

μ = 0.6

f_b = 95000 kpa

= 95000×10^3 pa

= 95000×10^3 N/m²

= 9500 N/cm²

f_s = 80000 kpa = 8000 N/cm²

f_t = 120000 kpa = 12000 N/cm²

C = 10

K = 280

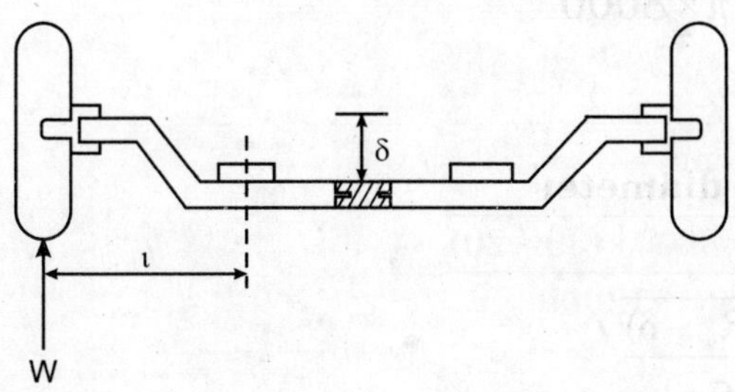

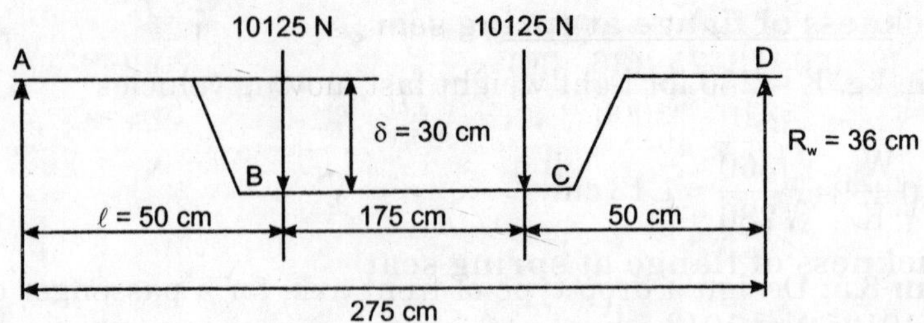

Design

(a) Section between spring pads

BM = Wℓ = 10125×50 = 506250 Ncm

Generally I-Section is preferred between the spring pads

We know that

$W\ell = \sigma_b z$

$506250 = 9500 \times 20.547 t^3$

t = 1.37 cm

Dimension of I-Section

d = 6t = 6×1.37 = 8.22 cm

b = 4.25t = 4.25×1.37 = 5.82 cm

w = 1.1t = 1.1×1.37 = 1.51 cm

c = 3.25t = 3.25×1.37 = 4.45 cm

h = 3.8t = 3.8×1.37 = 5.21 cm

(b) Section adjacent to knuckle head

We know,

$$d_s = \sqrt[3]{\frac{16\mu W R_w}{\pi \tau_s}}$$

$$= \sqrt[3]{\frac{16 \times 0.6 \times 10125 \times 36}{\pi \times 8000}}$$

d_s = 5.18 cm

(c) Spring bolt diameter

We know

$$d = \sqrt{\frac{2\mu W(R_w - \delta)}{\pi C \sigma_t}}$$

$$= \sqrt{\frac{2\times 0.6\times 10125\times (36-30)}{\pi\times 10\times 12000}}$$

$$= 0.439 \text{ cm}$$

(d) Thickness of flange at spring seat

$$W_s = 10125 \text{ N} = 1012.5 \text{ kg}$$

$$t_s = \sqrt{\frac{W_S}{K}} = \sqrt{\frac{1012.5}{280}} = 1.9 \text{ cm}$$

REAR AXLE

In the rear axle vehicles, the rear axle assembly supports from 50% to 80% of the vehicle weight and also driving the rear wheels at the same time.

The rear axle assembly can be divided roughly into the two axle shafts that drive the wheels, the differential that provides the necessary difference in the rotations between the left and right wheels, and the axle housing that encases these parts.

The rear axle shafts transfer the torque from the differential to the wheels. One end of each axle shaft is splined to the differential while the other end has the wheel bolted on.

REAR AXLE TYPES

The rear axle can be classified into three types depending on the bearing usage as follows: Semi-floating, three-quarter floating and full-floating.

Semi-floating type

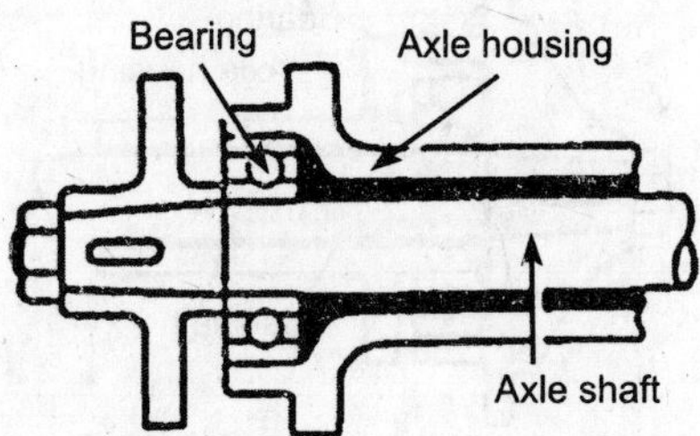

Fig.8.10: *Semi-floating axle*

The bearing is installed between the axle housing and axle shaft, and the wheel installed directly to the axle shaft. As a result, the axle shaft carries the entire vehicle load and also the lateral force created when the vehicle is turning. Since it is simple in construction, this type is used with most passenger cars.

Full-floating type

The bearings are located between the axle housing and wheel hub, with the wheel installed to the hub. In this type, the vehicle weight is supported entirely by the axle housing, the axle shaft serving only to drive the wheel. As there is no undue force acting on the axle shaft, this type is highly suited for transporting heavy loads and is therefore used in trucks.

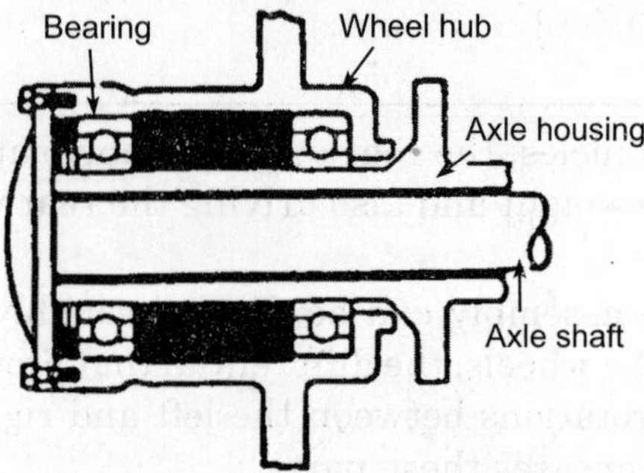

Fig.8.11: *Full floating axle*

Three-quarter floating type

There is only one set of bearing used between the axle housing and wheel hub. The wheel is installed directly to the axle shaft. Although most of the vehicle weight is supported by the housing at normal times, the lateral force that is created when the vehicle is turning will act on the axle shaft.

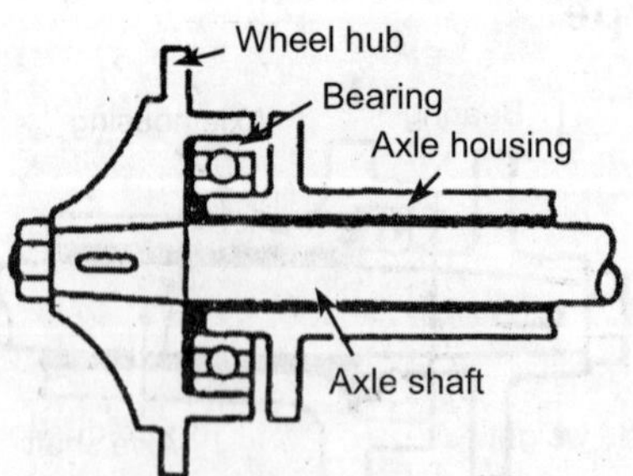

Fig.8.12: *Three-quarter floating axle*

A three-quarter floating axle was once very popular for cars and light commercial vehicles. However, due to availability of the compact, cheap and reliable semi floating axle, the three-quarter floating arrangement is rearly used today.

MATERIALS FOR REAR AXLES

The material chosen for axle shaft must ensure minimum unsprung weight. Hence, heat treated high grade alloy steel is used as material for axle shaft. Case hardening of axle shaft improves its resistance to torsional shear stress which is maximum at the surface.

Alloy steels of medium carbon content are normally used. After normalising and annealing the shafts are quenched in oil from a temperature, slightly above the critical range of steel and then tempered or drawn to give hardness, in the range of BHN 300. This is followed by machining operations, after which hardness is increased to about Brinnel 375 or 400 and tempered.

LOADS ON REAR AXLES

Fig.8.13 shows a line sketch of a simple axle shaft which is subjected to:

1. Shear stress due to the weight of the vehicle (Fig.8.13a).
2. Tensile and compressive stress due to cornering forces (Fig.8.13b).
3. Bending stress due to the weight of the vehicle (Fig.8.13c).
4. Torsional stress due to driving and braking torque (Fig.8.13d).

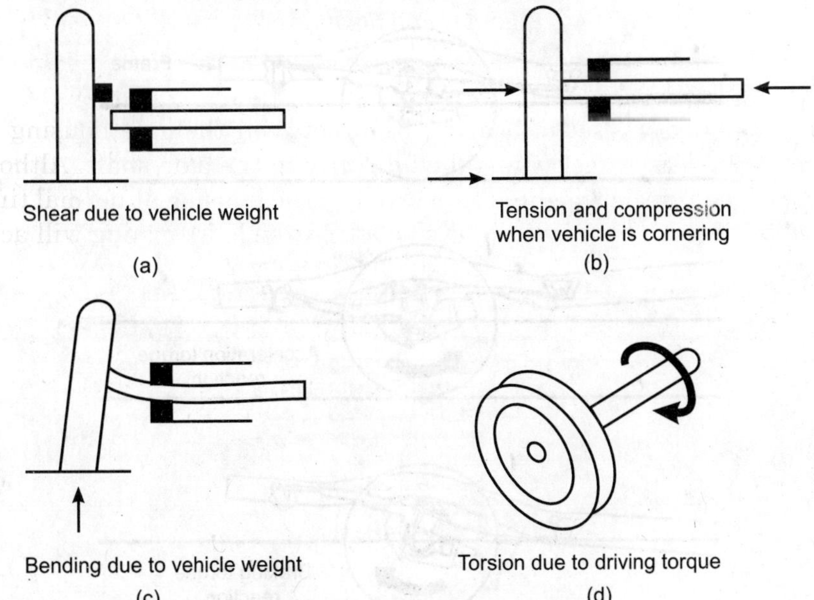

Fig.8.13: *Loads on rear axles*

FORCES ON THE REAR AXLE

The forces experienced by the rear axle are given as under :

Weight of the body

The weight of the body and the load is supported by the casing through the springs which are attached to the chassis frame. The casing is supported at its ends by the road wheels. The axle then acts as a beam and is subjected to a bending action as shown in the Fig.8.5, where W/2 are the load and weight of the body, applied to the casing through the each spring and R1 and R2 are the reactions supplied by the road wheels. The casing is made so strong to withstand the bending action without undue fixture.

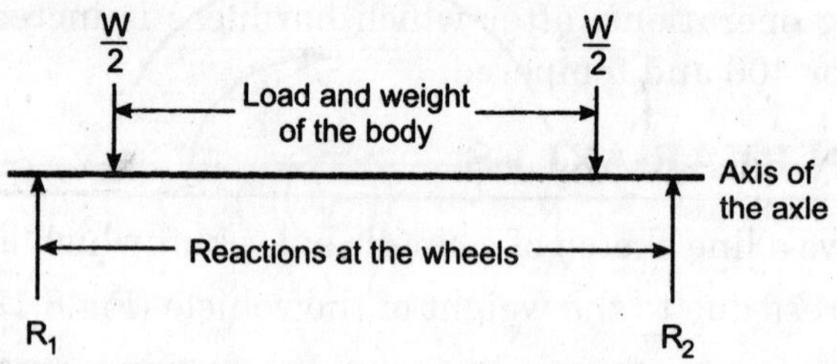

Fig.8.14: *Forces on rear axle due to load and weight of the body*

Torque reaction

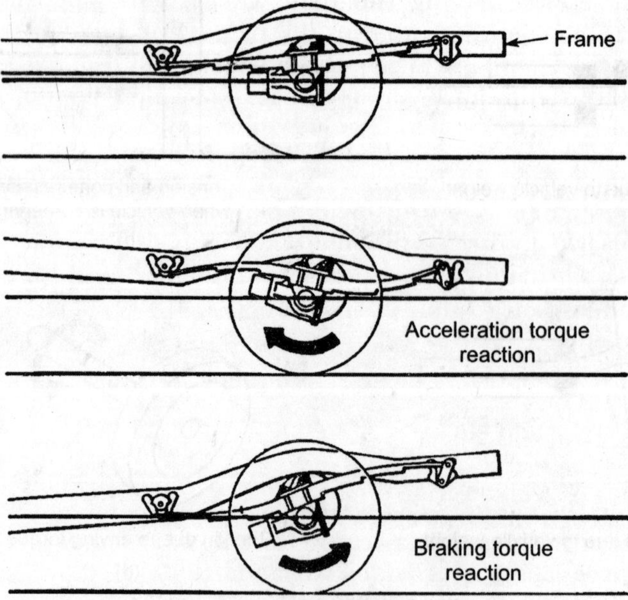

Fig.8.15: *Torque reaction*

The propeller shaft drives a torque to the pinion and the pinion will have to roll round the crown wheel carried in the casing. This tendency is also present when the vehicle is running. Reaction torque is equal and opposite to the driving torque applied to the road wheels. This phenomenon is called torque reaction. This opposition of the casing prevents bending of the propeller shaft.

Driving thrust

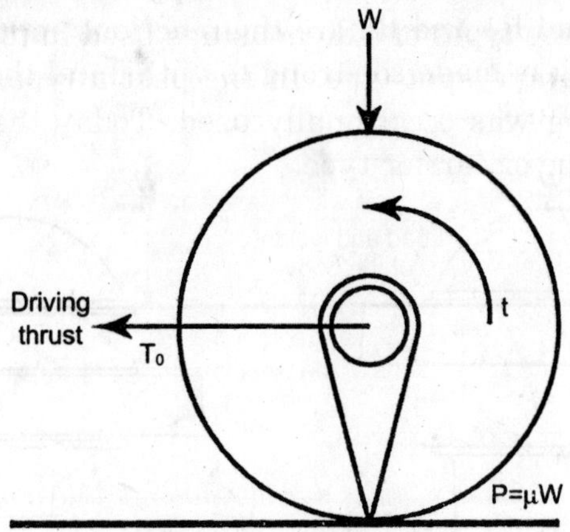

Fig.8.16: *Driving thrust*

The tractive effort, driving the road wheels with opposition of the casing has to be communicated to the frame and body of the vehicle. Therefore, some members or arms are attached to the axle casing and to the frame for this purpose. These members are known as "thrust members".

Braking thrust

When brakes are applied to the rear wheels, then the brake shoes tend to rotate with the drums and impose a braking torque on the axle casing in opposite direction to that of the driving-torque reaction. Similarly, the braking thrust on the axle casing is opposite in direction to the driving thrust.

Sideways forces

These forces on the wheels may be the reactions of the centrifugal force while turning the vehicle. The bending, shearing, and an end thrust in the axle case or in the axle may cause the sideways forces on the wheels. Therefore, axle casing is provided with additional reinforcement at the joints of the springs to avoid bending.

The axle rarely moves up and down parallel to the frame. Generally one wheel goes into a hole or over a bump while the other remains on the level. Then axle gains an angular position relative to the frame. Therefore, the connection between the various torque members, thrust members, springs and the frame must be such that the sideways forces may not induce undue stresses in the parts.

REAR AXLE HOUSING

The type of differential housing dictates the method that must be used to remove the final drive and differential assembly.

There are three types of axle housings used in the vehicles: banjo-type, carrier-type and split-type. In the past a type known as a split (trumpet) housing was occasionally used. Today the housing will be either a banjo-type or carrier-type.

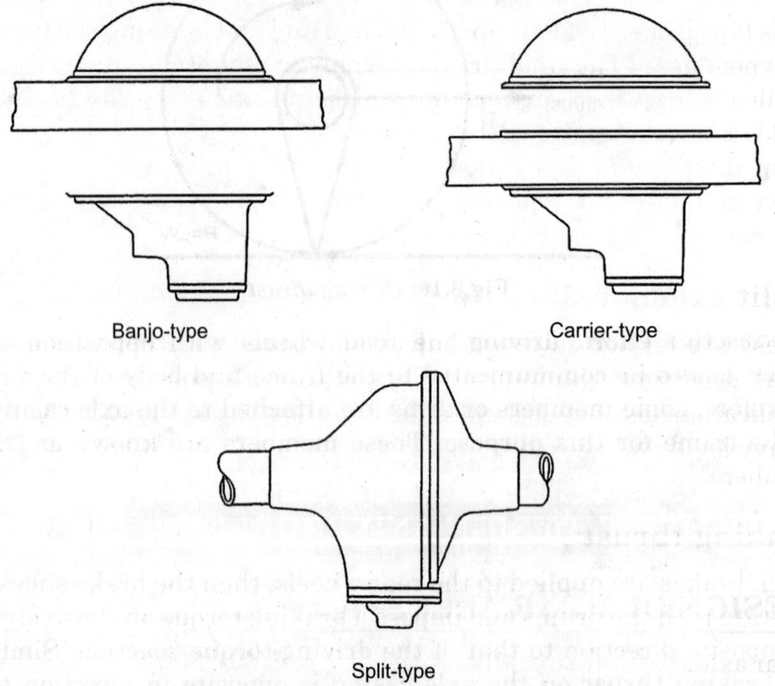

Fig.8.17: *Types of rear axle housings*

Banjo axle housing

The centre of the housing combined with the axle tube on one side, resembles a banjo, hence its name. The final drive assembly, which is mounted in detachable malleable iron housing, is secured by a ring of bolts to the axle housing. When this assembly is in position, the axle shafts are slid in from the road wheel end of the housing.

On some banjo axles a domed plate is bolted to the rear face of the housing. Removal of this plate allows the final drive gears to be inspected and in cases where the axle shaft is secured to the differential, the access point enables the axle shaft to be unlocked from the sun gear.

A lubricant level plug, set at a height about one-third up the crown wheel, is screwed into the domed cover or the final drive housing. The level is normally just below the axle tubes, so lubrication of the hub bearings is by splash caused by rotation of the crown wheel. Overfilling should be avoided because this swamps the oil seals and caused oil to enter the brakes.

When in operation the final drive gets hot, so some form of air vent is provided to release the pressure in the axle housing. This prevents oil being forced past the seals.

Carrier axle housing

This type is more rigid than a banjo housing and is often used to support a hypoid gear. The final drive assembly is mounted directly in a rigid malleable cast iron carrier, into which the axle tubes are pressed and welded. Extra rigidity is obtained by using reinforcing ribs that extend from the pinion nose to the main carrier casting. Access to the final drive and differential gears is by means of a domed plate at the rear of the housing for repair and maintenance.

Split axle housing

These are formed into two halves and bolted together to contain the final drive and differential. They are used more in heavy vehicle applications as they are more rigid in construction and can withstand heavy loads.

DESIGN AND PROBLEMS

DESIGN OF REAR AXLE

Rear axle shafts are usually subjected to combined torsional and bending loads. When the shaft is subjected to combined twisting moment and bending moment, then the shaft must be designed on the basis of the two moments simultaneously. Various theories have been suggested to account for the elastic failure of the materials when they are subjected to various types of combined stresses. The following two theories are important from the subject point of view:

1. Maximum shear stress theory or Guest's theory. It is used for ductile materials such as mild steel.
2. Maximum normal stress theory or Rankine's theory. It is used for brittle materials such as cast iron.

Let,

τ = Shear stress induced due to twisting moment (torsion)

σ_b = Bending stress (tensile or compressive) induced due to bending moment

DESIGN OF SEMI-FLOATING AXLE

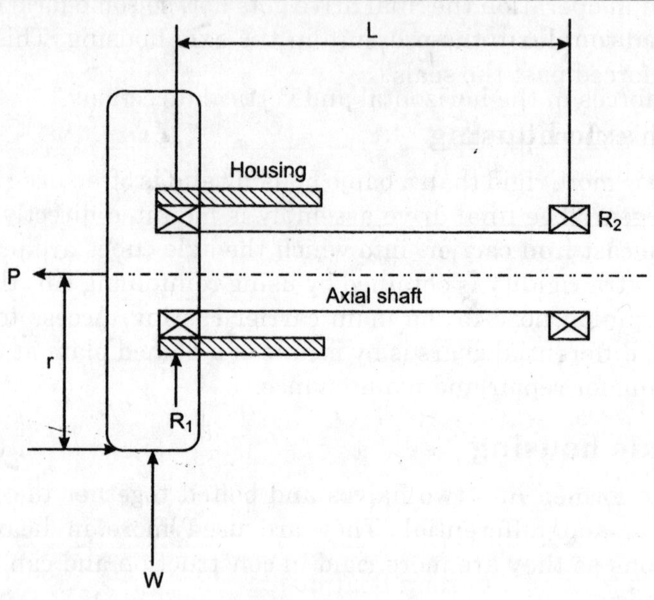

Bearing loads in semi-floating axle

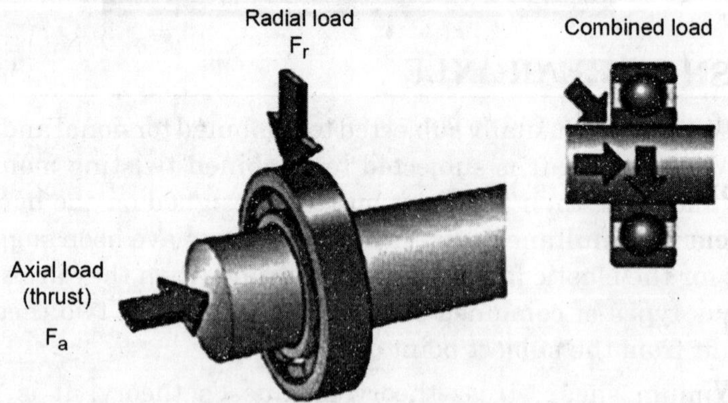

Bearings support a shaft or housing to permit their free motion about an axis of rotation. Load can be applied to bearings in either of two basic directions. Radial loads act at right angles to the shaft (bearing's axis of rotation). Axial (thrust) load acts parallel to the axis of rotation.

Bearing loads due to side thrust in a wheel in semi-floating axle:

F = Lateral force at the wheel rim

r = Effective wheel radius

L = Distance between the centers of wheel bearings

R_1 and R_2 = Radial reactions of the wheel bearing on the housing.

P = Thrust reaction of the bearing.

$\frac{r}{L} = 0.6$ for semi-floating axle

Consider forces in the horizontal and vertical directions.

Vertical direction, $R_1 = R_2$

Horizontal direction, F = P

Taking moments,

$$R_1 L = R_2 L = Fr$$

$$R_1 = R_2 = \frac{Fr}{L}$$

$$= 0.6\ F \qquad \left(\frac{r}{L} = 0.6\right)$$

$$= \frac{3}{5} F$$

∴ For a semi-floating axle,

$$P = F$$

$$R_1 = R_2 = \frac{3}{5} F$$

AXLE SHAFT DESIGN

Axle shafts are subjected to both bending and torsion.

Frictional force (F)

Co-efficient of friction

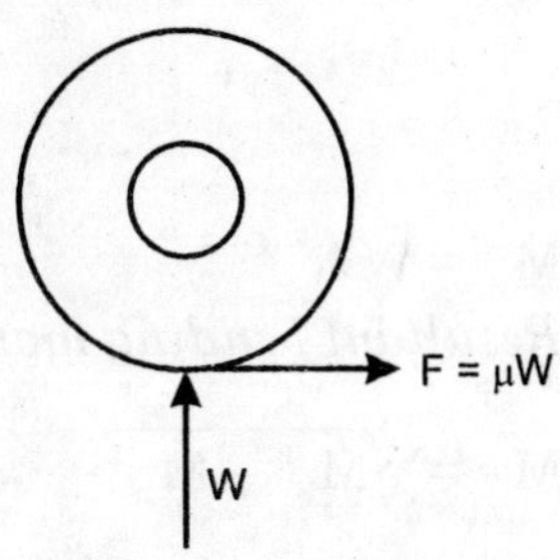

$$\mu = \frac{F}{W}$$

$$F = \mu W$$

Where,

μ = 0.6 for dry, rough road surface

W = Load supported on one wheel under fully loaded condition.

Frictional torque (T)

Frictional torque due to frictional force is given by,

Frictional torque = Frictional force × Effective wheel radius

$T = Fr = \mu Wr$

M: Bending moment

Bending moment due to frictional force at the outboard bearing is given by,

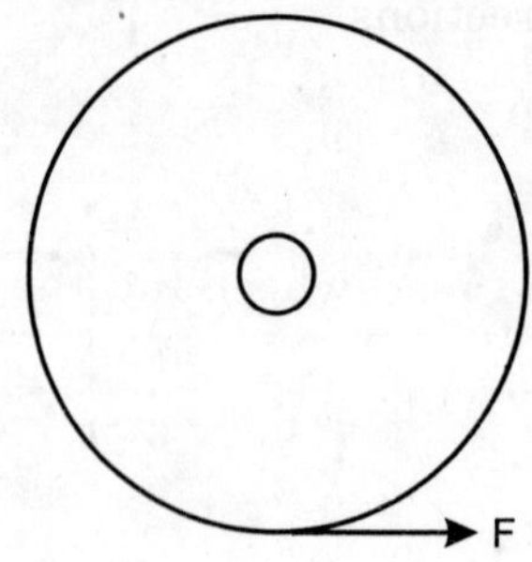

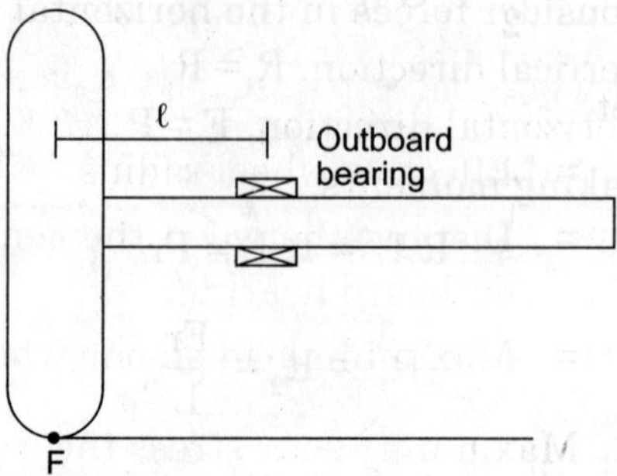

$$M_F = F\ell = \mu W\ell$$

Bending moment due to the load W on the axle at the center of the outboard bearing.

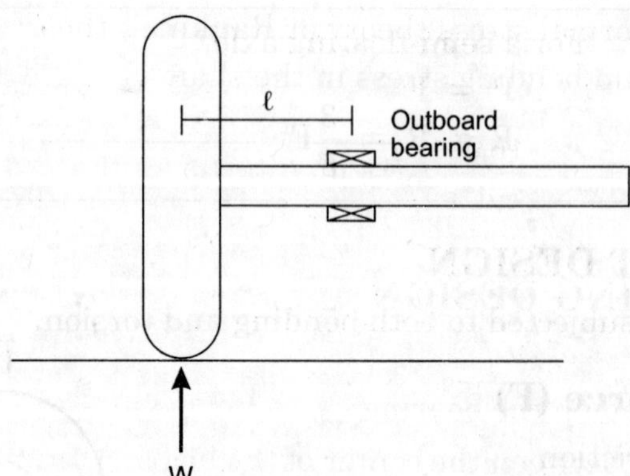

$M_W = W\ell$

Resultant bending moment is given by,

$$M = \sqrt{M_F^2 + M_W^2}$$

$$= \sqrt{(\mu W\ell)^2 + (W\ell)^2}$$

$= W\ell\sqrt{\mu^2 + 1}$

Driving torque on each wheel

$= \frac{T_W}{2}$

$= \frac{1}{2} T_e G \eta_t$

Total driving torque at the wheels (2WD)

$T_W = T_e \times GR \times AR \times \eta_t$ if losses are considered

$= T_e G \eta_t$

Where,

$G = GR \times AR$

η_t = Overall transmission efficiency

Let

r = Effective wheel radius

ℓ = Distance between the center plane of the road wheel and the outboard bearing

W = Maximum load on one wheel

i. Maximum shear stress theory or Guest's theory

The maximum shear stress in the shaft

$$\tau_{max} = \frac{16W}{\pi d^3}\sqrt{\mu^2(r^2 + \ell^2) + \ell^2}$$

ii. Maximum normal stress theory or Rankine's theory

The maximum bending stress in the shaft

$$\sigma_{b\,max} = \frac{16W}{\pi d^3}\left[\ell\sqrt{1+\mu^2} + \sqrt{\mu^2(r^2 + \ell^2) + \ell^2}\right]$$

AXLE HOUSING DESIGN

Let

W = Static load on the bearing

ℓ_s = Distance between the center of the bearing and the spring seat.

i. Maximum shear stress theory or Guest's theory

The maximum shear stress in the shaft

$$\tau_{max} = \frac{16}{\pi d^3}\sqrt{(\mu w r)^2 + (W_s \ell_s)^2 (1 + \mu^2)}$$

ii. Maximum normal stress theory or Rankine's theory

The maximum bending stress in the shaft

$$\sigma_{b\,max} = \frac{16}{\pi d^3}\left[W_s \ell_s \sqrt{1+\mu^2} + \sqrt{(\mu W r)^2 + (W_s \ell_s)^2 (1+\mu^2)}\right]$$

DESIGN OF THREE-QUARTER AXLE

AXLE SHAFT DESIGN

Let

r = Effective wheel radius

ℓ = Distance between the center plane of the road wheel and the outboard bearing

W = Maximum load on one wheel

i. Maximum shear stress theory or Guest's theory

The maximum shear stress in the shaft

$$\tau_{max} = \frac{16W}{\pi d^3}\sqrt{\mu^2(r^2+\ell^2)+\ell^2}$$

ii. Maximum normal stress theory or Rankine's theory

The maximum bending stress in the shaft

$$\sigma_{b\,max} = \frac{16W}{\pi d^3}\left[\ell\sqrt{1+\mu^2} + \sqrt{\mu^2(r^2+\ell^2)+\ell^2}\right]$$

AXLE HOUSING DESIGN

Let

W_s = Static load on the bearing

ℓ_s = Distance between the center of the bearing and the spring seat.

i. Maximum shear stress theory or Guest's theory

The maximum shear stress in the shaft

$$\tau_{max} = \frac{16}{\pi d^3}\sqrt{(\mu w r)^2 + (W_s \ell_s)^2 (1+\mu^2)}$$

ii. Maximum normal stress theory or Rankine's theory

The maximum bending stress in the shaft

$$\sigma_{b\,max} = \frac{16}{\pi d^3}\left[W_s \ell_s \sqrt{1+\mu^2} + \sqrt{(\mu W r)^2 + (W_s \ell_s)^2 (1+\mu^2)}\right]$$

DESIGN OF FULL-FLOATING AXLE

AXLE SHAFT DESIGN

In full-floating type, axle is subjected to pure torsion. Therefore, $\frac{M_t}{J} = \frac{\tau}{r}$, can be used to find 'd'.

Where,

M_t = Twisting moment (torsion)

J = Polar moment of inertia

= $\frac{\pi}{32}d^4$ for solid shaft

= $\frac{\pi}{32}[d_o^4 - d_i^4]$

d_0 = Outside diameter of the shaft

d_i = Inside diameter of the shaft

τ = Torsional shear stress

r = d/2 or $d_o/2$

AXLE HOUSING DESIGN

Let

r = Effective wheel radius

ℓ = Distance between the center plane of the road wheel and the spring seat

i. Maximum shear stress theory or Guest's theory

The maximum shear stress in the shaft

$$\tau_{max} = \frac{16W}{\pi d^3}\sqrt{\mu^2(r^2+\ell^2)+\ell^2}$$

ii. Maximum normal stress theory or Rankine's theory

The maximum bending stress in the shaft

$$\sigma_{b\,max} = \frac{16W}{\pi d^3}\left[\ell\sqrt{1+\mu^2} + \sqrt{\mu^2(r^2+\ell^2)+\ell^2}\right]$$

Problem 8.6: An engine develops 29.5kW at 2000 rpm when the torque developed is maximum. The bottom gear ratio is 3:1 and the back axle reduction is 4.5:1. The load on each driving axle is 7357.5N. When the car is fully loaded, diameter of road wheel over the tyre is 0.71 m and the co-efficient of adhesion between tyre and road is 0.6. If the permissible stress in the material of the shaft is not allowed to exceed 220725×10^4 Pa, find the diameter of the axle shaft.

Give data

P = 29.5 kW = 29.5×10^3 Watts

N = 2000 rpm

GR = 3

AR = 4.5

W = 7357.5 N

D = 0.71 m ⇒ r = 0.355 m

m = 0.6

t = 22072.5×10^4 pa = 22072.5×10^4 N/m²

Find

d

Solution

a. Axle shaft design

Using torsion formula, the minimum diameter can be calculated

$$\frac{M_t}{J} = \frac{\tau}{r}$$

$$\frac{M_t}{\frac{\pi}{32} d^4} = \frac{\tau}{\frac{d}{2}}$$

$$\frac{1901.5}{\frac{\pi}{32} d^4} = \frac{22072.5 \times 10^4}{\frac{d}{2}}$$

$d^3 = 4.3874 \times 10^{-5}$

d = 0.03526 m

= 35.26 mm @ 36 mm

Engine torque

$$P = \frac{2\pi \times 2000 \times T_e}{60}$$

$$29.5 \times 103 = \frac{2\pi \times 2000 \times T_e}{60}$$

T_e = 140.85 Nm

Torque transmitted to the wheels

Mt = Te × GR × AR

= 140.85 ×3 × 4.5

= 1901.5 Nm

Frictional torque due to frictional force is given by,

Frictional torque = Frictional force × Effective wheel radius

T = Fr

$1901.5 = F \times 0.355$

$F = 5356.33$ N

This frictional force develops horizontal shear stress and the weight on the axle develops vertical shear stress in the axle. The intensity of shear stress is zero at the surface and maximum at the center of the axle. The distribution is parabolic in nature.

Maximum horizontal shear stress = $\frac{4}{3}$ × Average shear stress due to frictional force

$$= \frac{4}{3} \times \frac{F}{A}$$

$$= \frac{4}{3} \times \frac{5356.3}{\frac{\pi}{4}(0.036)^2}$$

$$= 7016348.94 = 701.63 \times 10^4 \text{ N/m}^2$$

Maximum vertical shear stress = $\frac{4}{3}$ × Average shear stress due to the weight on the axle

$$= \frac{4}{3} \times \frac{W}{A}$$

$$= \frac{4}{3} \times \frac{7357.5}{\frac{\pi}{4}(0.036)^2}$$

$$= 9637715.99 = 963.77 \times 10^4 \text{ N/m}^2$$

Both together produce maximum stress at the center, which is too less in comparison to permissible stress given in the problem.

$$\tau_{cal} = 701.63\times10^4+963.77\times10^4 = 1665.4 \text{ N/m}^2 < 22072.5\times10^4 \text{ N/m}^2$$

Thus the shaft is quite safe in direct shear.

Problem 8.7: An engine develops 30 kW at 2000 rpm when the torque is maximum. The bottom gear ratio is 3:1 and the rear axle reduction is 4.5:1. The load on each driving axle is 0.75 tonne, when the car is fully loaded. Diameter of the road wheel over the tyre is 71 cm. The maximum permissible shear stress is 120 N/mm². The maximum permissible bending stress is 280 N/mm². If semi-floating axle is used, design axle shaft and axle housing of the value of inside diameter 0.8 times outside diameter and the distance between spring seat and center plane of wheel is 0.22m.

Given data

Semi-floating axle

P = 30 kW

N = 2000 rpm

GR = 3

AR = 4.5

W = 0.75 tonne = 0.75 × 1000 = 750 kg = 750 × 9.81 = 7357.5 N

D = 71 cm = 710 mm ⇒ r = 355 mm

t = 120 N/mm^2

σ_b = 280 N/mm^2

d_i = 0.8d_o

ℓ_s = 0.22 m = 220 mm

Find

d, d_i, d_o

Solution

a. Axle shaft design

Using torsion formula, the minimum diameter can be calculated

$$\frac{M_t}{J} = \frac{\tau}{r}$$

$$\frac{M_t}{\frac{\pi}{32}d^4} = \frac{\tau}{\frac{d}{2}}$$

$$\frac{1933.6 \times 10^3}{\frac{\pi}{32}d^4} = \frac{120}{\frac{d}{2}}$$

d^3 = 82064.74

d = 43.45 @ 44 mm

∴ minimum diameter d = 44 mm

Engine torque

$$P = \frac{2\pi N T_e}{60} \text{ Watts}$$

$$30 \times 10^3 = \frac{2\pi \times 2000 \times T_e}{60}$$

T_e = 143.23 Nm

Torque transmitted to the wheels

M_t = T_e × GR × AR

= 143.23 × 3 × 4.5

= 1933.6 Nm

= 1933.6 × 10^3 Nmm

At the bearing the equations for combined stresses can be used, assuming a diameter proportionately larger than the minimum and thus, both the allowable tensile and shear stresses in the shaft at the outboard bearing can be compared with the calculated values.

Let

d = 50 mm > 44 mm

ℓ =200 mm ($\ell < \ell_s$)

i. Maximum shear stress theory or Guest's theory

The maximum shear stress in the shaft

$$\tau_{Cal} = \frac{16W}{\pi d^3}\sqrt{\mu^2(r^2+\ell^2)+\ell^2}$$

$$= \frac{16\times 7357.5}{\pi 50^3}\left[200\sqrt{1+0.6^2}+\sqrt{0.6^2(355^2+200^2)+200^2}\right]$$

$= 94.68$ N/mm² $< \tau_{allow} = 120$ N/mm²

ii. Maximum normal stress theory or Rankine's theory

The maximum bending stress in the shaft

$$\sigma_{b\,max} = \frac{16W}{\pi d^3}\left[\ell\sqrt{1+\mu^2}+\sqrt{\mu^2(r^2+\ell^2)+\ell^2}\right]$$

$$= \frac{16\times 7357.5}{\pi 50^3}\left[200\sqrt{1+0.6^2}+\sqrt{0.6^2(355^2+200^2)+200^2}\right]$$

$= 164.60$ N/mm² $< \sigma_{ballow} = 280$ N/mm²

∴ d = 50 mm

b. Axle housing design

Let

d_i = 64 mm > d=50 mm

d_o = d_i/0.8 = 80 mm

i. Maximum shear stress theory or Guest's theory

The maximum shear stress in the shaft

$$\tau_{max} = \frac{16\,d_o}{\pi(d_o^4 - d_i^4)}\sqrt{(\mu wr)^2+(W_s\ell_s)^2(1+\mu^2)}$$

$$= \frac{16\times 80}{\pi(80^4-64^4)}\sqrt{(0.6\times 7357.5\times 355)^2+(7357.5\times 220)^2(1+0.6^2)}$$

$= 41.33$ N/mm² $< \tau_{allow} = 120$ N/mm²

ii. Maximum normal stress theory or Rankine's theory

The maximum bending stress in the shaft

$$\sigma_{b\,max} = \frac{16\,d_o}{\pi(d_o^{\,4}+d_i^{\,4})}\left[W_s\ell_s\sqrt{1+\mu^2}+\sqrt{(\mu Wr)^2+(W_s\ell_s)^2(1+\mu^2)}\right]$$

$$= \frac{16\times 80}{\pi(80^4+64^4)}\left[7357.5\times 220\sqrt{1+0.6^2}+\sqrt{(0.6\times 7357.5\times 355)^2+(7357.5\times 220)^2(1+0.6^2)}\right]$$

$= 73.13$ N/mm² $< \sigma_{ballow} = 280$ N/mm²

$\therefore d_i = 64$ mm

$d_o = 80$ mm

Problem 8.8: Design a full floating rear axle from the following data: Maximum engine torque = 196 Nm; Rear axle ratio = 5:1; Gear ratios = 4:1, 2.5:1, 1.4:1, & 1:1; Permissible shear stress for axle shaft materials = 13734 N/mm²; The diameter of the shaft alone is to be found out.

Given data

Full-floating rear axle

T_e = 196 Nm

AR = 5

GR_1 = 4:1 ; GR_2=2.5:1 ; GR_3=1.4:1 ; GR_4=1:1

τ_{allow} = 13734 N/m²

Find

d

Solution

In full-floating type, axle is subjected to pure torsion. Therefore,

$\frac{M_t}{J} = \frac{\tau}{r}$, can be used to find 'd'.

a. First gear

$M_t = T_e \times GR_1 \times AR$

$= 196 \times 4 \times 5$

$= 3920$ Nm $= 3920 \times 10^3$ Nmm

$$\frac{3920\times 10^3}{\frac{\pi}{32}d^4} = \frac{13734}{\frac{d}{2}}$$

$d^3 = 1453.64$

d = 11.32 @ 12 mm

b. Second gear

$M_t = T_e \times GR_2 \times AR$

$= 196 \times 2.5 \times 5$

$= 2450 \text{ Nm} = 2450 \times 10^3 \text{ Nmm}$

$$\frac{2450 \times 10^3}{\frac{\pi}{32} d^4} = \frac{13734}{\frac{d}{2}}$$

$d^3 = 908.52$

d = 9.68 @ 10 mm

c. Third gear

$M_t = T_e \times GR_3 \times AR$

$= 196 \times 1.4 \times 5$

$= 1372 \text{ Nm} = 1372 \times 10^3 \text{ Nmm}$

$$\frac{1372 \times 10^3}{\frac{\pi}{32} d^4} = \frac{13734}{\frac{d}{2}}$$

$d^3 = 508.77$

d = 7.98 @ 8 mm

d. Top gear

$M_t = T_e \times GR_4 \times AR$

$= 196 \times 1 \times 5$

$= 980 \text{ Nm} = 980 \times 10^3 \text{ Nmm}$

$$\frac{980 \times 10^3}{\frac{\pi}{32} d^4} = \frac{13734}{\frac{d}{2}}$$

$d^3 = 363.41$

d = 7.13 @ 7 mm

Steering

STEERING

The steering system serves to allow changing the direction the direction in which the vehicle is moving.

FUNCTIONS

The function of the steering system is to:

- help in swinging the wheels to the left or right.
- maintain the correct effort needed to turn the wheels.
- absorb a major part of the road shocks thereby preventing them to get transmitted to the hands of the driver.

REQUIREMENTS

The main requirements of a steering system are:

1. Excellent maneuverability
2. Proper steering effort
3. Smooth recovery
4. Minimum transmission of shock from road surface

COMPONENTS

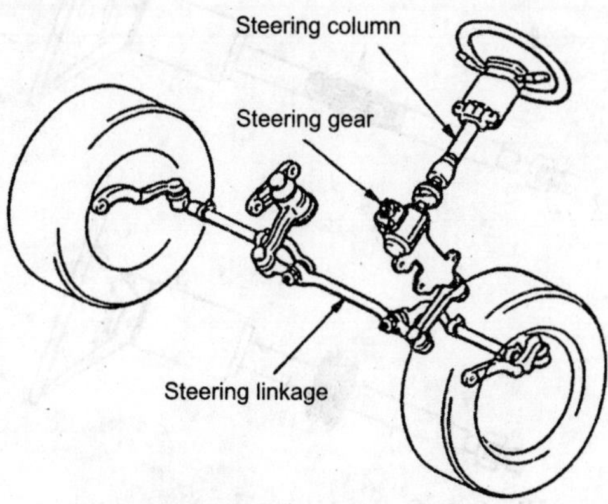

Fig.9.1: *Steering components*

The steering system generally consists of three main sections:

1. Steering column
2. Steering gearbox
3. Steering linkages

OPERATION

Turning the steering wheel turns the main shaft connected to the steering wheel. The steering gear attached to the end of main shaft changes the rotational motion of main shaft to oscillating motion of the pitman arm or back and forth movement of the rack. Since the pitman arm or rack is connected to the front wheels by linkage, one wheel is pushed and the other pulled, allowing each wheel to rotate around the king pin or ball joint.

STEERING COLUMN

A solid or hollow main shaft whose one end is attached to the steering wheel and the other end to a steering driving gear is housed inside a steel-column, is called steering column. The main types of the steering columns are named as under:

1. Rigid (or Integral type) steering column.
2. Collapsible (or energy absorbing) steering column.
3. Tilt steering column.
4. Tilt and telescopic steering column.
5. Steering column with anti-theft lock.

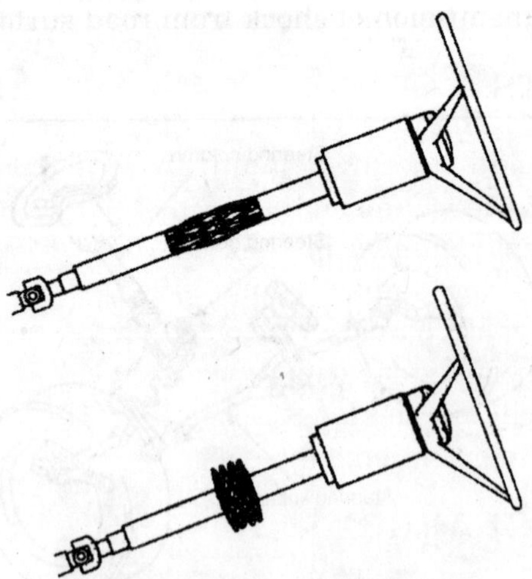

Fig.9.2: *Collapsible steering column*

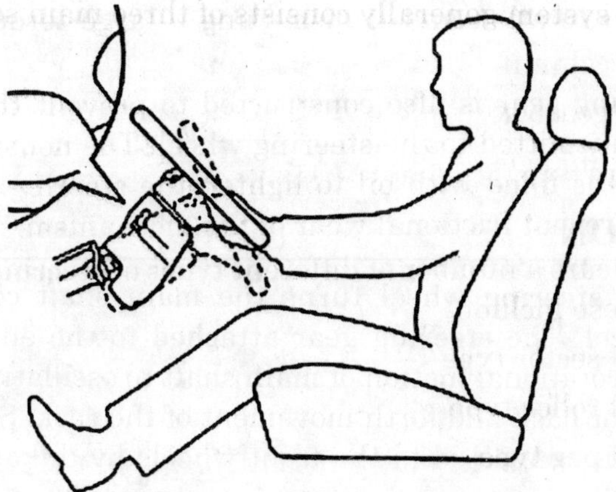

Fig.9.3: *Tilt steering column*

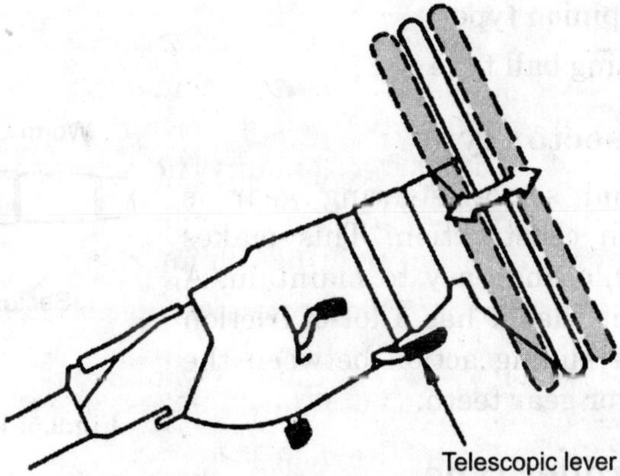

Fig.9.4: *Telescopic steering column*

The steering column can be designed to collapse on impact. In the event of collision, the driver is exposed to the risk of injury from the steering wheel and steering column. In addition to the collapsible or impact-absorbing mechanism, the main steering shaft on some vehicles may also contain a number of steering control systems. For example, the steering lock mechanism which locks the main shaft completely; the tilt steering mechanism, which allows the driver to adjust the vertical position of the steering wheel; the telescopic steering, which lets the driver change the length of the steering shaft freely to obtain an optimal driving position; etc.

STEERING GEARBOX

The steering gearbox serves to change the direction of steering wheel movement and to lighten the steering force required by reducing the steering wheel rotation and transmitting a large torque to the front wheels.

The steering gear is also constructed to prevent the road shocks from being transmitted to the steering wheel. The housing is bolted to the frame and is filled with oil to lighten the steering wheel turning effort and to prevent frictional wear of the mechanism.

Over the years a number of different types of steering gearbox have been used. These include:

1. Worm and sector type
2. Worm and roller type
3. Worm and peg type
4. Worm and wheel type
5. Screw and nut type
6. Rack and pinion type
7. Recirculating ball type

Worm and sector type

The worm and sector steering gear is very simple in construction. This makes it cheap to build and easy to maintain. A disadvantage is that it has a lot of friction because of the sliding action between the worm and sector gear teeth.

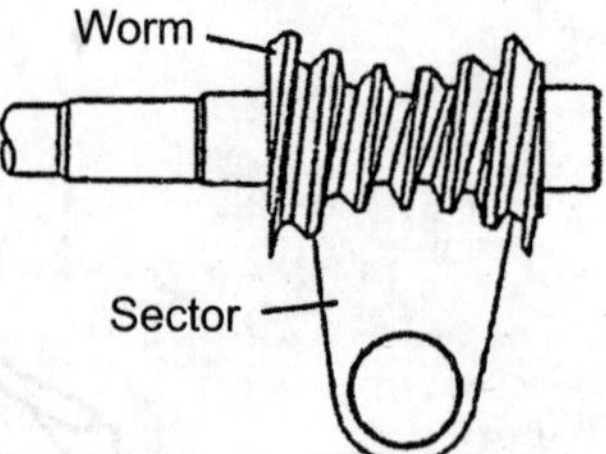

Fig.9.5: *Worm and sector type*

Worm and roller type

The worm and roller steering gear is much like the worm and sector, but the sliding friction is changed to rolling friction so that less effort is required to turn the steering wheel. When the worm gear is turned, the roller follows along the worm gear teeth and turns the pitman arm clockwise or counterclockwise.

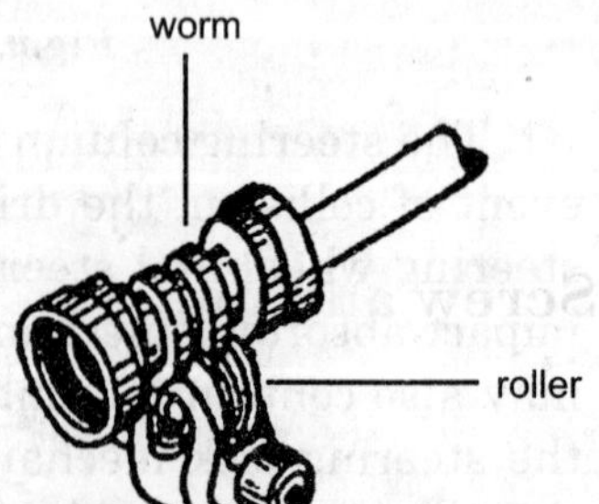

Fig.9.6: *Worm and roller type*

Worm and peg type

This type of steering gearbox is used widely in lighter cars. A cone shaped peg travels along the worm gear as the steering shaft is rotated. Peg is mounted on a lever which drives the drop arm via a shaft.

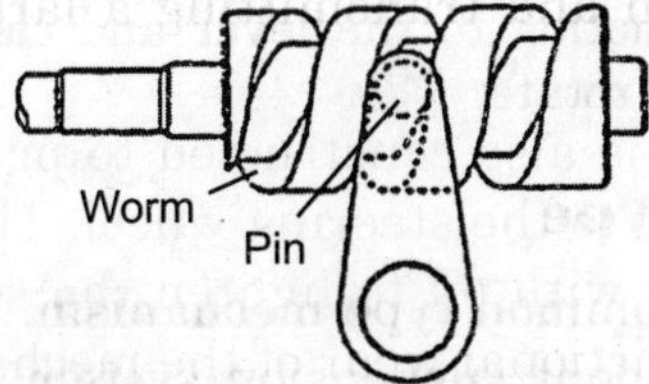

Fig.9.7: *Worm and peg type*

Worm and wheel type

The worm which is keyed on to a steering shaft has a mesh with the worm wheel. The outer end of the worm wheel is fixed to a drop arm which is having a ball end to connect the side rod. The side rod is connected to the steering arm which is fixed to the stub axles. When driver rotates the steering wheel then drop arm moves either backward or forward direction. This motion results in motion of the stub axles.

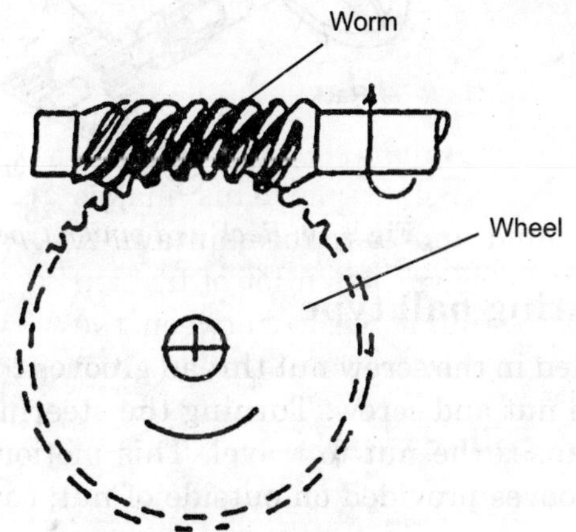

Fig.9.8: *Worm & wheel type*

Screw and nut type

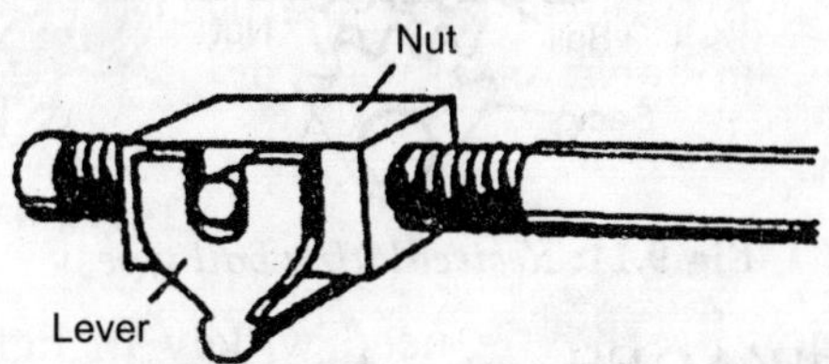

Fig.9.9: *Screw and nut type*

Threads are cut on the lower end of main shaft and a nut screwed into it. Nut end is provided with a protrusion which fits into a lever installed on the housing. Turning the steering wheel moves the nut up and down, imparting a swinging motion to the lever and causing the pitman arm connected to the lever to rotate.

Rack and pinion type

It is very simple and common type mechanism. This type is very well suitable in an independent suspension system. The rack and pinion steering gear, used mostly on smaller cars, has pinion connected to the lower end of the steering main shaft. The pinion is meshed with the rack. When the pinion is given rotary motion with the steering wheel, then the rack slides in either side. This sliding motion of the rack is used through the track rods to turn the wheels in desired side.

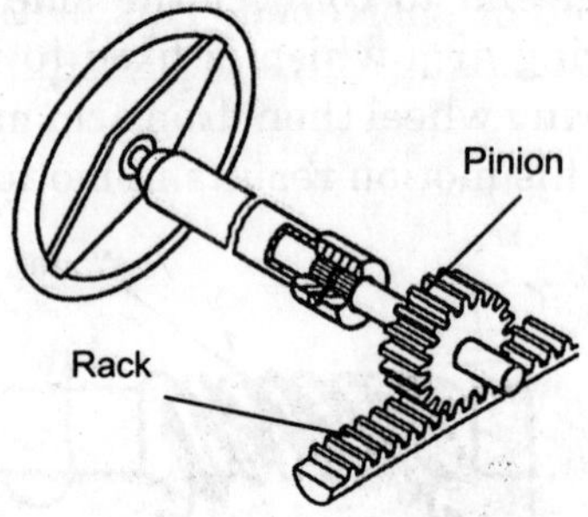

Fig.9.10: *Rack and pinion type*

Recirculating ball type

Balls are filled in the screw nut thread grooves to form a rolling contact between the nut and screw. Turning the steering wheel circulates the balls and causes the nut to travel. This motion is transferred to the sector by grooves provided on outside of nut, causing the pitman arm to rotate.

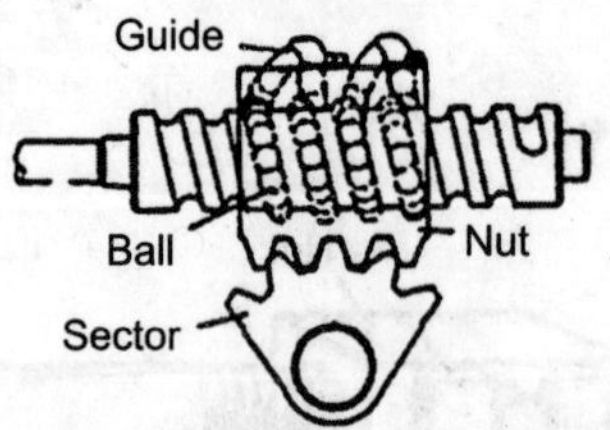

Fig.9.11: *Recirculating ball type*

STEERING LINKAGES

The steering linkage for rigid axle suspension is an assemblage of rods and arms that transfer the power from the steering wheel to the front wheels. The construction of steering linkage differs on whether the front suspension is of the rigid axle type or independent suspension type.

Rigid axle type

The steering linkage is comprised of the pitman arm, drag link, knuckle arm and tie rod. A typical construction is as shown in Fig. Ball joints are used to connect the links. These ball joints allow free movement at the connections so as not to impose a strain on the links.

Tie rod ends are screwed into both ends of the tie rod. One end is provided with right hand threads and the other end with left hand threads. Thus, by turning these tie rod ends, the overall length of the tie rod can be lengthened or shortened. This method is used to adjust the toe-in.

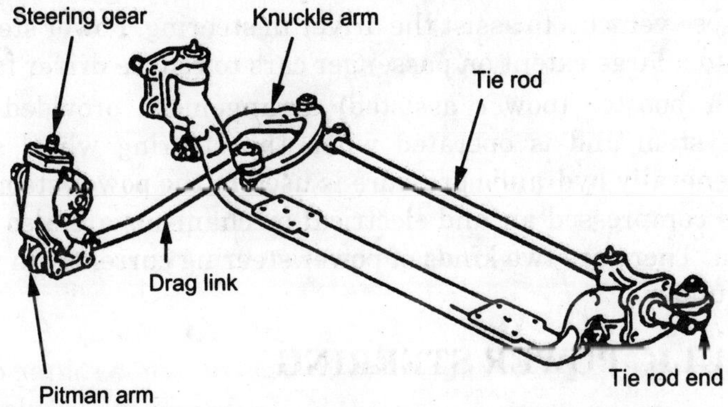

Fig.9.12: *Steering linkage for rigid axle suspension*

Independent suspension type

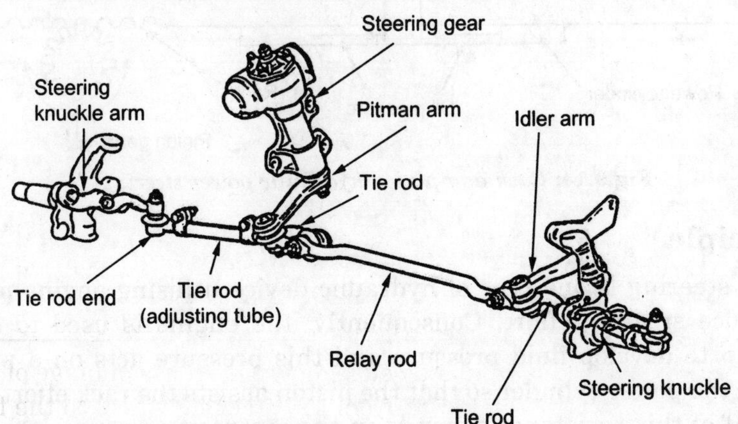

Fig.9.13: *Steering linkage for independent suspension*

In the case of independent suspension type, the wheels move up and down independently so that the distance between the knuckle arms changes. Therefore, if the wheels are connected together with a single tie rod, the toe-in would change each time the wheels move up and down. For this reason, two tie-rods are used and linked to the relay rod as shown in Fig.9.13. An adjusting tube is provided in each tie-rod to allow making toe in adjustments. The adjusting tube has left handed threads on one end and right handed threads at the other to enable adjusting the tie rod length by turning the tube. One of the tie rods has about the same length as the lower suspension arm.

POWER STEERING

Large vehicles require considerably more force to steer due to the greater load on the front wheels. Thus, power steering is equipped on some of these vehicles to assist the driver in steering. Power steering is also used to a large extent on passenger cars to reduce driver fatigue.

It is a booster (power assisted) arrangement provided in the steering system and is operated when the steering wheel shaft is turned. Generally hydraulic pressure is used in the power steering but sometimes compressed air and electrical mechanisms are also used in the system. There are two kinds of power steering currently in use and are given under.

HYDRAULIC POWER STEERING

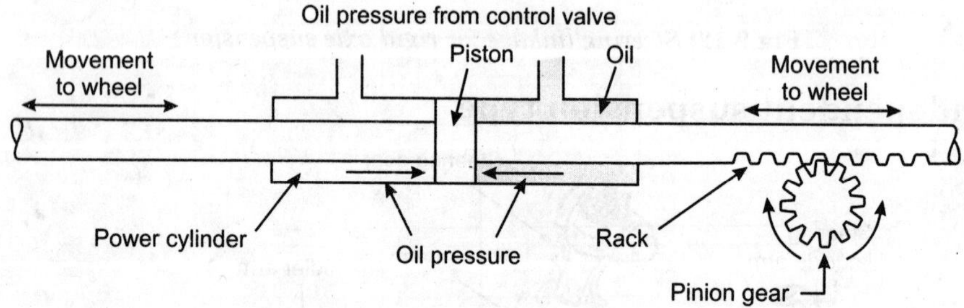

Fig.9.14: *Rack-and-pinion hydraulic power steering*

Principle

Power steering is one type of hydraulic device utilising engine power to reduce steering effort. Consequently, the engine is used to drive a pump to develop fluid pressure and this pressure acts on a piston within the power cylinder so that the piston assists the rack effort. The amount of this assistance depends on the extent of pressure acting on the piston. Therefore, if more steering force is required, the pressure must be raised. The variation in the fluid pressure is accomplished by a control valve which is linked to the steering main shaft.

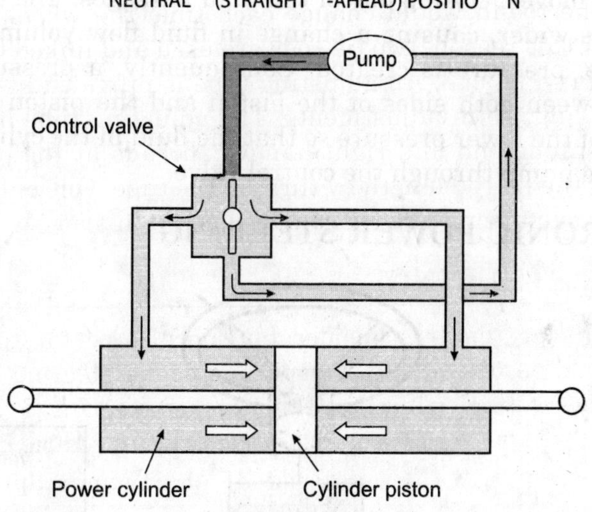

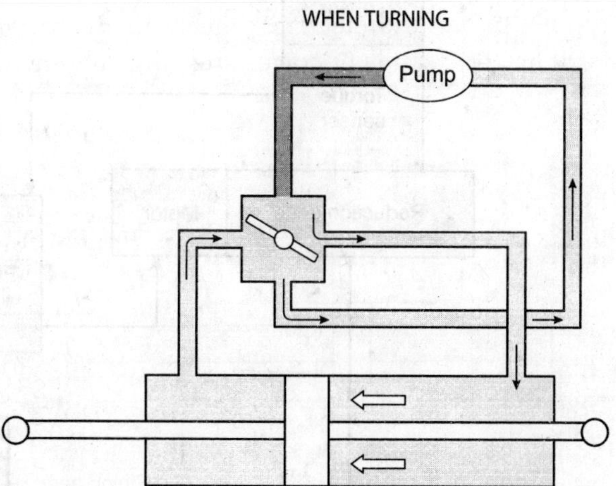

Fig.9.15: *Principle of hydraulic power steering*

Neutral (straight-ahead) position

Fluid from the pump is sent to the control valve. If the control valve is in the neutral position, the fluid will flow through the control valve into the relief port and back to the pump. At this time, hardly any pressure is created and because the pressure on the cylinder piston equal on both sides, the piston will not move in either direction.

When turning

When the steering main shaft is turned in either direction, the control valve also moves, closing one of the fluid passages. The other passage then opens wider, causing a change in fluid flow volume and, at the same time, pressure is created. Consequently, a pressure difference occurs between both sides of the piston and the piston moves in the direction of the lower pressure so that the fluid in the cylinder is forced back to the pump through the control valve.

ELECTRONIC POWER STEERING

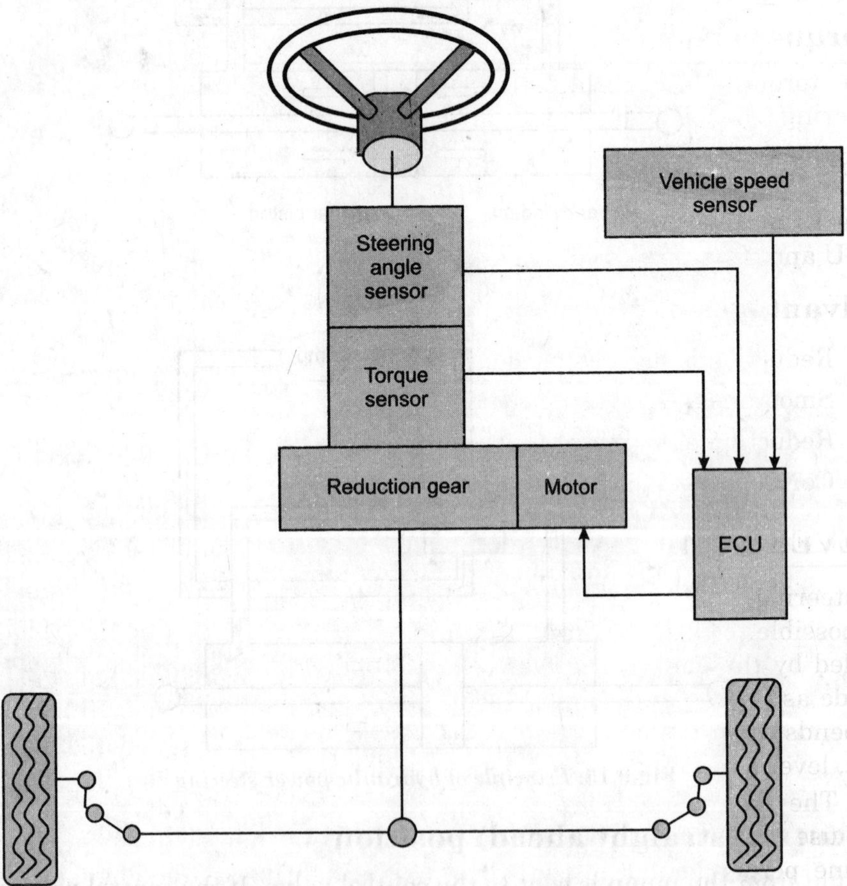

Fig.9.16: *Electronically controlled power steering*

The electronically controlled power steering improves steering ability and performance. This power steering reduces the steering effort when driving at low speeds and to provide a corrective action when driving at high speeds.

It consists of wheel speed sensor, a steering sensor (torque, angular velocity), an electronic control unit, a drive unit and a motor. Signal output from each sensor are input to ECU and necessary steering assistance is calculated and applied by the drive unit to control the operation of the motor.

Power steering ECU

The power steering ECU calculates assisting power based on steering torque signals from the torque sensor and vehicle speed signals from the skid control ECU. For vehicles without ABS, the power steering ECU receives vehicle speed signals from the speedometer.

Torque sensor

The torque sensor detects the steering effort generated when the steering wheel is turned and converts it to an electrical signal.

EPS motor

The EPS motor is activated by the current from the power steering ECU and generates torque to assist the steering effort.

Advantages of electronic power steering

1. Reduction of driver's burden
2. Smoothness in steering operation
3. Reduction of kick back
4. Correction of steering reaction forces

REVERSIBILE AND IRREVERSIBLE STEERING

A steering gear is defined as reversible if by turning of stub axles it is possible to turn the steering wheel by which the road shocks being feeled by the driver. Therefore the passenger- car steering gears are made as nearly nonreversible. This characteristic of the steering gear depends upon the design of the worm or cam and its meshing roller, nut, lever and segment etc.

The nonreversible property in the steering gear is obtained by the use of worm gear having thread angle less than six degrees with a line perpendicular to the worm axis. Slightly greater angle than this results semi-reversible of the steering gear. The main advantage of the non-reversible gear is that the driver does not feel any shock but disadvantage of it is that the steering operation becomes heavy. Therefore looking to straight motion of the vehicle without undue stresses on the mechanism the semi-reversible type of steering gear is preferred.

WHEEL ALIGNMENT

The purpose of the wheel alignment is to position the front wheels and steering mechanism to reduce the tyre wear, road shocks, wheel wobbles and providing ease in steering control. It provides directional stability and the vehicle can run straight down a road under all conditions and at all loads. The other factors those effect the wheel alignment are as frame alignment, position of the rear axle, spring condition and condition of the shock absorbers. Therefore for smooth running of the vehicle, it is very necessary to consider and understand the factors which influence the stability and control of it. The balance of wheels, inflation of tyres and brake adjustment are also considerable in cases of wheel wobble, change of rolling radius and to pull the vehicle to one side respectively.

FRONT END GEOMETRY

Steering geometry is the term manufacturers use to describe steering and wheel alignment. The fundamental angles or specifications that are required for a proper wheel alignment are caster, camber, steering axis inclination, toe-in, toe-out on turns, tracking, and scrub radius. The factors of the front end geometry are considered for the steering ease, steering stability, tyre wear and the riding qualities of the vehicle. The front end geometry depends upon the following factors:

CAMBER

The camber is the tilt of the front wheels from the vertical plane. When the tilt is outward at the top then the camber is positive and when the tilt is inward at the top the camber is called negative. The wheels are given a positive camber to start so that when the vehicle is loaded and rolling along on road, they automatically come to a vertical position. The positive camber is shown in the Fig.9.17 which is measured in degrees from the vertical plane. The camber angle should not exceed 2°, while running any amount of camber, positive or negative tends to cause more rapid tyre wear because under this condition most of the load is shared on one side of the tyre. Therefore it is clear that the tyre life will be maximum when the camber angle in running condition is zero with average load.

Similarly unequal camber in the front wheels causes the vehicle to roll in the direction of the wheel having the greater camber. With excessive camber a tyre does not come in correct contact with the road which causes it to wear only on the side.

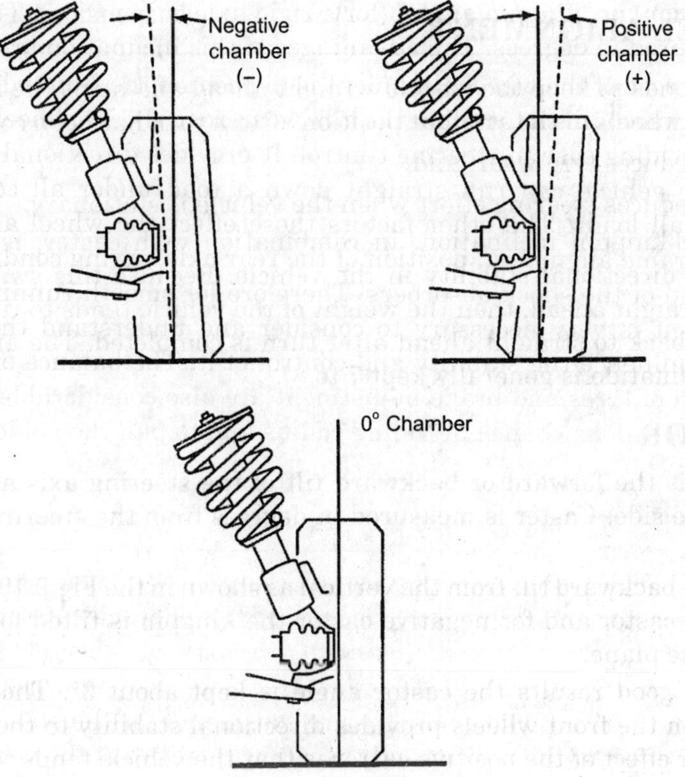

Fig.9.17: *Camber*

KINGPIN INCLINATION

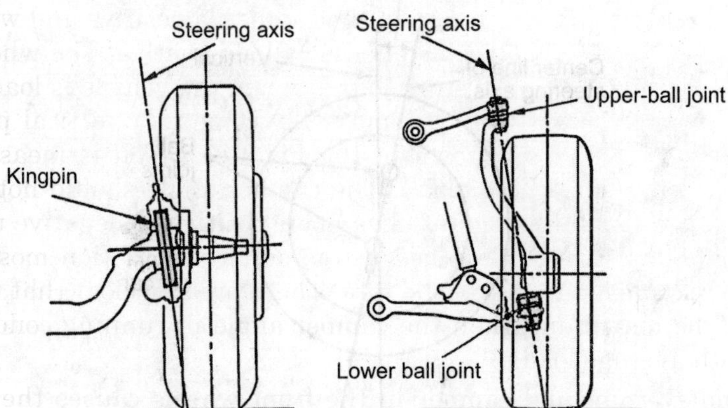

Fig.9.18: *KPI and SAI*

The steering axis is tilted inward as viewed from the front of the car and is called the "steering axis inclination (SAI)" (for an independent suspension) or "kingpin angle" (for a rigid axle suspension). This angle is measured in degrees. The advantages of this inclination are:

1. It tends to keep the wheels straight ahead and helps the return of the wheels to the straight position after a turn has been completed.
2. It reduces tyre wear, and
3. It reduces steering effort when the vehicle is stationary.

The kingpin inclination, in combination with caster, is used to provide directional stability in the vehicle. Because this swing away from straight ahead, then the weight of the vehicle tends to return the wheels back to straight ahead after turn is completed. The amount of this inclination is generally kept 7 to 8°.

CASTOR

Caster is the forward or backward tilt of the steering axis as viewed from the side. Caster is measured in degrees from the steering axis to vertical.

The backward tilt from the vertical as shown in the Fig.9.19 is called positive castor and for negative castor the kingpin is tilted forward in the same plane.

For good results the castor angle is kept about 3°. The positive castor on the front wheels provides directional stability to the vehicle. Another effect of the positive castor is that the vehicle tends to roll out or lean out on turns. But with negative castor the vehicle tends to bank or lean in on turns.

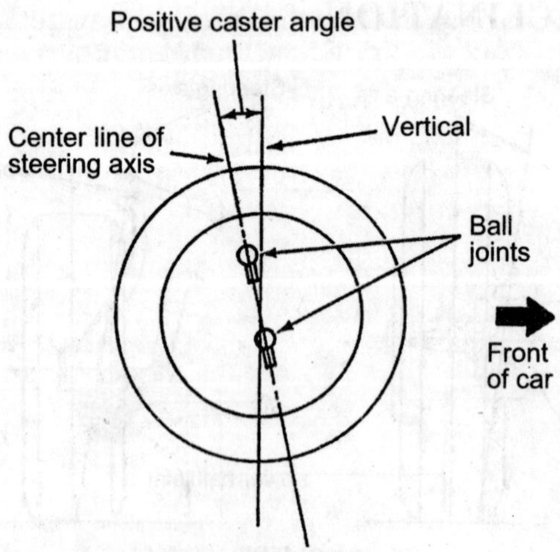

Fig.9.19: *Castor*

With positive castor when the right turn is made, the left side of the vehicle is lowered and the right side of the vehicle is lifted, at this position the positive castor helps the centrifugal force in rolling out the vehicle which is not desired. With negative castor, the left side of the vehicle would lift during a right turn while the right side of the vehicle would drop. Thus the negative castor tends to roll in, the vehicle.

The positive castor, causes a tendency in the wheels to toe in because the body of the vehicle is lifted as the front wheels toe out and it is lowered when the toe in. With negative castor, the wheels would try to toe out.

The positive castor provides directional stability and tries to keep the wheels straight ahead. But more amount of positive castor increases the effort required to steer. Therefore modern heavy-duty trucks, tend to have a negative castor which makes steering easier.

TOE-ANGLE (TOE-IN OR TOE-OUT)

It is the difference in distance between the front wheels of a car as measured at the front and at the rear as viewed from the top when the vehicle is stationary. Toe angle is usually expressed by a distance (B-A).

Toc-in

Toe-in is amount by which the front wheels are set closer together at the front than at the rear. The amount of toe-in generally does not exceed 3 mm. The purpose of it is to move the wheels perfectly straight ahead under running condition of the vehicle, to stabilize steering, to prevent side slipping and excessive wear of tyres. Due to rolling resistance the small deflections come in the wheel support system, when the vehicle is moving. But the toe-in of the wheels in standing position tends to roll them parallel when the vehicle is moving.

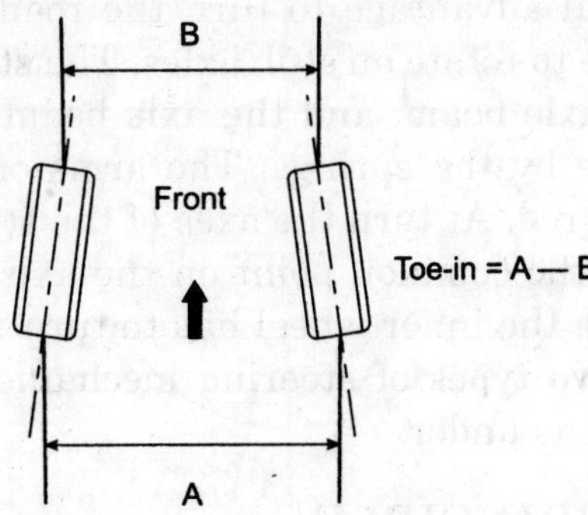

Fig.9.20: *Toe-in*

Toe-out on turns

The wheels may be set closer at the rear than at the front in which case the difference of the distances between the front wheels at the front and at the rear is called toe-out.

The steering system of a vehicle is designed to turn the inside wheel through a larger angle than the outside wheel when making a turn. This action causes the front wheels to toe-out on turns.

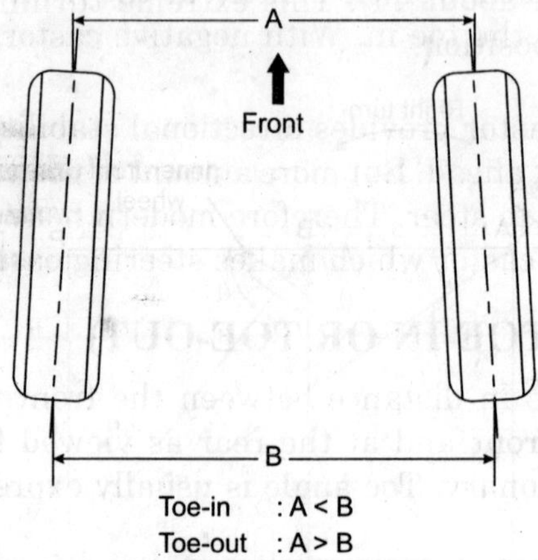

Fig.9.21: *Toe-out*

Therefore in this way the steering angle of the inner wheel may has a maximum value of about 44°. This extreme turning position of the wheel is called lock position.

STEERING MECHANISMS

The main function of the steering system of a vehicle is to convert the rotary movement of the steering wheel into angular turn of the front wheels. The system also multiplies the effort applied at the steering wheel by mechanical advantage to turn the road wheels easily. The front wheels are free to rotate on stub axles. The stub axles are pivoted at the ends of the axle beam, and the axle beam is connected to the frame of the vehicle by the springs. The arms of the stub axles are coupled to the track rod. At turn the axes of the stub axles of the front wheels intersect at the common point on the axis of the rear wheels. Under this condition the inner wheel has to turn more than the outer wheel. There are two types of steering mechanisms in general uses which are explained as under.

ACKERMANN PRINCIPLE

Referring Fig.9.22, the front wheels are steered to make the left side turn. The inner wheel turns at an angle $\theta°$ and the outer wheel turn at an angle $\varphi°$ with the axis of the front axle beam. The axes of the front wheels meet the rear wheel axis at the point P, which is called the instantaneous centre of the vehicle. Thus it is clear that the inside wheel is required to turn to a greater angle than the outer wheel. This permits the inner wheel to follow a shorter radius than the outer wheel.

Therefore in this way the steering angle of the inner wheel may has a maximum value of about 44°. This extreme turning position of the wheel is called lock position.

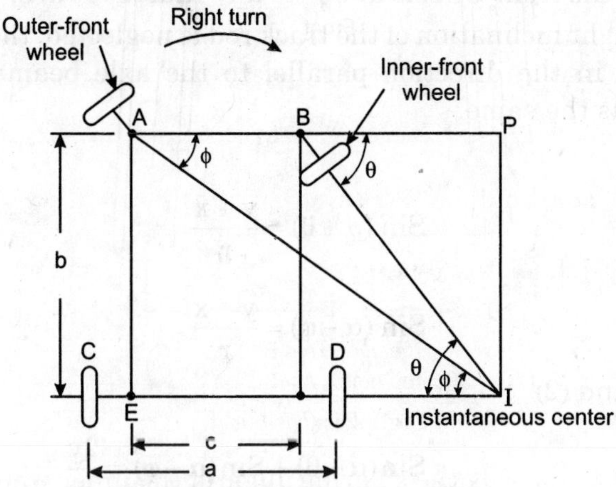

Fig.9.22: *Ackermann principle*

ACKERMANN MECHANISM

The difference in movement of the inner and outer wheels is obtained by inclining the steering arms. In the straight–ahead position, the steering arms are arranged so that their centre lines, if produced, would meet on the centre line of the vehicle on or near the nose of the rear axle. The track rod is therefore shorter than the distance between the kingpin axes if it lies behind the axle beam.

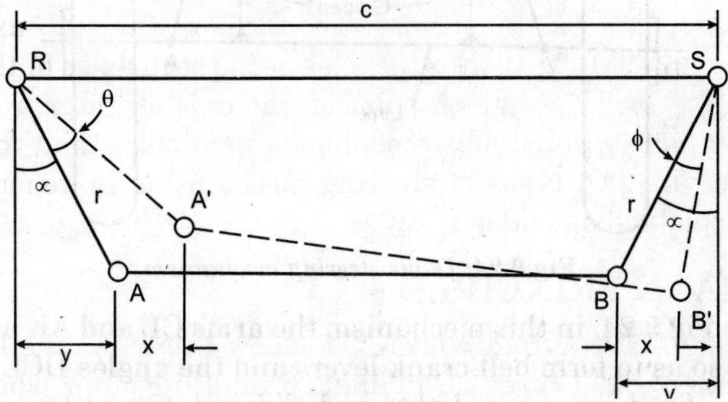

Fig.9.23: *Ackermann linkage*

When the vehicle is travelling in the straight–ahead position, the track rod is parallel to the axle beam but, as it moves the steering arms, the inner end moves nearer to the axle than the outer end (as indicated by the dotted lines in Fig.9.23). The result of this action is that the inner steering arm and road wheel are turned through a greater angle than the outer arm and wheel. It should be noted that the difference between these two angles is not constant but is automatically corrected to maintain the front wheels at 90° to any radius of turn.

If the slight inclination of the track rod is neglected, the movements of A and B in the direction parallel to the axle beam (RS) can be considered as the same.

Then

$$\text{Sin}\,(\alpha + \theta) = \frac{y + x}{r} \qquad \text{....... (1)}$$

$$\text{Sin}\,(\alpha - \varphi) = \frac{y - x}{r} \qquad \text{....... (2)}$$

Adding (1) and (2)

$$\text{Sin}\,(\alpha + \theta) + \text{Sin}\,(\alpha - \varphi) = \frac{2y}{r} \qquad \text{Sin}\,\alpha = \frac{y}{r}$$

$$\text{Sin}\,(\alpha + \theta) + \text{Sin}\,(\alpha - \varphi) = 2\,\text{Sin}\,\alpha$$

DAVIS STEERING MECHANISM

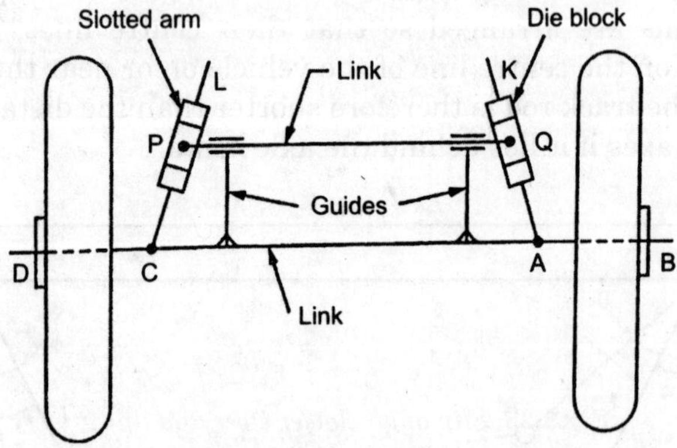

Fig.9.24: *Davis steering mechanism*

Referring Fig.9.24, in this mechanism the arms CL and AK are fixed to the axles so as to form bell-crank levers and the angles DCL, BAK are equal. The fixed arms are slotted and slide relative to two die-blocks.

The die blocks are pivoted to the link PQ, which is supported in two guides so that it can be moved parallel to the link CA. The sliding of the link PQ to the left or to the right effects the steering of the vehicle. The vehicle moves along a straight path when the steering arms CL and AK are inclined at the equal angle to the centre line of the vehicle.

STEERING CHARACTERISTICS

The steering characteristics of the vehicle depend on the design, conditions under which it operates, tyre pressures and the loads carried by it under different conditions.

When a vehicle moves in a circular path the centre of that path does not coincide with the point of intersection of the wheel axes. Under this condition the centrifugal force acts on the wheels which are provided with slip angle. Therefore the value of the slip angle has important effects on the steering characteristics of the vehicle. The cases arise under the different slip angle are as given under.

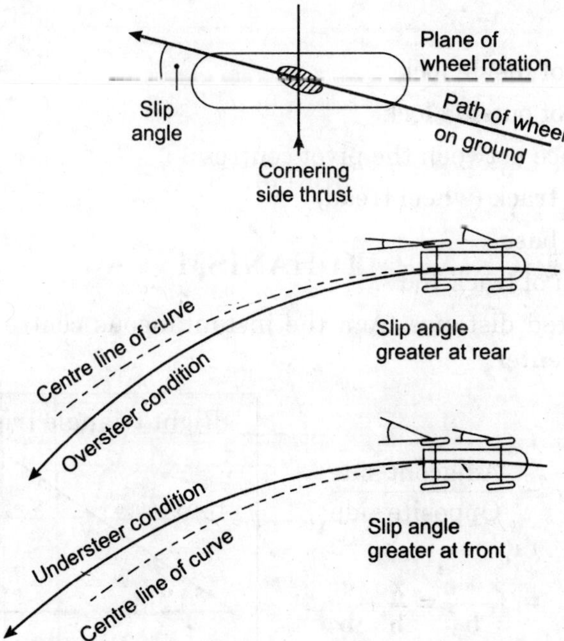

Fig.9.25: *Slip angle, lower steer and under steer*

Over steer

Refer Fig.9.25, if the slip angle for the rear wheels is greater than for the front wheels then the radius of the turn is decreased. Therefore the vehicle will turn more sharply into the curve and the steering wheel has to be turned back to correct this tendency. Therefore to keep the vehicle on the right path the wheels should be steered a little less than is theoretically desired. This condition of cornering is called oversteer.

Understeer

Refer Fig.9.25, if the slip angle is greater for the front wheels than for the rear, the radius of the turn is increased. Therefore on turn the vehicle will tend to straighten out and the steering wheel will generally have to be turned a little more to counteract this tendency. This condition of cornering is called understeer.

DESIGN AND PROBLEMS

DESIGN OF STEERING SYSTEM

CONDITION FOR TRUE ROLLING

Let,

q = Angle of inside lock

f = Angle of outside lock

c = Distance between the pivot centres

a = Wheel track (wheel tread)

b = Wheel base

d = Length of track rod

x = Projected distance from the instantaneous centre to the inner pivot center.

From ∇IEA,

$$\text{Cot } \phi = \frac{\text{Adjacent side}}{\text{Opposite side}}$$

$$= \frac{x+c}{b} = \frac{x}{b} + \frac{c}{b}$$

$$= \text{Cot } \theta + \frac{c}{b}$$

$$\text{Cot } \phi - \text{Cot } \theta = \frac{c}{b}$$

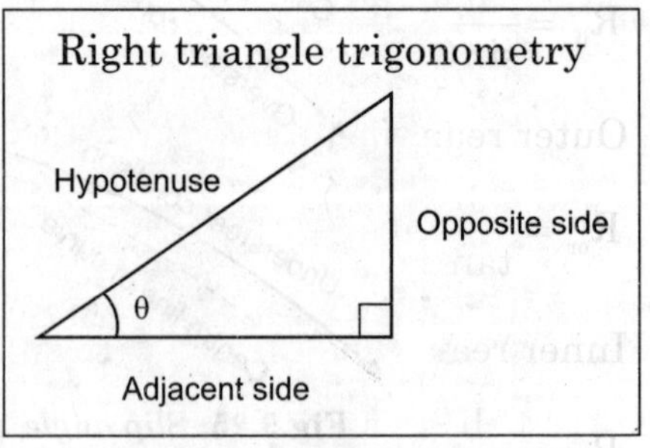

From ∇IFB

$$\text{Cot } \theta = \frac{x}{b}$$

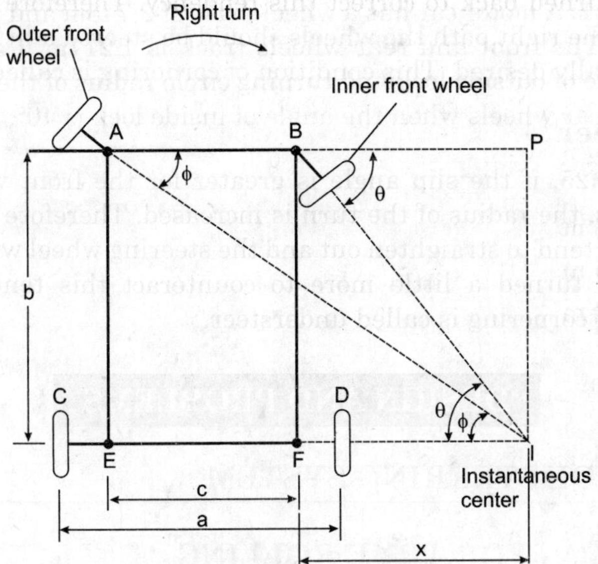

TURNING CIRCLE RADIUS

The turning circle radius of different wheels without any lateral slip can be found using the following expressions.

Outer front wheel

$$R_{of} = \frac{b}{\sin\varphi} + \frac{a-c}{2}$$

Inner front wheel

$$R_{if} = \frac{b}{\sin\theta} - \frac{a-c}{2}$$

Outer rear wheel

$$R_{or} = \frac{b}{\tan\varphi} + \frac{a-c}{2}$$

Inner rear wheel

$$R_{ir} = \frac{b}{\tan\theta} - \frac{a-c}{2}$$

From SAE method turning radius can be found by

$$R_{of} = \left\{\left(-\frac{b}{\sin\theta}\right)^2 + c^2 + \frac{2bc}{\tan\theta}\right\}^{\frac{1}{2}} + \frac{a-c}{2}$$

Problem 9.1: A motor car has a wheel–base of 2.743m and pivot centre of 1.065m. The front and rear wheels track is 1.217m. Calculate the correct angle of outside lock and turning circle radius of the outer front and inner rear wheels when the angle of inside lock is 40°.

Give data

$b = 2.743$ m

$c = 1.065$ m

$a = 1.217$ m

$\theta = 40°$

Find

R_{of}, R_{ir} and ϕ

Solution

We know that, condition for true rolling

$$\text{Cot}\,\phi - \text{Cot}\,\theta = \frac{c}{b}$$

$$\text{Cot}\,\phi - \text{Cot}\,40° = \frac{1.065}{2.743}$$

$$\phi = 32.4°$$

Outer front wheel

$$R_{of} = \frac{b}{\sin\varphi} + \frac{a-c}{2} = \frac{2.743}{\sin 32.4} + \frac{1.217-1.065}{2} = 5.2 \text{ m}$$

Inner rear wheel

$$R_{ir} = \frac{b}{\tan\theta} - \frac{a-c}{2} = \frac{2.743}{\tan 40} - \frac{1.217-1.065}{2} = 3.2 \text{ m}$$

Problem 9.2: A track has a pivot pins 1.37m apart, the length of each track arm is 0.17m and the track rod is behind front axle and 1.17m long. Determine the wheel base which will give true rolling for all wheels when the car is turning so that the inner wheel stub axle is 60° to the center line of the car. A geometrical construction may be used.

Note: θ is indirectly given.

Given

$c = 1.37$ m

r = 0.17 m

d = 1.17 m

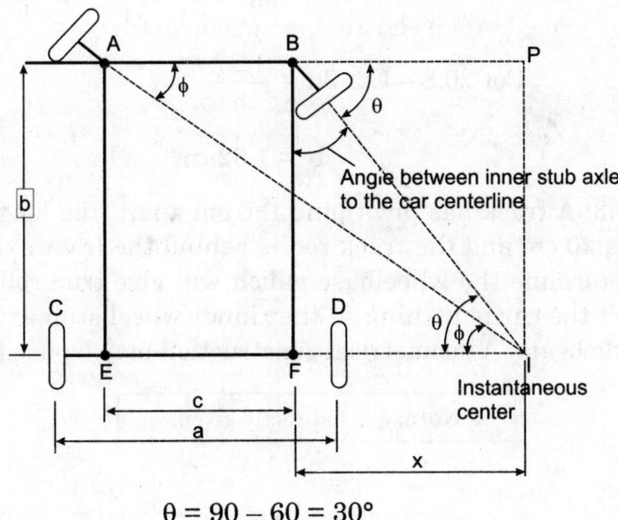

∴ $\theta = 90 - 60 = 30°$

Find

α, ϕ, b

Solution

$$\text{Sin}\alpha = \frac{\text{Opposite side}}{\text{Hypotenuse}} = \frac{c-d}{2r} = \frac{1.37-1.17}{2\times 0.17} = 0.5882$$

$$\alpha = 36.03°$$

We know that

$$\text{Sin}(\alpha+\theta) + \text{Sin}(\alpha-\varphi) = 2\,\text{Sin}\,\alpha$$

$$\text{Sin}(36.03+30) + \text{Sin}(36.03-\phi) = 2\text{Sin}(36.03)$$

$$\text{Sin}(66.03) + \text{Sin}(36.03-\phi) = 2\text{Sin}(36.03)$$

$$0.9137 + \text{Sin}(36.03-\phi) = 1.1764$$

$$\text{Sin}(36.03-\phi) = 1.1764 - 0.9137$$

$$\text{Sin}(36.03-\phi) = 0.2627$$

$$36.03-\phi = \text{Sin}^{-1}(0.2627)$$

$$36.03-\phi = 15.23$$

$$\phi = 20.8°$$

Now

$$\text{Cot } \phi - \text{Cot } \theta = \frac{c}{b}$$

$$\text{Cot } 20.8 - \text{Cot } 30 = \frac{1.37}{b}$$

$$\mathbf{b = 1.52\ m}$$

Problem 9.3: A track has pivot pins 136 cm apart, the length of each track arm is 20 cm and the track rod is behind the front axle and 126 cm long. Determine the wheelbase which will give true rolling for all wheels when the car is turning so that inner wheel stub axle is 30° to the front axle beam. A geometrical construction may be used.

Note: θ is indirectly given.

Given

c = 136 cm

r = 20 cm

d = 126 cm

θ = 30°

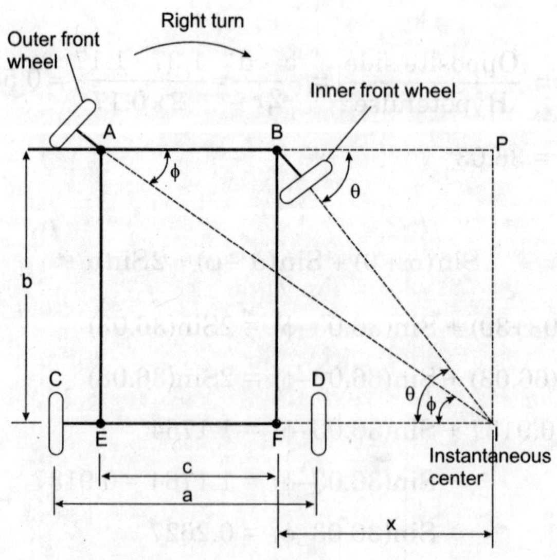

Find

α, ϕ, b

Solution

$$\text{Sin}\alpha = \frac{\text{Opposite side}}{\text{Hypotenuse}} = \frac{c-d}{2r} = \frac{136-126}{2\times 20} = 0.25$$

$$a = 14.48°$$

We know that

$$\text{Sin}(\alpha+\theta) + \text{Sin}(\alpha-\varphi) = 2\text{Sin}\,\alpha$$

$$\text{Sin}(14.48+30) + \text{Sin}(14.48-\phi) = 2\text{Sin}(14.48)$$

$$\text{Sin}(44.48) + \text{Sin}(14.48-\phi) = 2\text{Sin}(14.48)$$

$$0.7006 + \text{Sin}(14.48-\phi) = 0.5$$

$$\text{Sin}(14.48-\phi) = -0.2005$$

$$14.48-\phi = \text{Sin}^{-1}(-0.2005)$$

$$14.48-\phi = -11.56$$

$$\phi = 26.04°$$

Now

$$\text{Cot}\,\phi - \text{Cot}\,\theta = \frac{c}{b}$$

$$\text{Cot}\,26.04 - \text{Cot}\,30 = \frac{136}{b}$$

$$\mathbf{b = 432.26\ cm}$$

Problem 9.4: The distance between the kingpins of a car is 1.3 m. The track arms are 0.1525 m long and the length of track rod is 1.2 m for a track of 1.42 m and a wheel base of 2.85 m. Find the radius of curvature of the path followed by the near side front wheel at which correct steering is obtained when the car is turning to the right.

Given

c = 1.3 m

r = 0.1525 m

d = 1.2 m

a = 1.42 m

b = 2.85 m

Find

R_{if}

Solution

We know that, condition for true rolling

$$\text{Cot}\,\phi - \text{Cot}\,\theta = \frac{c}{b}$$

$$\text{Cot}\,\phi - \text{Cot}\,\theta = \frac{1.3}{2.85} = 0.456$$

We also know that,

$$\text{Sin}(\alpha + \theta) + \text{Sin}(\alpha - \varphi) = 2\,\text{Sin}\,\alpha$$

$$\text{Sin}\,\alpha = \frac{c - d}{2r} = \frac{1.3 - 1.2}{2 \times 0.1525}$$

$$\alpha = 19.14°$$

Sin(19.14+θ) + Sin(19.14–ϕ) = 2Sin(19.14)

By trail and error method, θ can be found.

Assume θ=30°

Sin(19.14+30) + Sin(19.14–ϕ) = 2Sin(19.14)

0.7593 + Sin(19.14–ϕ) = 0.6558

ϕ = 25.17°

For θ=30° and ϕ=25.17°

Cot 25.17 – Cot 30 = 0.3959 < **0.456**

Assume θ=35°

Sin(19.14+35) + Sin(19.14–ϕ) = 2Sin(19.14)

ϕ = 28.45°

For θ=35° and ϕ=28.45°

Cot 28.45 – Cot 35 = 0.4175 < **0.456**

Assume θ=40°

Sin(19.14+40) + Sin(19.14–ϕ) = 2Sin(19.14)

ϕ = 30.83°

For θ=40° and ϕ=30.83°

Cot 30.83 – Cot 40 = 0.4837 > **0.456**

∴ θ lies between 35° and 40°

Assume θ=36°

Sin(19.14+36) + Sin(19.14–ϕ) = 2Sin(19.14)

ϕ =28.62 °

For θ=36° and ϕ=28.62°

Cot 28.62 – Cot 36 = 0.456 = **0.456**

∴ **θ=36 and ϕ=28.62°**

Radius of curvature of the path followed by the near side front wheel is

$$R_{if} = \frac{b}{\sin\theta} - \frac{a-c}{2}$$

$$= \frac{2.85}{\sin 36} - \frac{1.42-1.3}{2} = 4.77 \text{ m}$$

Problem 9.5: A vehicle with the wheel base = 2.1m and front wheel track = 1.22m is provided with Ackermann steering system. The distance from the centre plane of each front wheel to the nearest kingpin is 0.11m while taking a turn the inner wheel is deflected through a maximum angle of 42 deg. Calculate the corresponding deflection of the outer front wheel, assuming that all wheels are in true rolling motion. Also find the turning radius of the outer front wheel

Given

b = 2.1 m

a = 1.22 m

θ = 42°

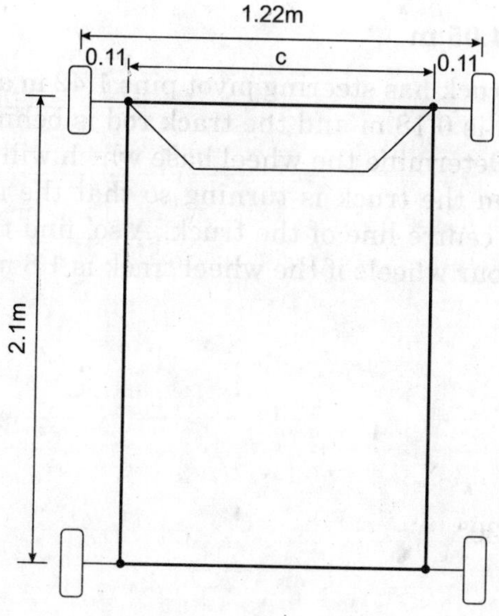

Find

ϕ, R_{of}

Solution

$$c = 1.22 - (2\times0.11) = 1 \text{ m}$$

We know that, condition for true rolling

$$\text{Cot}\,\phi - \text{Cot}\,\theta = \frac{c}{b}$$

$$\boxed{\text{Cot}\,\theta = \frac{1}{\tan\theta}}$$

$$\text{Cot}\,\phi - \text{Cot}\,42 = \frac{1}{2.1}$$

$$\text{Cot}\,\phi - \text{Cot}\,42 = 0.476$$

$$\text{Cot}\,\phi - 1.11 = 0.476$$

$$\text{Cot}\,\phi = 1.586$$

$$\boldsymbol{\phi = 32.2^\circ}$$

Turning circle radius of outer front wheel

$$R_{of} = \frac{b}{\sin\phi} + \frac{a-c}{2}$$

$$= \frac{2.1}{\sin 32.2} + \frac{1.22-1}{2}$$

$$= \mathbf{4.05\ m}$$

Problem 9.6: A truck has steering pivot pins 1.42 m apart. The length of each track arm is 0.18 m and the track rod is behind the front axle and 1.18m long. Determine the wheel base which will give true rolling of all wheels when the truck is turning so that the inner wheel stub axle is 60° to the centre line of the truck. Also, find the turning circle radius of all the four wheels if the wheel track is 1.8 m.

Give data

$c = 1.42$ m

$r = 0.18$

$d = 1.18$

$\theta = 90^\circ - 60^\circ = 30^\circ$

$a = 1.8$ m

Find

b, R_{if}, R_{of}, R_{ir}, R_{or}

Solution

$$\text{Sin}\alpha = \frac{c-d}{2r} = \frac{1.42-1.18}{2\times0.18} = 0.667$$

$$\alpha = 41.8°$$

We know that

$$\text{Sin}(\alpha+\theta) + \text{Sin}(\alpha-\varphi) = 2\,\text{Sin}\,\alpha$$

$$\text{Sin}(41.8+30) + \text{Sin}(41.8–\phi) = 2\text{Sin}(41.8)$$

$$\text{Sin}(71.8) + \text{Sin}(41.8–\phi) = 2\text{Sin}(14.48)$$

$$0.94 + \text{Sin}(41.8–\phi) = 1.33$$

$$\text{Sin}(41.8–\phi) = 0.38$$

$$41.8–\phi = \text{Sin}^{-1}(0.38)$$

$$41.8–\phi = 22.3$$

$$\mathbf{\phi = 19.46°}$$

Now

$$\text{Cot}\,\phi - \text{Cot}\,\theta = \frac{c}{b}$$

$$\text{Cot } 19.46 - \text{Cot } 30 = \frac{1.42}{b}$$

$$2.83 - 1.732 = \frac{1.42}{b}$$

$$\mathbf{b = 1.29\ m}$$

Turning circle radius,

Outer front wheel

$$R_{of} = \frac{b}{\sin\varphi} + \frac{a-c}{2} = \frac{1.29}{\sin 19.46} + \frac{1.8-1.42}{2} = \frac{1.29}{\sin 19.46} + 0.19 = 4.05\ \text{m}$$

Inner front wheel

$$R_{if} = \frac{b}{\sin\theta} - \frac{a-c}{2} = \frac{1.29}{\sin 30} - 0.19 = 2.39\ \text{m}$$

Outer rear wheel

$$R_{or} = \frac{b}{\tan\varphi} + \frac{a-c}{2} = \frac{1.29}{\tan 19.46} + 0.19 = 3.83 \text{ m}$$

Inner rear wheel

$$R_{ir} = \frac{b}{\tan\theta} - \frac{a-c}{2} = \frac{1.29}{\tan 30} - 0.19 = 2.04 \text{ m}$$

STEERING BOX RATIO (MOVEMENT RATIO)

The ratio between the number of turns of the steering wheel (in degrees) to the turn of the drop arm (in degree) is known as steering box gear ratio or movement ratio.

For worm and roller type, worm and peg type, worm and nut type, etc.

$$\text{Movement ratio} = \frac{360° \times \text{No. of turns of steering wheel}}{\text{No. of degrees turned through by drop arm}}$$

For worm and wheel type

$$\text{Movement ratio} = \frac{\text{No. of teeth on worm wheel}}{\text{No. of starts on worm}}$$

For rack and pinion type

In order the pinion to make one revolution, the input steering wheel moves through $2\pi R$ and output rack moves through $2\pi r$.

$$\text{Movement ratio} = \frac{2\pi R}{2\pi r} = \frac{R}{r}$$

Or

$$\text{Movement ratio} = \frac{2\pi R}{2\pi r} = \frac{2\pi R}{tp}$$

$2\pi r = tp$

Where,

R = Steering wheel radius

r = Pinion pitch circle radius

t = Number of pinion teeth

p = Linear or circular pitch

When there is no friction,

$$\text{Movement ratio} = \frac{W}{E}$$

Where,

W = Load on the rack (resistance on the rack)

E = Input steering wheel effort by two hands

This reduction ratio is called steering gear ratio and is usually from 14:1 and 24:1 for manual steering but the power steering reduces ratios on an average by 20 per cent. The low gear ratios produce fast steering, while the high ratios produce slow steering.

STEERING BOX TORQUE

Let,

F = Force exerted by each hand of driver on the steering wheel

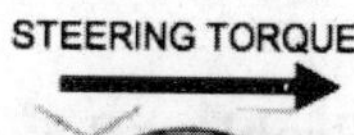

d = Diameter of steering wheel

T_s = Torque applied to the steering wheel

= F.d

G_s = Gear ratio of steering box

η_s = Efficiency of gearing of steering box

T_t = Torque transmitted to the drop arm shaft

= $F.d.G_s.\eta_s$

In practice, the torque exerted through the steering wheel by the driver is multiplied by 10 to 20 times or even more.

TORQUE RATIO

The torque ratio of the steering system is given by,

$$= \frac{\text{Output torque}}{\text{Input torque}}$$

$$= \frac{\text{Torque transmitted to the drop arm shaft}}{\text{Torque applied to steering wheel}}$$

$$= \frac{T_{\text{drop arm}}}{T_{\text{steering wheel}}}$$

EFFICIENCY OF STEERING BOX

The percentage efficiency of the steering mechanism is calculated as,

$$\eta = \frac{\text{Torque ratio}}{\text{Gear ratio}} \times 100$$

Problem 9.7: Calculate the gear ratio of a steering box which requires 1.25 turns of the steering wheel to move the drop arm through 30° from the central position to full lock in one direction.

$$\text{Gear ratio} = \frac{360° \times \text{No. of turns of steering wheel}}{\text{No. of degrees turned through by drop arm}}$$

$$= \frac{360° \times 1.25}{30}$$

$$= 15:1$$

Problem 9.8: A steering box has a gear ratio of 16:1 and an efficiency of 85%. Assume the driver to exert a force of 60N at the rim of a wheel 0.4 m in diameter, determine the torque at the drop arm shaft.

Given

$G_s = 16:1$

$\eta_s = 85\% = 0.85$

$F = 60$ N

$d = 0.4$ m

Find

T_t

Solution

Torque on the drop arm shaft

$$T_t = F.d.G_s.\eta_s$$

$$= 60 \times 0.4 \times 16 \times 0.85$$

$$= 326.4 \text{ Nm}$$

Problem 9.9: The gear ratio of a steering box is 14:1. When the driver applies a force of 25 N with each hand on the steering wheel of 0.38 m diameter, the torque transmitted to the droparm shaft is 110 Nm. Determine the percentage efficiency of the steering mechanism.

Given

Gear ratio = 14:1

F = 25 N

d = 0.38 m

Torque transmitted to the droparm shaft=110 Nm

Find

η

Solution

Torque applied to the steering wheel

$$= F.d$$

$$= 25 \times 0.38$$

$$= 9.5 \text{ Nm}$$

$$\text{Torque ratio} = \frac{T_{\text{drop arm}}}{T_{\text{steering wheel}}} = \frac{110}{9} = 11.58:1$$

Percentage efficiency of the steering mechanism

$$\eta = \frac{\text{Torque ratio}}{\text{Gear ratio}} \times 100$$

$$= \frac{11.58}{14} \times 100$$

$$= 82.71\%$$

Problem 9.10: A rack and pinion steering has a pinion of 16 mm pitch circle diameter. What effort must be applied on the 320 mm diameter steering wheel to overcome a resistance of 500 N experienced transversely on the rack?

Given

d = 16 mm ⇒ r = 8 mm

D = 320 mm ⇒ R = 160 mm

W = 500 N

Find

E

Solution

$$\text{Movement ratio} = \frac{R}{r} = \frac{160}{8} = 20:1$$

When there is no friction,

$$\text{Movement ratio} = \frac{W}{E}$$

$$20 = \frac{500}{E}$$

$$E = 25 \text{ N}$$

Problem 9.11: A rack and pinion steering gear has a 5 teeth pinion of 10 mm pitch. If an effort of 15 N is applied by each hand on the 350 mm diameter steering wheel, determine (a) movement ratio, (b) the force transmitted to the tierods.

Given

T = 5

P = 10 mm

Effort applied by each hand = E/2 = 15 N ⇒ E = 30 N

D = 350 mm ⇒ R = 175 mm

Find

MR, W

Solution

$$\text{Movement ratio} = \frac{2\pi R}{tp} = \frac{2\pi \times 175}{5 \times 10} = 22:1$$

When there is no friction,

$$\text{Movement ratio} = \frac{W}{E}$$

$$22 = \frac{W}{30}$$

$$W = 660 \text{ N}$$

Suspension System

The suspension system is the mechanism located between the body or frame and the wheels. The job of this system is to maximise the contact between the tyres and the road surface, to provide steering stability with good handling.

A vehicle with a solid suspension or no suspension would bounce off the ground when the tyres hit a bump. If the tyres are off the ground, even for a fraction of a second, loss of control is possible. The suspension is also serving to cushion the road shocks and make riding more comfortable.

FUNCTIONS

The functions of the suspension in vehicles are:

1. To provide the comfort and safeguard the occupants from road shocks.
2. To prevent the transmission of the road shocks to the vehicle components.
3. To give the stability to the vehicle in case of rolling or pitching, while in motion.
4. To provide the particular height to body structure and to bear the torque and braking reactions.

REQUIREMENTS

The main functional requirements of a suspension system are:

1. Minimum deflection consistent with required stability
2. Low maintenance operating costs
3. Minimum weight
4. Strong and durable
5. Minimum tyre wear

SPRUNG AND UNSPRUNG WEIGHTS

The weight of the body, engine, transmission, etc., which is supported by suspension springs is called the sprung weight. On the other hand, the wheels and axles and other parts which are not supported by springs, make up the unsprung weight.

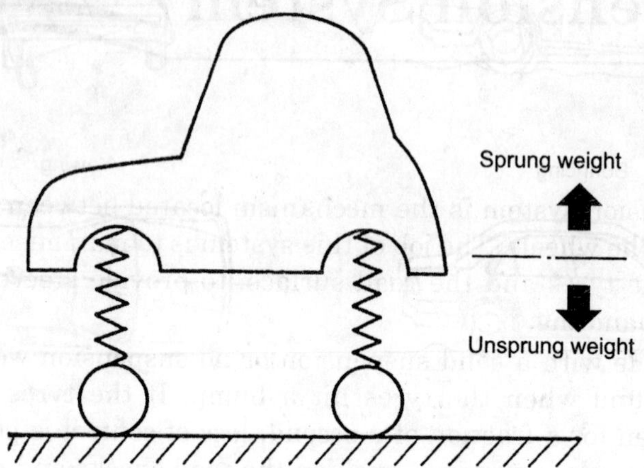

Fig.10.1: *Sprung weight and unsprung weight*

If the unsprung weight is less, the riding comfort is better because of the tendency of the body to be jolted decreases.

OSCILLATION OF THE SPRUNG AND UNSPRUNG WEIGHTS

Oscillation and jolting of the sprung and unsprung parts of the vehicle have a particularly great effect on riding comfort. This oscillation and jolting can be classified as follows:

OSCILLATION OF SPRUNG WEIGHT

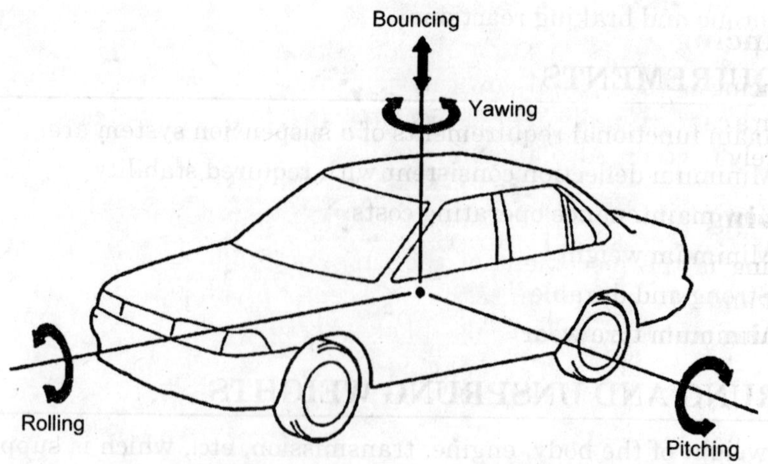

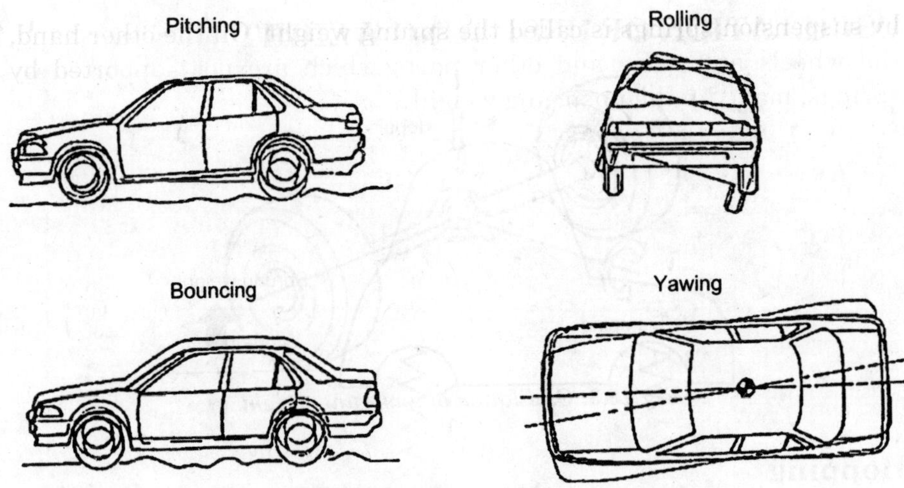

Fig.10.2: *Oscillation of sprung weight*

Pitching

Pitching is the up-and-down oscillation, in relation to the car's centre of gravity, of the front and back of the car. This happens especially when the car goes over large ruts or bumps in the road or when driving over an unpaved road which is rough and full of potholes. Also, pitching occurs more easily in vehicles with softer (easily compressed) springs than in those with harder springs.

Rolling

When turning or when driving on a bumpy road, the springs on one side of the vehicle expand, while those on the other side contract. This results in body rolling in the lateral (side-to-side) direction.

Bouncing

Bouncing is the up-and-down movement of the auto body as a whole. When a car is running at high speeds on an undulating surface, bouncing is likely to occur. Also, it occurs easily when the springs are soft.

Yawing

Yawing is the movement of the car's longitudinal centerline to the right and left, in relation to the car's center of gravity. On roads where pitching occurs, yawing is also likely to occur.

OSCILLATION OF UNSPRUNG WEIGHT

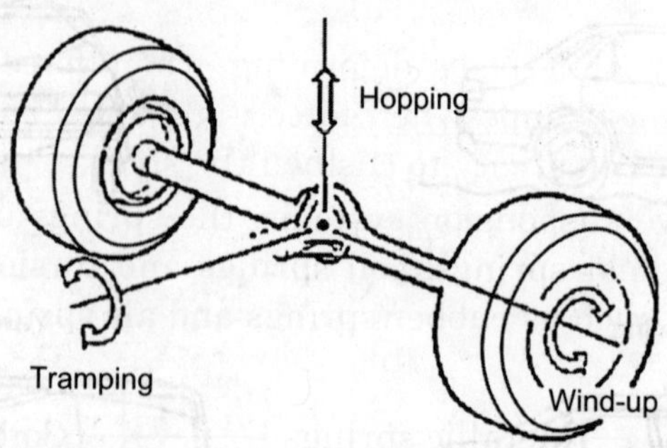

Fig.10.3: *Oscillation of unsprung weight*

Hopping

Hopping is the up-and-down bouncing of the wheels which usually occurs on corrugated roads while driving at medium and high speeds.

Tramping

Tramping is the up-and-down oscillation in opposite directions of the left and right wheels, causing the wheels to skip over the road surface. This occurs more easily in vehicles with rigid axle suspension.

Wind-up

Wind-up is the phenomenon in which the leaf springs attempt to wind themselves around the axle due to the driving torque.

COMPONENTS OF A SUSPENSION SYSTEM

It consists of the following principle components:

1. Springs, which neutralize the shocks from the road surface.
2. Shock absorbers (dampers), which act to improve riding comfort by limiting the free oscillation of the springs.
3. Stabilizer (sway or anti-roll bar), which prevents lateral swaying of the car.
4. A linkage system, which acts to hold the above components in place and to control the longitudinal and lateral movements of the wheels.

SUSPENSION SPRINGS

In the suspension system, the springs perform the very important function of connecting the frame to the axles and by cushioning the impacts from the road prevent them from being directly transmitted into the body.

The property of a body deforming when force is applied and resuming its former shape when force is removed is called elasticity. The spring utilises elasticity to cushion the shocks.

In automotive suspension systems, the springs used are metallic springs, such as leaf springs, coil springs and torsion bars and non-metallic springs, such as rubber springs and air springs.

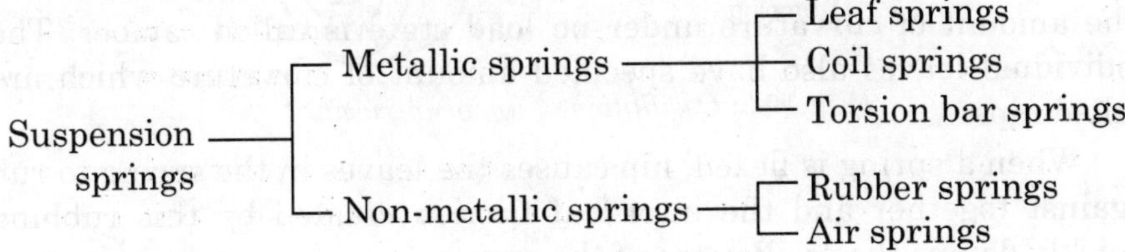

Spring materials

Most springs are made of a tempered steel alloy known as spring steel, usually chrome silicon or chrome-vanadium alloy. Tempering is a process of heating and cooling metal under controlled conditions, which increases the resilience of the metal. Resilience is the ability of the metal to return to or spring back to its original shape after being twisted or compressed.

LEAF SPRING

Leaf springs are the suspension system for light commercial vehicles (LCV) and heavy goods vehicles (HGV), helping them to take the load and absorb the weight of the vehicle on the axle.

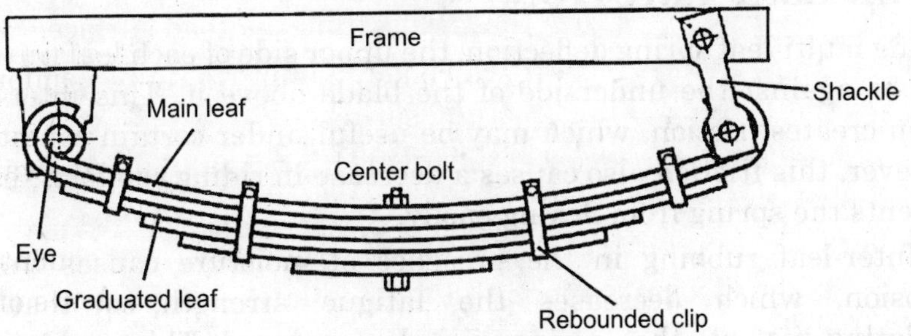

Fig.10.4: *Leaf spring*

At present, leaf springs are mostly made from silicon manganese steels. The leaf spring is made up of a series of 3 to 10 narrow steel plates each 3 to 6 mm thick and of graduated lengths stacked together such as shown in Fig.10.4. The leaves are piled one on top of the other starting from the shortest leaf and riveted or bolted together at the center. The longest leaf (main leaf) is rolled at both ends to form spring eyes to allow installing the spring to the frame. Leaf spring is formed into elliptic shape to raise the elasticity. Today leaf springs are still used in heavy commercial vehicles such as vans and trucks, SUVs, and railway carriages.

CAMBER AND NIP

The amount of curvature under no load state is called camber. The individual leaves also have specified amount of curvature which are called nips.

When a spring is flexed, nip causes the leaves in the spring to rub against together and the inter-leaf friction created by this rubbing quickly damps the oscillations of the spring.

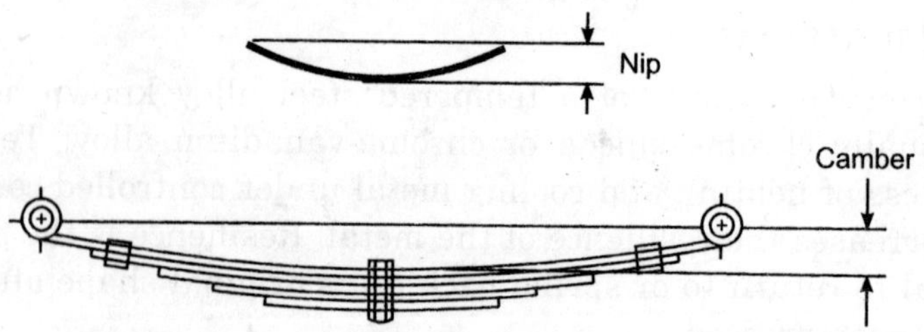

Fig.10.5: *Nip and Camber*

When the spring rebounds, nip prevents gaps from occurring between each of the leaves, thus preventing dirt and sand, etc., from penetrating between the leaves and causing wear.

INTER-LEAF FRICTION

During multi-leaf spring deflection, the upper side of each leaf tip slides or rubs against the underside of the blade above it. This inter-leafs action creates friction, which may be useful under certain conditions. However, this friction also causes a decrease in riding comfort, since it prevents the spring from flexing easily.

Inter-leaf rubbing in the presence of moisture causes fretting corrosion, which decreases the fatigue strength, so that the oscillating life of the spring is also reduced. This problem can be reduced to some extent by applying phosphate paint between the blades. Also by fitting a thin layer of lead or anti-friction pads (silencer pads) between the blades (Fig.10.6), the friction in the interference and hence fretting can be reduced.

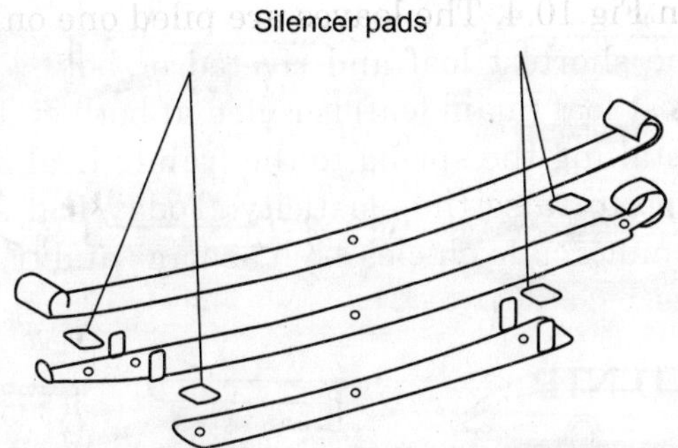

Fig.10.6: *Anti-friction pads*

GRADUATED LEAVES

The leaf at the top has maximum length. The length gradually decreases from the top leaf to the bottom leaf. To ensure a constant stress throughout the spring the leaves are graduated in length.

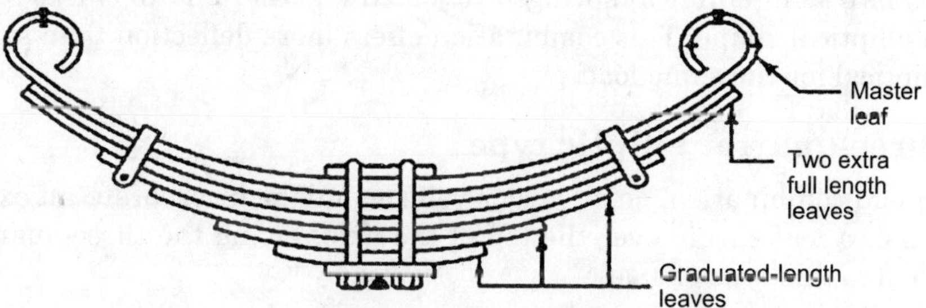

Fig.10.7: *Graduated leaves*

CHARACTERISTICS OF LEAF SPRING

- Since the springs themselves have adequate rigidity to hold the axle in the proper position, it is not necessary to use linkages for this.
- They control their own oscillation through inter-leaf friction.
- They have sufficient durability for heavy-duty use.
- Due to inter-leaf friction, it is difficult for them to absorb the minute vibrations from the road surface. Therefore, leaf springs are generally used for large commercial vehicles which carry heavy loads.

CONFIGURATIONS OF LEAF SPRING

Laminated leaf-spring configurations may be classified by the portion of an ellipse they represent i.e. fully elliptic, three-quarter elliptic, semi-elliptic, quarter elliptic, etc.

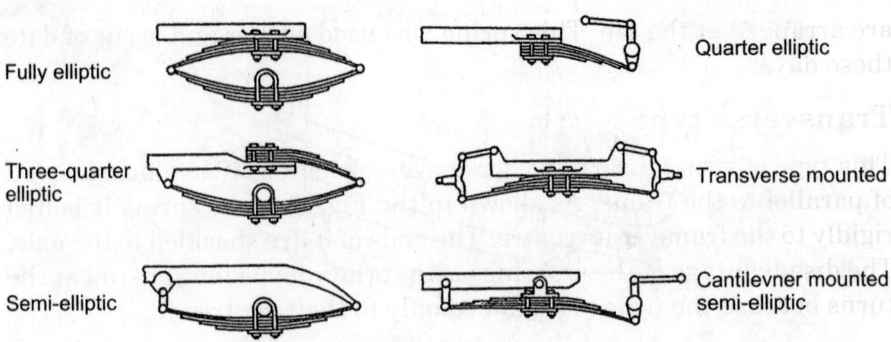

Fig.10.8: *Configurations of leaf spring*

Fully-elliptic type

Elliptical" or "full elliptical" leaf springs referred to two circular arcs linked at their tips.

Two semi-elliptical springs are joined together end to end to form an elliptical shape. This combination offers more deflection than semi-elliptical for the same load.

Three-quarter elliptic type

An odd combination, some larger vehicles might incorporate an extra half of a leaf spring over the top of the axle, giving the three-quarter name to this type of shape.

It is a combination of semi-elliptical and quarter elliptical to form a three-quarter elliptical spring.

Semi-elliptic type

The most commonly used leaf spring today is the half or semi-elliptic spring. It is so called because its shape is like a semi-ellipse. Ends are supported and load is applied at the center. It is used for both front and rear suspensions of heavy duty vehicles such as van, bus and lorry. This type of spring is used at front side of the vehicle with any one of the ends shackled with the frame, while the other end is attached with the simple pin. This spring is fitted longitudinally to the chassis.

Quarter-elliptic type

The quarter-elliptic springs is used on small sports cars where a compact short spring is preferred.

This shape is just half of the semi-elliptical spring. It is cantilever type spring, which is fixed from its one end and the other end is shackled or pivoted to the axle. The short leaves in this type of spring are arranged at the top. This spring was used earlier and is out of date these days.

Transverse type

This type of spring is arranged transversely across the vehicle instead of parallel to the frame. As shown in the Fig.10.8, the spring is bolted rigidly to the frame at its centre. The ends of it are shackled to the axle. The disadvantage of these springs is that the vehicle tends to roll at the turns because the frame is clamped only to their centres.

Cantilever mounted type

The cantilever mounted semi-elliptic spring has been used in some cars such as the Jaguar for the rear suspension. The central pivot of this spring extends the effective spring length. The spring lies parallel and very close to the chassis, so that a compact and effective suspension is achieved.

Of all these types and shapes, the standard semi-elliptical multi leaf spring is the most used and trusted system, with additional strain taken by a full elliptic where needed.

SHACKLES

A swinging shackle accommodates the alteration in spring length. This does not alter the suspension and steering geometry as the spring deflects and the various forces act on the system. This also reduces wear and noise.

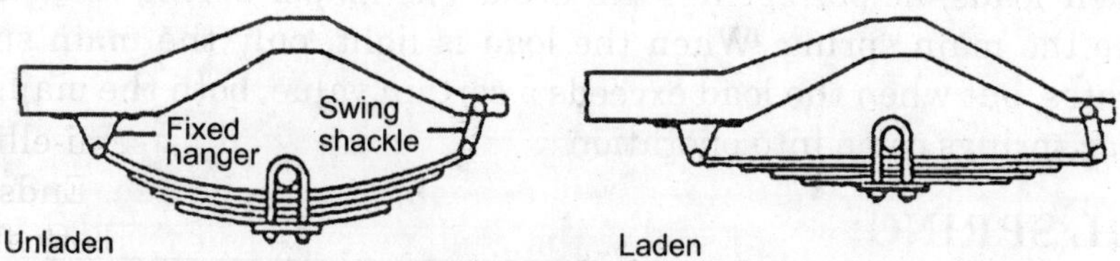

Fig.10.9: *Shackle*

Shackle is the flexible link which connects between leaf spring rear eye and frame. When the vehicle comes across a road bump, the wheel moves up, leading to deflection of the spring. This changes the length between the spring eyes. If both the ends are fixed, the spring will not be able to accommodate this change of length. So, to accommodate this change in length shackle is provided at one end, which gives a flexible connection.

FORCES SUPPORTED BY A LEAF SPRING

Leaf springs are capable of exerting large forces within comparably small spaces. The leaf springs carry lateral loads, brake torque, driving torque etc., in addition to shocks.

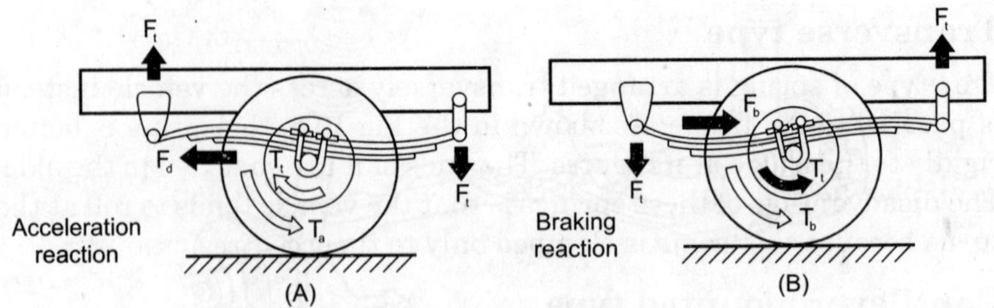

Fig.10.10: *Springs carry torque reactions*

HELPER SPRING

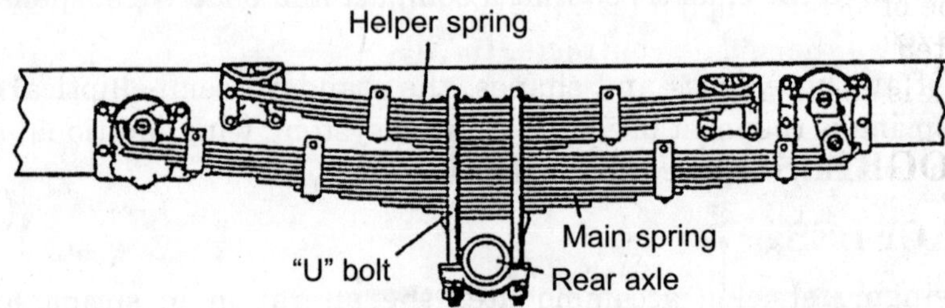

Fig.10.11: *Helper spring*

In trucks and many other vehicles which undergo great fluctuations in their loads, helper springs are used. The helper spring is installed above the main spring. When the load is light, only the main spring operates, but when the load exceeds a certain value, both the main and helper springs come into operation.

COIL SPRING

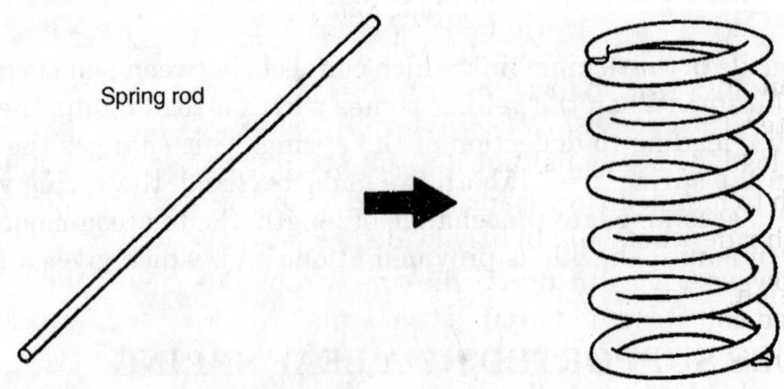

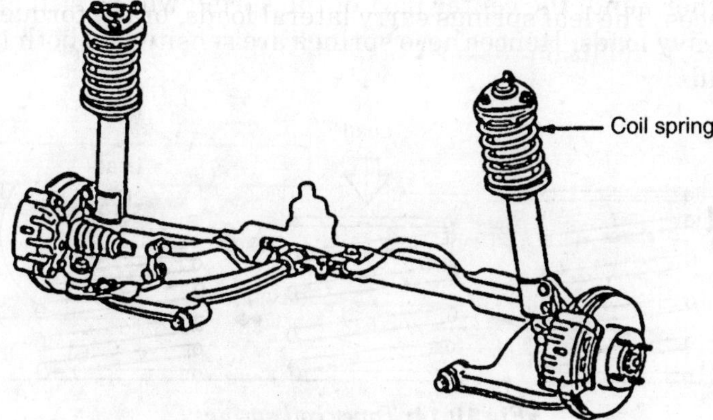

Fig.10.12: *Coil springs*

Coil springs are made from rods of special spring steel formed into the shape of a coil. When a load is placed on a coil spring, the entire rod is twisted as the spring contracts. In this way, the energy is stored and shock is cushioned.

PROGRESSIVE COIL SPRINGS

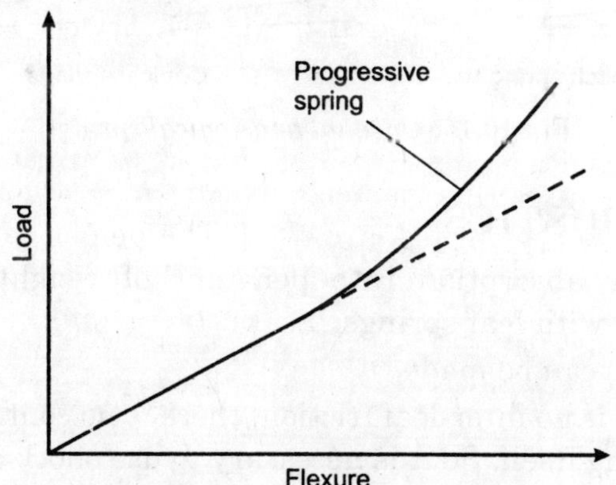

Fig.10.13: *Progressive coil springs*

If a coil spring is made from a rod of spring steel having a uniform diameter, the entire spring will flex uniformly in proportion to changes in the load. This means that if a soft spring is used, it will not be stiff enough to handle heavy loads, while if a hard spring is used, it will give a rough ride when only lightly loaded.

However, if a spring rod having a constantly changing diameter (tapered coil spring) is used, as shown above in Fig.10.14, the ends of the spring will have a lower spring constant than the center. Under light loads, the ends of the spring will contract and absorb road shock.

On the other hand, the center part of the spring will be stiff enough to handle heavy loads. Hence these springs are sensitive to both light and heavy loads.

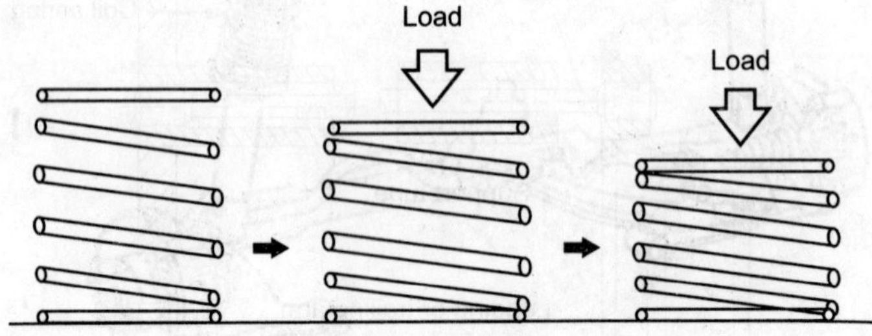

Fig.10.14: *Taper coil springs*

The following types of springs can also be used to achieve the same effect.

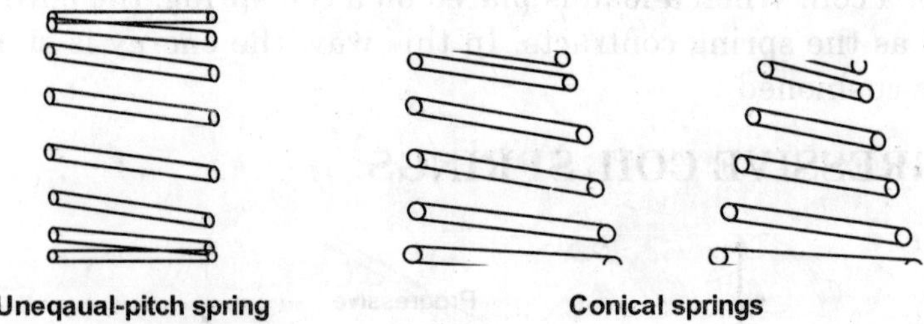

Fig.10.15: *Unequal and conical springs*

CHARACTERISTICS

- The energy absorption rate per unit of weight is greater in comparison with leaf springs.
- Soft springs can be made.
- Since there is no inter-leaf friction, there is no control of oscillation by the spring itself. So it is necessary to use shock absorbers along with them.
- Since there is no resistance to lateral forces, linkage mechanisms to support the axle (suspension arms, lateral control rod, etc.) are necessary.

TORSION BAR SPRING

A torsion bar spring is a spring-steel rod that uses its torsional elasticity to resist twisting. One end of the torsion bar is anchored to the frame or other structural member of the body and the other end to suspension arm that is subjected to torsional load.

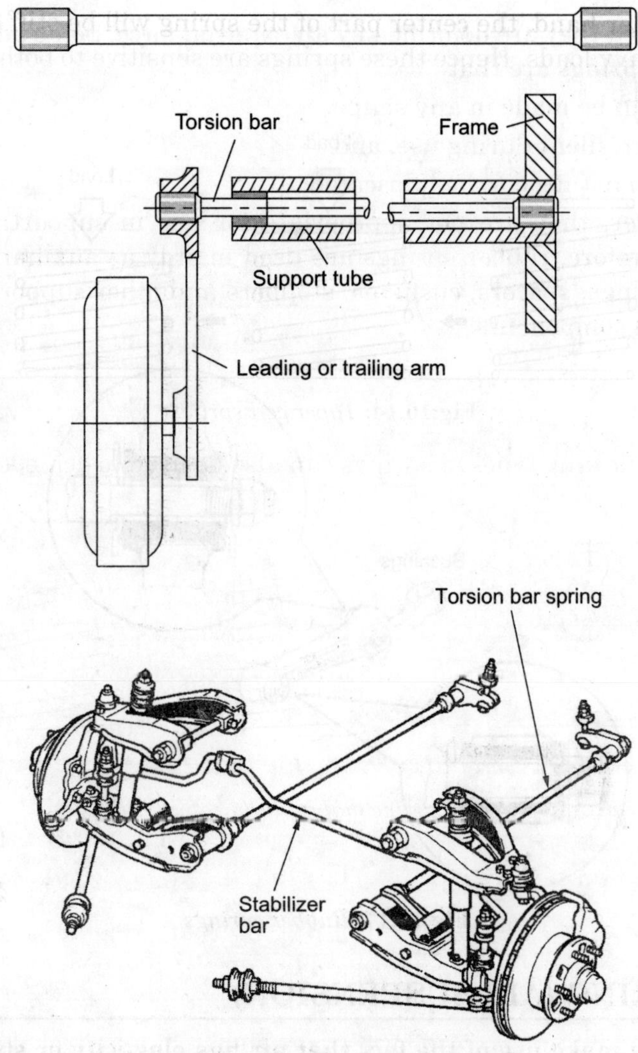

Fig.10.16: *Torsion bar spring*

CHARACTERISTICS

- Since the energy absorption rate per unit weight is great as compared to other springs, the suspension can be lightened.
- The layout of the suspension system is simplified.
- As with coil springs, torsion bar springs do not control oscillation, so it is necessary to use shock absorbers along with them.

RUBBER SPRING (RUBBER SUSPENSION)

Rubber springs absorb oscillations through the generation of internal friction when they are deformed by an external force. Some advantages of rubber springs are that

- they can be made in any shape.
- they are silent during use, and
- they do not need to be lubricated.

However, they are not appropriate for use in supporting heavy loads. Therefore, rubber springs are used mainly as auxiliary springs or as bushings, spacers, cushions, stoppers and other supports for the suspension components.

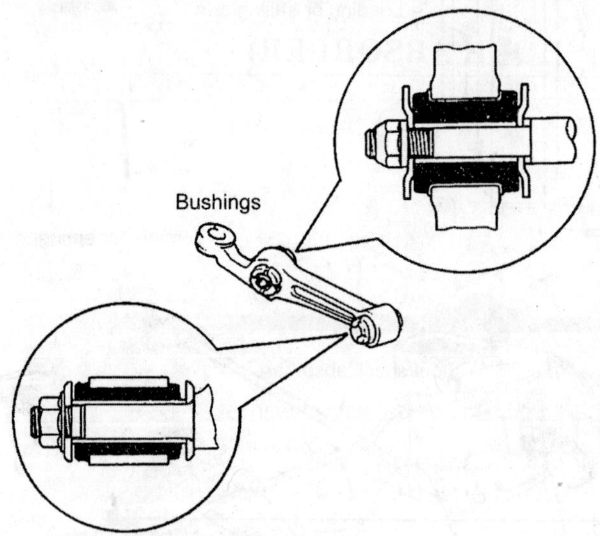

Fig.10.17: *Rubber springs*

AIR SPRING (AIR SUSPENSION)

Air springs make use of the fact that air has elasticity or springiness when compressed. Air springs have the following characteristics:

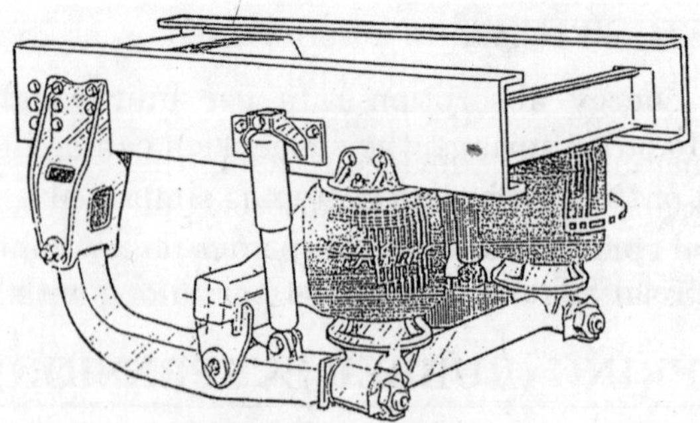

Fig.10.18: *Air suspension*

- They are extremely soft when the vehicle is not loaded, but their spring constant can be increased as the load is increased by increasing the air pressure inside the chamber. This provides optimum riding comfort both when the vehicle is lightly loaded and when it is fully loaded.
- The height of the vehicle can be kept constant, even if the load changes, by adjusting the air pressure.

However, in air suspensions using air springs, devices for controlling the air pressure and compressors for compressing air, etc, are necessary, so the suspension becomes complex.

DAMPER (SHOCK ABSORBER)

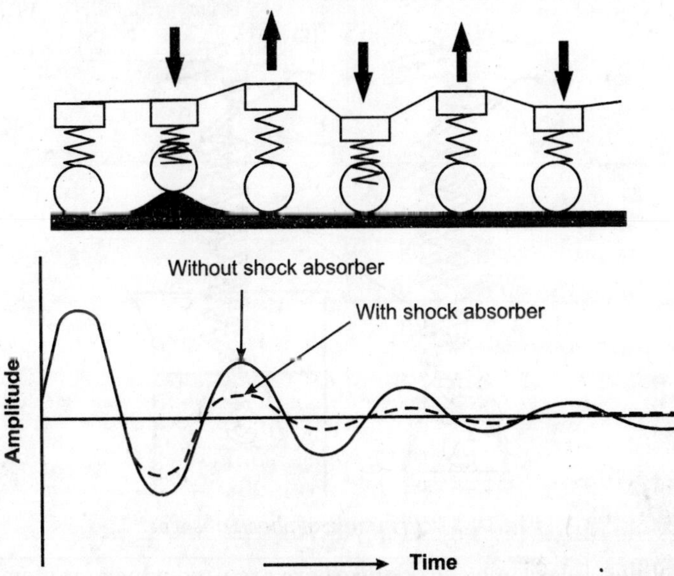

Fig.10.19: *Need for damper*

When a car is subjected to shocks from the road surface, the suspension springs compress and expand to absorb those shocks. However, because springs have the characteristic of continuing to oscillate, and because it often takes a long time for this oscillation to stop, riding comfort will be poor unless some means is provided to damp this oscillation. This is the job of the shock absorber. The shock absorbers not only absorb the excess oscillation of the springs, thus improving riding comfort, but they also give the tyres better road-holding characteristics and improve steering stability.

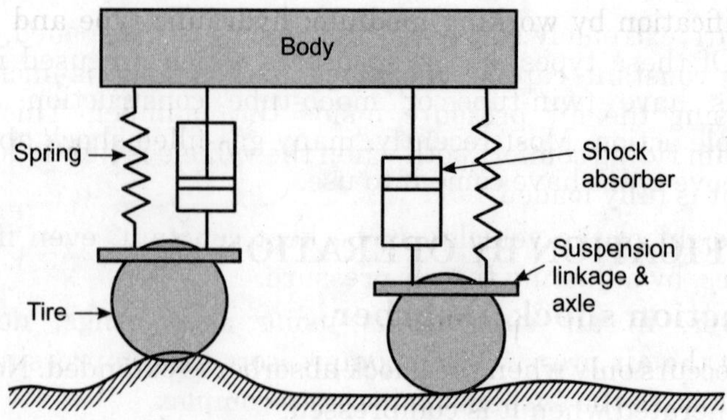

Fig.10.20: *Location of shock absorber*

PRINCIPLE OF OPERATION

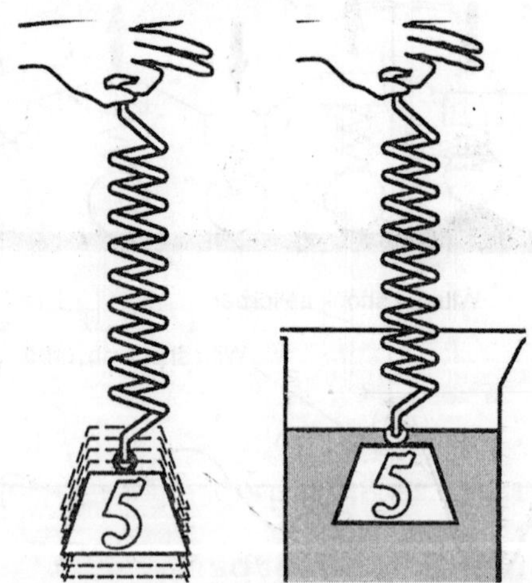

Fig.10.21: *Principle of shock absorber*

In automobiles, telescopic shock absorbers are used which employ a special fluid, called shock absorber fluid, as the working medium. In this type of shock absorber, damping force is generated by the flow resistance caused by the fluid being forced through an orifice (small hole) by the movement of a piston.

TYPES OF TELESCOPIC SHOCK ABSORBERS

Shock absorbers are classified according to their construction and operation as follows:

1. Classification by operation: single-action type and multiple-action type
2. Classification by construction: mono-tube type and twin-tube type

3. Classification by working medium: hydraulic type and gas-filled type Of these types, shock absorbers which are used in current models have twin-tube or mono-tube construction and have multiple action. Most recently, many gas-filled shock absorbers of the above types have come into use.

CLASSIFICATION BY OPERATION

Single-action shock absorber

Damping occurs only when the shock absorber is extended. No damping force is generated when it is compressed.

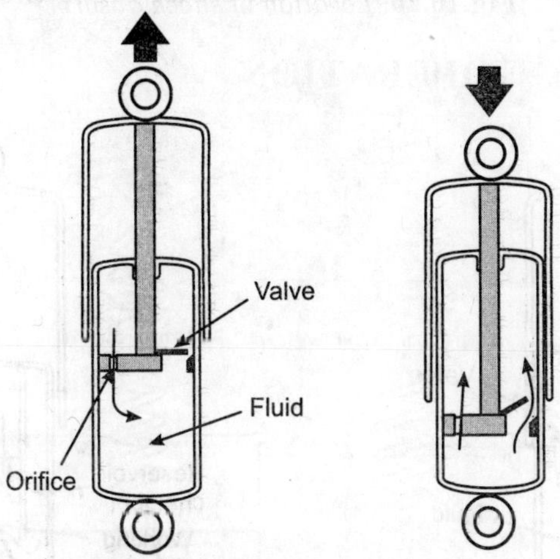

Fig.10.22: *Single-action shock absorber*

Multiple-action shock absorber

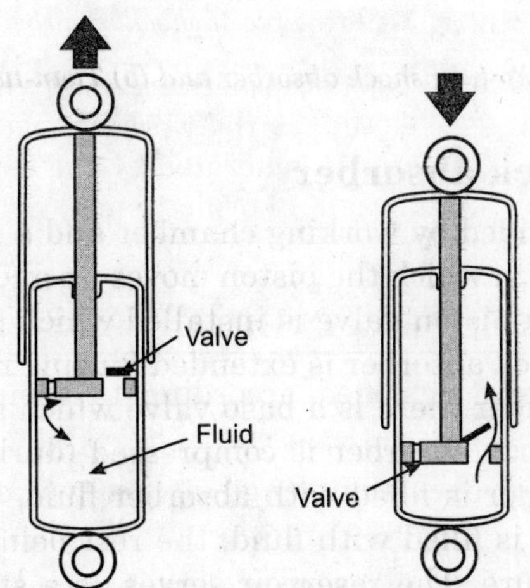

Fig.10.23: *Double-action shock absorber*

Damping occurs both when the shock absorber is extended and when it is compressed. Currently, most shock absorbers used on cars are of this type.

CLASSIFICATION BY CONSTRUCTION

Mono-tube shock absorber

This is a shock absorber with only a single cylinder. The mono-tube shock absorber does not have a base valve. Instead, all of the control during compression and extension takes place at the piston. The advantage of this type is it may run cooler since the working tube is exposed to the air.

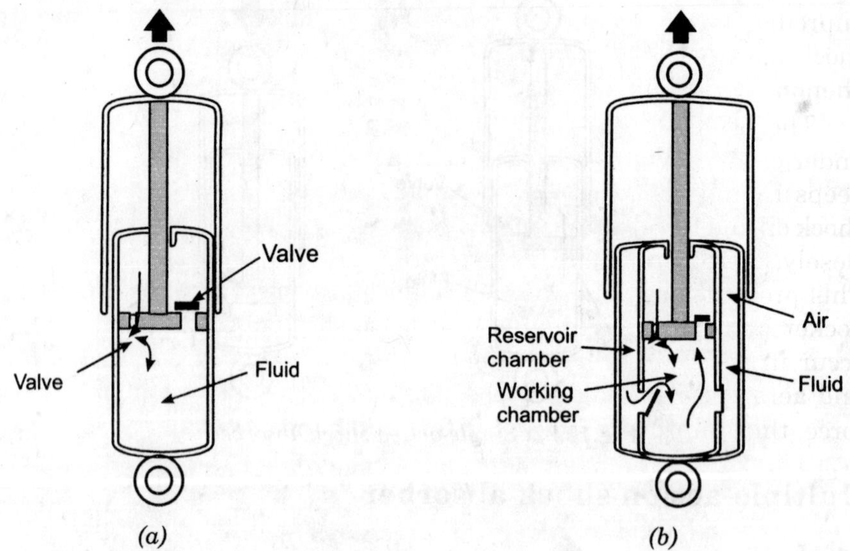

Fig.10.24: *(a) Mono-tube shock absorber and (b) Twin-tube shock absorber*

Twin-tube shock absorber

The cylinder is divided by working chamber and a reservoir chamber. There is a cylinder in which the piston moves up and down. At the base of the piston rod, a piston valve is installed which generates damping force when the shock absorber is extended (during rebounding). At the bottom of the cylinder there is a base valve which generates damping force when the shock absorber is compressed (during bounding). The inside of the cylinder is filled with absorber fluid, but only 2/3 of the reservoir chamber is filled with fluid, the rest being filled with air at atmospheric pressure. The reservoir serves as a storage tank for the fluid entering and leaving the cylinder.

CLASSIFICATION BY WORKING MEDIUM

Hydraulic shock absorber

This is an ordinary shock absorber which uses only shock absorber fluid as a working medium.

Gas-filled shock absorber

Gas shock absorbers represent an advance over traditional shocks. At high vehicle speeds, the suspension will be cycling at a tremendous rate, thereby forcing the piston within the shock absorber to move at a tremendous rate as well. When this happens, the oil within a regular shock absorber gets air bubbles forced into it can cause foam of oil and air to form. The oil will flow through the orifices of the piston at unpredictable rates and decreases the performance of any standard shock absorber. The addition of nitrogen under pressure limits this phenomenon and increases efficiency.

The principle gas used is nitrogen (an inert gas), which is kept under either low pressure (3–6 bar) or high pressure (20–30 bar). This keeps the oil from aerating because nitrogen does not mix with the shock oil and forces the oil molecules to stay packed together much more closely, thereby preventing the oil from getting any air bubbles within. This prevents the generation of abnormal noise due to cavitation (air pockets) and aeration (mixing of air with shock absorber fluid) that can occur in shock absorbers using only fluid. Minimisation of cavitation and aeration also makes it possible to obtain a more stable damping force, thus improving riding comfort and handling stability.

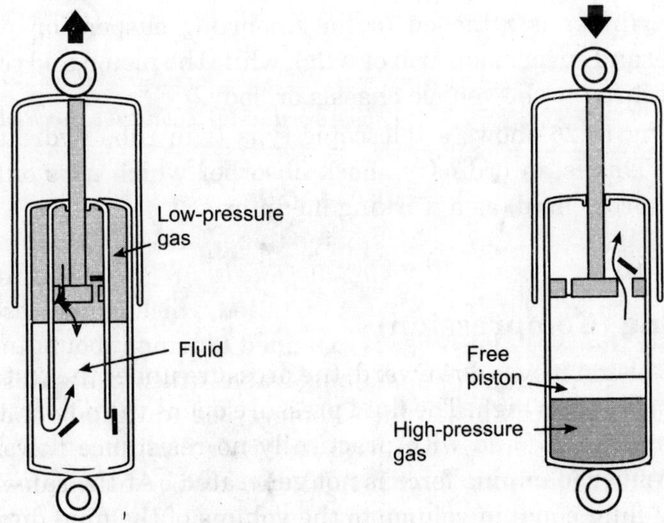

Fig.10.25: *Gas-filled shock absorbers*

WORKING OF TELESCOPIC SHOCK ABSORBER

Dampers are usually of the telescopic type. The telescopic damper consists of a steel cylinder that contains a piston attached to a connecting rod. The piston divides the cylinder into two chambers, the upper and lower. The space between the cylinder and the outer casing is the hydraulic fluid reservoir.

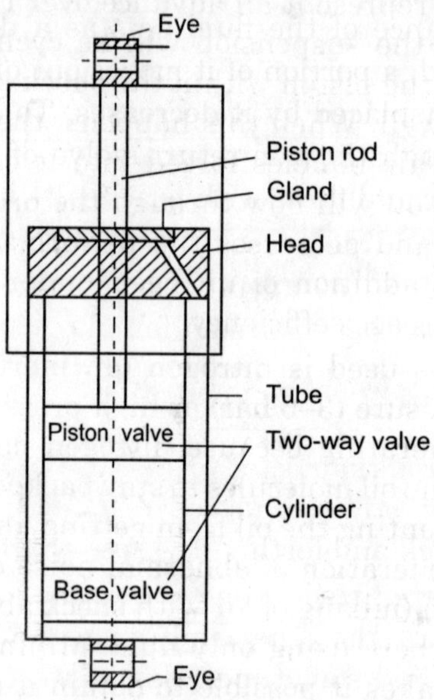

Fig.10.26: *Telescopic shock absorber*

The cylinder is attached to the unsprung suspension parts (the roadwheel and suspension arm or axle), while the piston and connecting rod are secured to the vehicle chassis or body.

The Fig.10.26 shows a telescopic type twin tube hydraulic shock absorber. This is an ordinary shock absorber which uses only a fluid (shock absorber fluid) as a working medium.

Operation

Bounding (Compression)

When the piston moves downward, the pressure under the piston (lower chambers) becomes high. The fluid pressure opens the non-return valve of the piston valve, and with practically no resistance flows into the upper chamber (damping force is not generated). At the same time, an amount of fluid equal in volume to the volume of the fluid displaced by the piston rod as it is pushed in to the cylinder, is forced through the leaf valve in the base valve and flows into the reservoir chamber. It is at this time that the flow resistance generates damping force.

Rebounding (Expansion)

When the piston rod is moving upward, the pressure above the piston becomes high and the fluid inside the upper chamber opens the leaf valve in the piston valve and flows into the lower chamber. At this time, the flow resistance of the fluid acts as a damping force. Since the rod moves upward, a portion of it moves out of the cylinder, so the volume of the fluid displaced by it decreases. To compensate for this, the fluid passes through the non-return valve of the base valve from the reservoir chamber and flows with practically no resistance into the lower chamber.

Although telescopic dampers are compact and require no maintenance, they must be replaced if found to be leaking or defective.

RADIUS ROD

When the wheels are rods subjected to shocks due to road irregularities, obstacles, braking, etc., the radius rods prevent the lower suspension arms from moving back and forth.

The radius rods are fitted at one end to the lower suspension arm by a bolt or nut and at the other end to the body or frame member, via cushions. Radius rod sometimes also called as strut rod or brake reaction rod or drag rod.

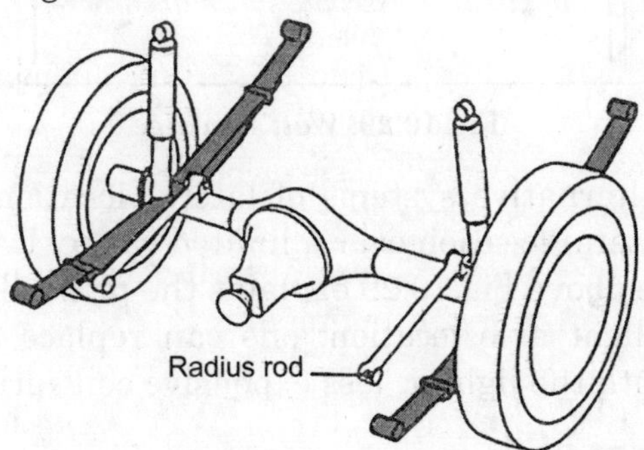

Fig.10.27: *Radius rods*

PANHARD ROD

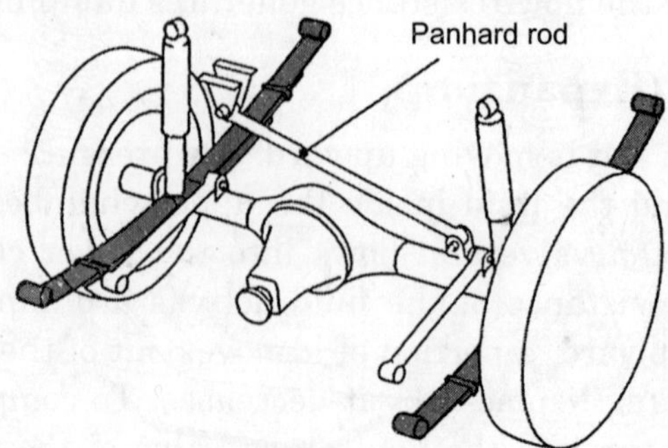

Fig.10.28: *Panhard rod*

In order to bear forces in the lateral direction and to help position the axle so that it is not displaced in the lateral direction, a lateral control rod is used, mainly on the rear suspension.

The lateral control rod is mounted transversely between the axle and a structural member of the body.

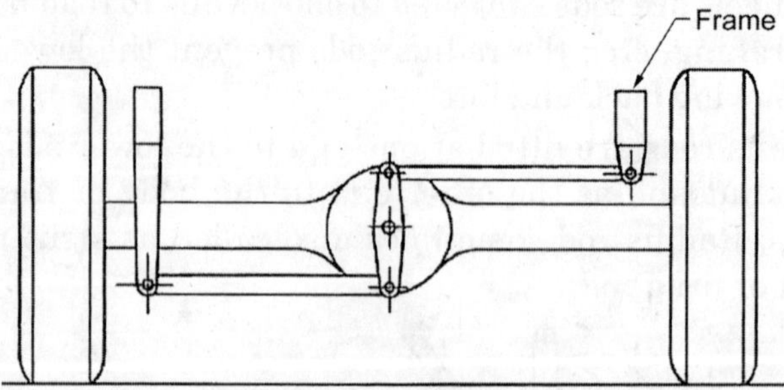

Fig.10.29: *Watt's linkage*

There are alternative systems of lateral location. A popular one giving perfect lateral location over a limited range, is that invented by James watt. The above Fig.10.29 explains the principle of this linkage. With such excellent axle location, one can replace the heavy semi-elliptic spring with the lighter, less expensive coil spring.

STABILIZER BAR

The stabilizer bar is designed to prevent excessive body roll in the lateral direction (side-to-side) when the vehicle rolls on turns and to improve stability while driving over rough road surfaces. It also improves the traction of the tyres. The other names for stabilizer bar are sway bar and anti-roll bar.

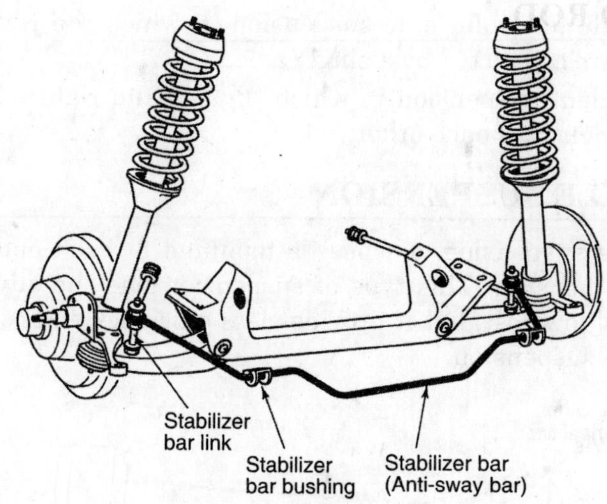

Fig.10.30: *Stabilizer bar*

The stabilizer bar is normally fitted at both ends to the lower suspension arms via rubber cushions or ball joints and linkages. The center section is fitted to the vehicle frame or body at two points via rubber bushings.

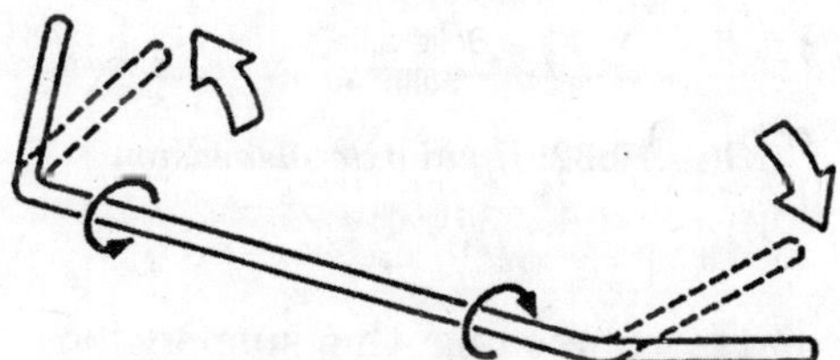

Fig.10.31: *Stabilizer bar in action*

Operation

When turning or when running on a bumpy road, the springs on one side of the vehicle expand, while those on the other side contract. This results in body rolling in the lateral (side-to-side) direction. If stabilizer bar is provided, one end of the stabilizer bar is twisted up while the other end is twisted down. The stabilizer bar is twisted as a result and tends to resist the torsion. It is the resistance that reduces the vehicle roll and maintains the body as level as possible.

TYPES OF SUSPENSION

Many different suspension systems are in use. Most of them provide an acceptable degree of roadability and riding comfort. Suspensions can be divided into two types, based on the differences in their construction:

- Rigid axle (or) solid axle suspension in which the right and left wheels are connected by a single axle.
- Independent suspension in which the left and right wheels move independently of each other.

RIGID AXLE SUSPENSION

In rigid axle suspension, a wheel is mounted at each end of a solid axle or axle housing. This type of suspension has the advantages of strength and low cost, but it provides less efficiency and comfort than independent suspension.

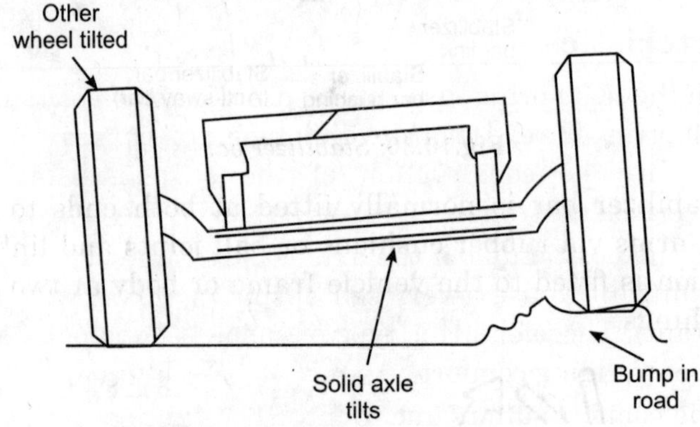

Fig.10.32: *Rigid axle suspension*

Advantages

- The number parts composing the suspension are less and the construction is simple. Therefore, maintenance is simple.
- It is durable enough for heavy-duty use.
- When turning there is little tilting of the body.
- There is little change in the alignment due to the up-and-down movements of the wheels. Therefore, there is less tyre wear.

Disadvantages

- Since the unsprung weight is great, riding comfort is poor.
- Since the movements of the left and right wheels mutually influence one another, vibration and oscillation occur easily.

TYPES OF RIGID AXLE SUSPENSION

PARALLEL LEAF SPRING TYPE

This type of suspension is used for the front suspension and the rear suspensions in commercial vehicles.

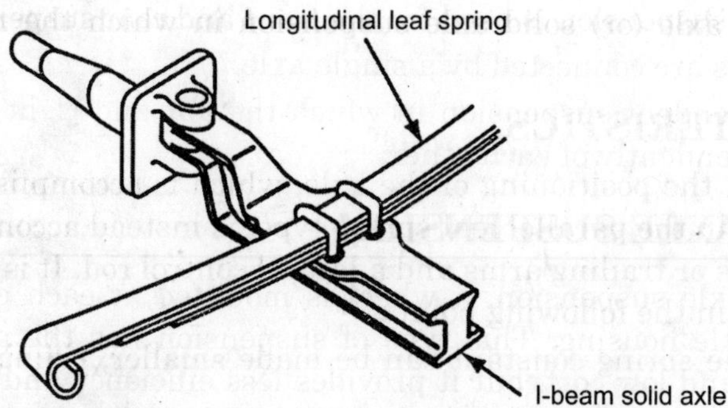

Fig.10.33: *Parallel leaf spring suspension*

Characteristics

- Since the leaf spring also acts as a linkage for positioning the axle (holding in place), separate linkages are generally unnecessary. Therefore the construction of the suspension is simple but comparatively strong.
- Since the positioning of the axle is carried out by the leaf springs, it is difficult to use a very soft spring (a spring with a low spring constant). Therefore, this type of suspension is not very good with respect to riding comfort.
- Riding comfort suffers due to inter-leaf friction in the leaf springs. Acceleration and braking torque tend to causes wind-up and vibration and wind-up in turn causes rear-end squat and nose-diving.

Leading arm and trailing arm types with lateral rods

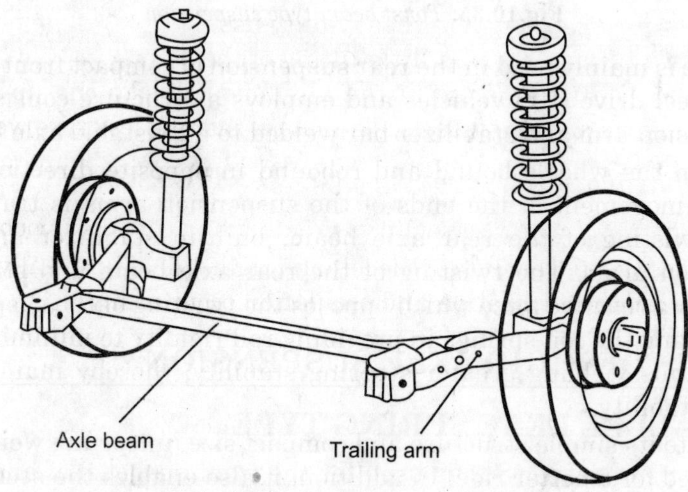

Fig.10.34: *Trailing arm suspension*

This type of suspension is used for the front and rear suspension of the SUVs.

CHARACTERISTICS

In this type, the positioning of the axle, which is accomplished by the leaf springs in the parallel leaf spring type, is instead accomplished by leading arms or trailing arms and a lateral control rod. It is superior to leaf springs in the following points.

- Since the spring constant can be made smaller, riding comfort is good.
- Since the trailing arm's rigidity is high, it is difficult for wind-up to occur.

Trailing arm with twist-beam type

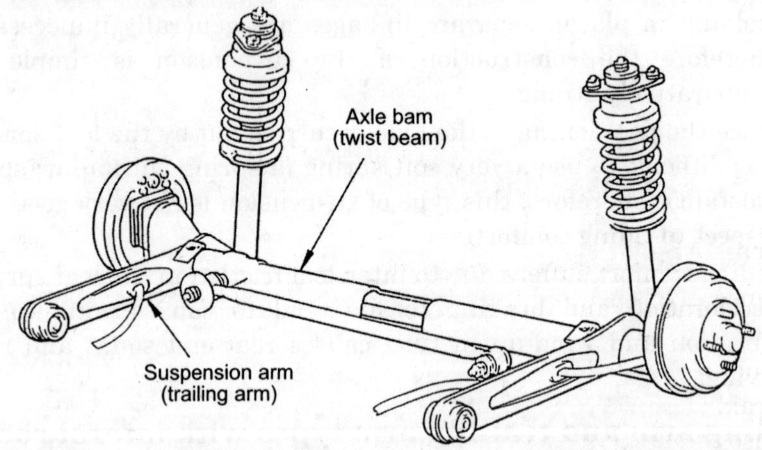

Fig.10.35: *Twist beam type suspension*

This type is mainly used in the rear suspension of compact front-engine front-wheel drive (FF) vehicles and employs a structure consisting of a suspension arm and stabilizer bar welded to a twistable axle beam.

When the wheels bound and rebound in opposite directions, the twisting movement of the ends of the suspension arms is translated into a twisting of the rear axle beam, built-in stabilizer and rear suspension arms. The twisting of the rear axle beam and stabilizer generates a reactive force which opposes the twisting of the suspension arms, aiding the coil springs in providing roll rigidity to minimize body rolling, thus yielding greater steering stability, thereby maintaining vehicle stability.

Due to its simple structure and compact size, unsprung weight can be reduced for a better ride. In addition, it also enables the securing of a luggage space.

FOUR-LINK TYPE

This type is used for the rear suspension. It provides the best riding comfort of all the rigid-axle suspensions.

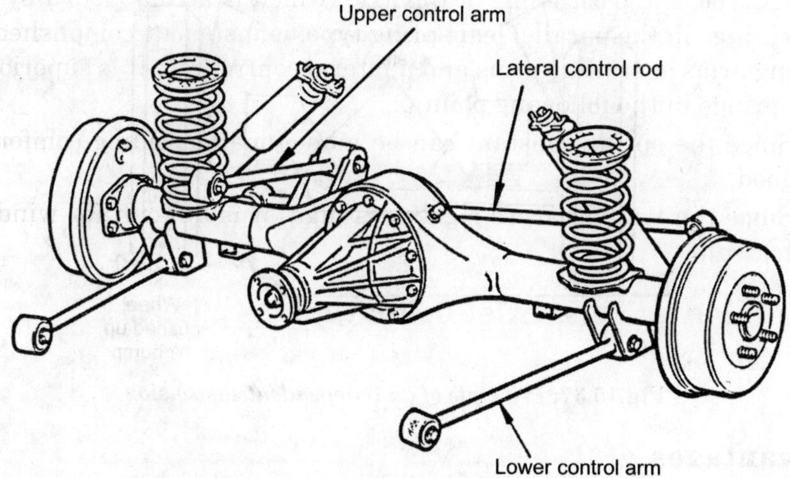

Fig.10.36: *Four-link suspension*

Characteristics

- Since axle positioning is accomplished by a linkage, soft springs can be used and riding comfort is good.
- Due to the geometrical layout of the linkages, nose diving during braking and rear-end squat during acceleration are prevented. Also the up-and-down movement of the front end of the differential during bounding and rebounding can be minimized by making the upper control arms shorter than the lower control arms. This in turn means that the floor over the nose of the differential can be lowered, allowing more passenger compartment space.
- Use of coil springs minimizes friction in the suspension, so minute shocks from the road surface can be absorbed and riding comfort can be improved.
- Generally, when this suspension system is used in the rear, the linkage is located forward of the axle, and as a result, the capacity of the trunk can be larger on passenger cars.

INDEPENDENT SUSPENSION

Independent suspension systems provide a separate mounting for each wheel. There is no connection between the left and right wheels. Therefore, the linear motion of one wheel is not transmitted to the other, and there is no tilting of an axle, that in turn, might tilt the entire car.

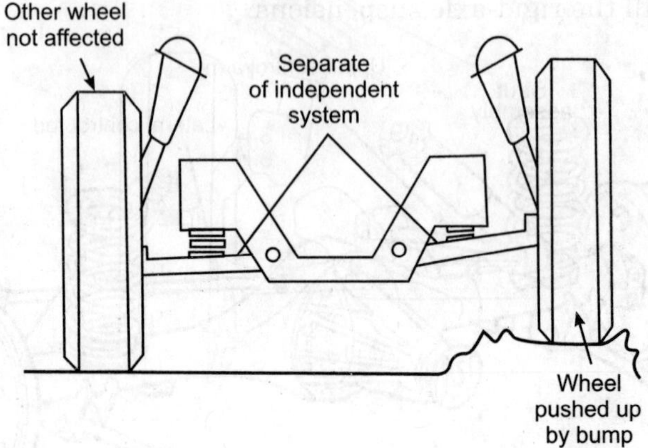

Fig.10.37: *Principle of an independent suspension*

Advantages

- The unsprung weight can be kept lo and the road-holding characteristics of the wheels are good, so riding comfort and handling stability are good.
- In independent suspensions, the springs only support the body. They do not help to position the wheels (this is being done by the linkages). This means that softer springs can be used.
- Since there is no axle connecting the left and right wheels, the floor and the engine mounting position can be lowered. This means that the vehicle's centre of gravity will be lowered, and the passenger compartment and luggage room can be made larger.

Disadvantages

- The construction is complex.
- Tread and alignment change with the up-and-down motions of the wheels.

TYPES OF INDEPENDENT SUSPENSION

There are various types of independent suspension.

MACPHERSON STRUT TYPE

This is the most widely used independent suspension system for the front suspension of small and medium sized cars. This type is also used as the rear suspension of front-engine front wheel drive cars.

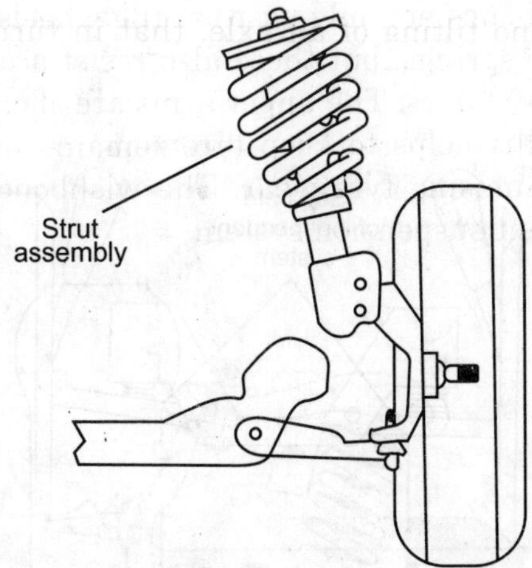

Fig.10.38: *Macpherson strut suspension*

In this type only lower wishbones are used. A strut containing shock absorber and the spring carries also the stub axle on which the wheel is mounted. The wishbone is hinged to the cross member and positions the wheel as well as resists acceleration, braking and side forces.

Characteristics

- The construction of the suspension is relatively simple.
- Since there is a small number of parts, it is light, so unsprung weight can be reduced.
- Since the space taken up by the suspension is small, the usable space in the engine compartment can be increased.
- Since the distance between suspension support points is great, there is little disturbance of the front wheel alignment due to installation error or part manufacturing error. Therefore, except for toe-in, alignment adjustments are ordinarily unnecessary.

DOUBLE-WISHBONE

Double-wish bone type is used widely for the front suspension of small trucks and for the front and rear suspensions of passenger cars. This type is also referred to as short-and-long arm (SLA) suspension. It consists of upper and lower wishbone arms pivoted to the frame members. The vehicle weight is transmitted from the body and cross-member to the coil spring through which it goes to the lower wishbone member.

The wishbone arms are like the chicken wishbone or letter V in shape, because of which the system is so called. Because of this

V-shape, the wishbones not only position the wheels and transmit the vehicle load to the springs, but these also resist acceleration, braking and cornering (side) forces. The upper arms are shorter in length than the lower ones. This helps to keep tyre remains in a more constant position, thus minimising tyre wear. The wishbone type is the most popular independent suspension system.

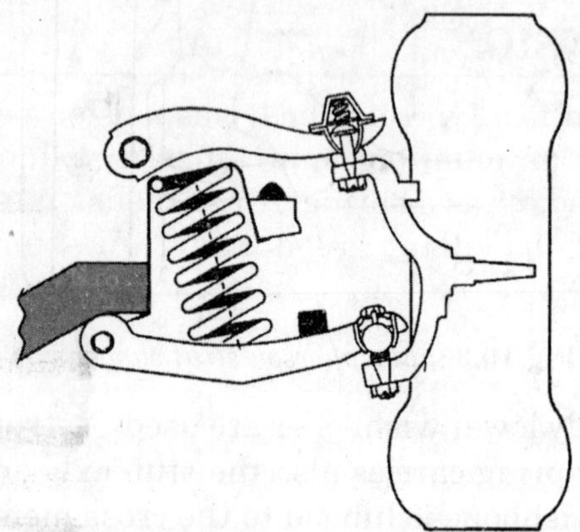

Fig.10.39: *SLA suspension*

Characteristics

In this type, the upper control arm is shorter than the lower control arm. During wheel jounce and rebound travel, the upper control arm moves in a shorter arc than the lower arm. This action moves the top of the tyre in and out slightly, but the bottom of the tyre remains in a more constant position.

SEMI-TRAILING ARM TYPE

This type is used for the rear suspension in a few models.

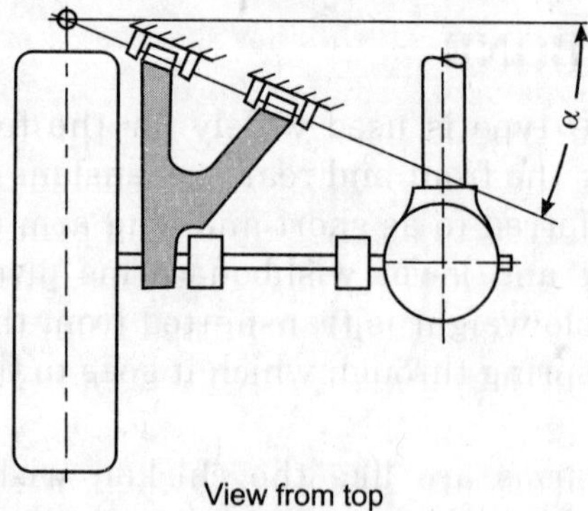

Fig.10.40: *Semi-trailing arm suspension*

Characteristics

With this suspension, the amount by which the toe angle and camber change (due to the up-and-down motion of the wheels) can be controlled by varying the length of each arm and setting the arm installation angle and angle of axis sway at the design stage, in order to determine the handling characteristics of the car.

AIR SUSPENSION

Air suspension is a type of vehicle suspension powered by an engine driven or electric air pump or compressor. This pump pressurises the air, using compressed air as a spring. Air suspension replaces conventional steel springs. If the engine is left off for an extended period, the car will settle to the ground. The purpose of air suspension is to provide a smooth ride quality and in some cases self-leveling.

Operation

The air suspension comprises a compressor supplying air to an air tank, a leveling valve to maintain a pressure a fixed distance between the chassis frame and ground and four air bags. Each air bag is filled with compressed air and supports the weight of the vehicle. The air bag is placed between the chassis frame and the axle beam. As the load is applied to the frame, the air bags get compressed, actuating the leveling valves (control actuators). Air from the tank fills the compressed air bags and hence raises the level of the frame. The supply of air is automatically cut as soon as the frame reaches a predetermined level. The air from the air bag gets released as soon as the load on the chassis frames decreases. The leveling valve again does this.

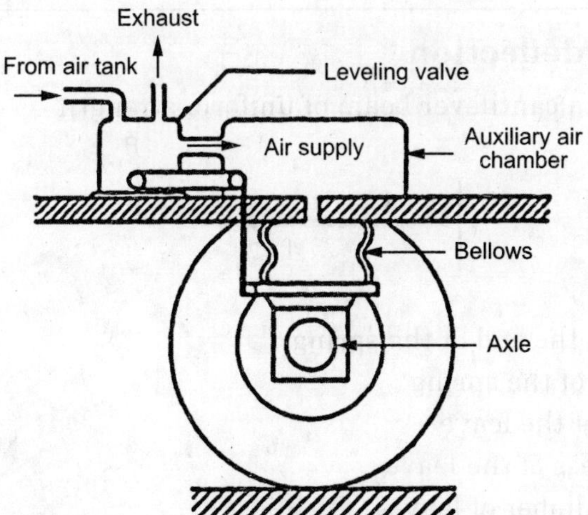

Fig.10.41: *Layout of air suspension system*

Air pressure is supplied to the air-bags from the air braking system which may be as high as 8 bar, but the supply pressure is reduced before it is introduced to the air-bag to around 5 bar when the vehicle is fully laden and decreasing to about 2.5 bar in the unladen state. Rolling-lobe air bags can operate at bounce frequencies as low as 80 cycles per minute fully laden, rising to roughly 100 cycles per minute when unladen.

Advantages

Air spring suspensions have become popular with heavy goods vehicles, coaches and buses in recent years, due to the following reasons:

- Maintain a constant frequency of vibration whether the vehicle is laden or unladen.
- Constant frame height is maintained.
- It gives smooth and comfort ride of the vehicle.
- The stiffness of the system increases with the increase of deflection.
- It lowers stresses imposed on the vehicle because high peak axle loads are eliminated.
- Generates less noise than steel leaf springs.
- Air spring bag life is as good as or even better than for steel leaf springs.
- Maintenance is minimal and replacement units are easy to fit.

DESIGN AND PROBLEMS

DESIGN OF LEAF SPRINGS

Stress and deflection

i. Spring as a cantilever beam of uniform strength

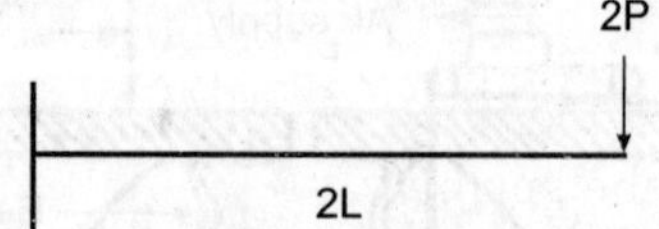

Let,

2P	Load at the end of the spring
2L	Length of the spring
b	Width of the leave
t	Thickness of the leave
n	Total number of leaves
E	Young's modulus (Modulus of elasticity)

$$\text{Bending stress } \sigma_b = \frac{6\,P\,L}{n\,b\,t^2}$$

$$\text{Deflection of the spring } y = \frac{6\,P\,L^3}{E\,n\,b\,t^3}$$

ii. Semi-elliptical leaf spring (Simply supported type without extra full length leaves)

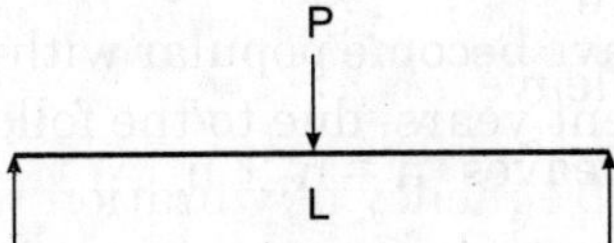

Let,

P Load at the centre of the spring

L Length of the spring

b Width of the leave

t Thickness of the leave

n Total number of leaves $n = n_e + n_g$

n_e Number of full length leaves

n_g Number of graduated leaves

E Young's modulus (Modulus of elasticity)

$$\text{Bending stress at the center } \sigma_b = \frac{3\,P\,L}{n\,b\,t^2}$$

$$\text{Deflection of the spring } y = \frac{3\,P\,L^3}{E\,n\,b\,t^3}$$

Due to inter-leaf friction, deflection of the springs becomes somewhat smaller giving.

$$\text{Deflection of the spring } y = \frac{2.85\,P\,L^3}{E\,n\,b\,t^3}$$

iii. Semi-elliptical leaf spring (Simply supported type with extra full length leaves and bands/U-bolts)

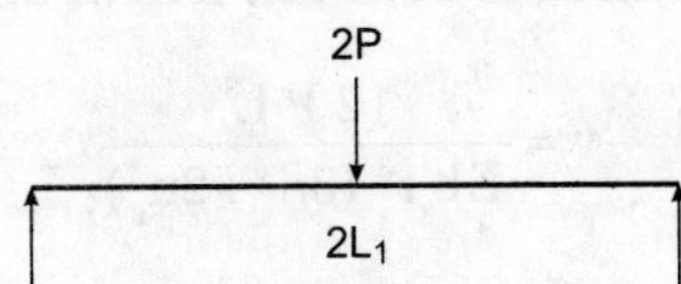

Let,

2P	Total load at the center of the spring
$2L_1$	Total length of the semi-elliptic spring (Span of the spring)
2L	Effective length of the spring
L	Length of the spring as a cantilever beam
ℓ	Width of band or distance between centers of U-bolts (ineffective length of the spring)
b	Width of the leave
t	Thickness of the leave
n	Total number of leaves $n = n_e + n_g$
n_e	Number of full length leaves
n_g	Number of graduated leaves
E	Young's modulus (Modulus of elasticity)

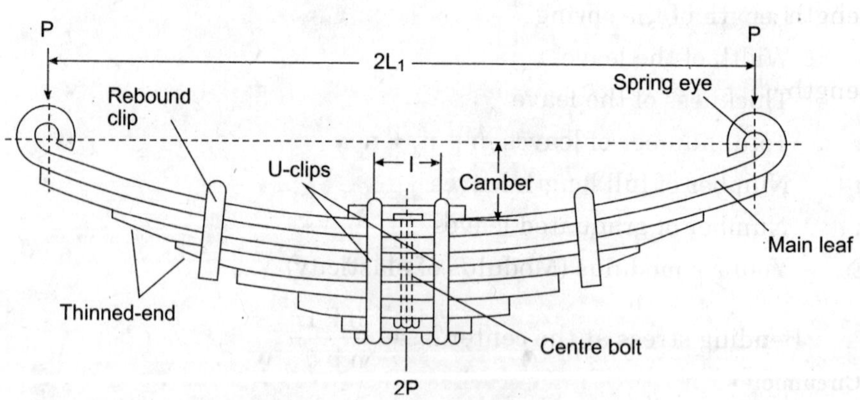

Bending stress induced in graduated leaves

$$\sigma_{bg} = \frac{12\,P\,L}{b\,t^2\,(3n_e + 2n_g)}$$

Bending stress induced in full length leaves

$$\sigma_{be} = \frac{18\,P\,L}{b\,t^2\,(3n_e + 2n_g)}$$

Deflection which is common to both full length and graduated leaves

$$y = \frac{12\,P\,L^3}{E\,b\,t^3\,(3n_e + 2n_g)}$$

Length of leaf spring leaves

Let 'n' be the total number of leaves including master leaf, The number of leaves not having eyes are 'n-1'.

For a leaf spring having two full length leaves including one master leaf, the length of any leaf starting from smallest leaf is obtained as:

$$\text{Length of } r^{th} \text{ leaf} = \frac{\text{Effective length}}{n-1} \times r + \text{Ineffective length}$$

Where, 'r' varies from 1 to (n–1)

$$\text{Length of smallest (first) leaf} = \frac{2L}{n-1} \times 1 + \ell$$

$$\text{Length of second leaf} = \frac{2L}{n-1} \times 2 + \ell$$

$$\text{Length of third leaf} = \frac{2L}{n-1} \times 3 + \ell$$

Length of nth leaf which is a master leaf = Length of full length leaf + Length of spring eye

$$= \frac{2L}{n-1} \times n - 1 + \ell + \pi(d+t)2$$

$$= 2L + 1 + \pi\,(d+t)2$$

$$= 2L_1 + \pi\,(d+t)2$$

Circumference of a circle is the actual length around the circle.

∴ Circumference of a spring eye

= πd where d-inside diameter of eye

= $\pi(d+t)$ if thickness of a leaf is considered,

Effective length of the spring

If $2L_1$ is the span of the spring, effective length of the spring will be 2L.

- When band is used, $2L = 2L_1 - \ell$
- When U-bolts are used, $2L = 2L_1 - \frac{2}{3}\ell$
- If band or U-bolts are not given, $2L = 2L_1$

Nip (curvature of each leaf)

The initial gap that should be provided between the full length and graduated leaves before the band or clip bolts load is applied.

$$C = \frac{2PL^3}{Enbt^3}$$

Camber (Initial radius)

An initial curvature or camber to the leaf springs is given so that they are substantially straight when under their full static load.

$$R = \frac{L^2}{8Z} + \frac{Z}{2}$$

Where, Z = Perpendicular distance between the line joining the centres of the eyes of a semi-elliptic spring and the inner side of the main leaf.

= Camber

= y

Load on clip bolts

The load on the clip bolts or band required to close the gap is,

$$P_b = \frac{2Pn_e n_g}{n\,(3n_e + 2n_g)}$$

Case 1 Cantilever type 2P 2L	$\sigma_b = \frac{6\,P\,L}{n\,b\,t^2}$ $y = \frac{6\,P\,L^3}{E\,n\,b\,t^3}$	Load and length given in the problem as 2P 2L
Case 2 Semi-elliptical leaf spring (Simply supported type without extra full length leaves) P L	$\sigma_b = \frac{3\,P\,L}{n\,b\,t^2}$ $y = \frac{2.85\,P\,L^3}{E\,n\,b\,t^3}$	Load and length given in the problem as P L

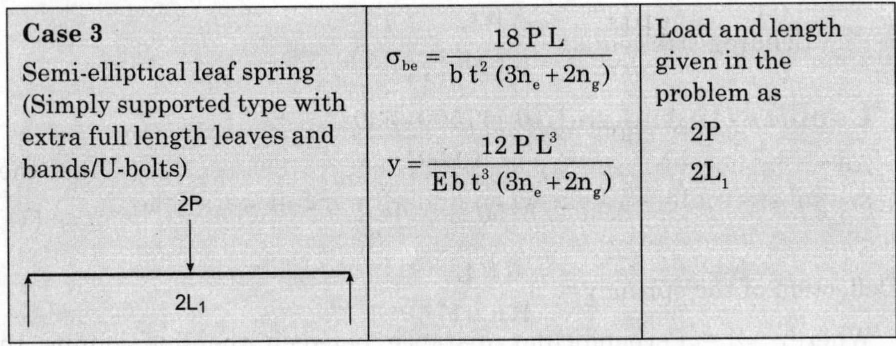

Case 3 Semi-elliptical leaf spring (Simply supported type with extra full length leaves and bands/U-bolts)	$\sigma_{be} = \dfrac{18\,P\,L}{b\,t^2\,(3n_e + 2n_g)}$ $y = \dfrac{12\,P\,L^3}{E\,b\,t^3\,(3n_e + 2n_g)}$	Load and length given in the problem as 2P $2L_1$

Problem 10.1: Design a leaf spring for the following specifications:

Total load = 140 kN

Number of springs supporting the load = 4

Maximum number of leaves = 10

Span of the spring = 1000 mm

Permissible deflection = 80 mm

Take young's modulus as 200 kN/mm² and allowable bending stress as 600 N/mm².

Given data

Considering cantilever type leaf spring,

Total load = 140 kN = 140,000 N

Number of springs supporting the load = 4

$n = 10$

$2L = 1000$ mm $\quad \therefore L = 500$ mm

$y = 80$ mm

$E = 200$ kN/mm² $= 200 \times 10^3$ N/mm²

$\sigma_b = 600$ N/mm²

Solution

Considering cantilever type leaf spring,

Total load on each spring

$$2P = \frac{\text{Total load on springs}}{\text{No of springs supporting total load}} = \frac{140000}{4}$$

$$= 35000 \text{ N}$$

$$\therefore \quad P = 17500 \text{ N}$$

$$\text{Bending stress } \sigma_b = \frac{6\,P\,L}{n\,b\,t^2}$$

$$600 = \frac{6 \times 17500 \times 500}{10\,b\,t^2}$$

$$bt^2 = 8750 \qquad \text{....... (1)}$$

$$\text{Deflection of the spring } y = \frac{6\,P\,L^3}{E\,n\,b\,t^3}$$

$$80 = \frac{6 \times 17500 \times 500^3}{200 \times 10^3 \times 10 \times b\,t^3}$$

$$bt^3 = 82031 \qquad \text{....... (2)}$$

Dividing (2) by (1)

$$\frac{bt^3}{bt^2} = \frac{82031}{8750}$$

$$t = 9.37 \text{ mm}$$

From equation (1)
$bt^2 = 8750$
$b\,(9.37)^2 = 8750$
$b = 99.66$ mm

From equation (2)
$bt^3 = 82031$
$b\,(9.37)^3 = 82031$
$b = 99.71$ mm

Taking larger value,

$$b = 99.71 \text{ mm}$$

Problem 10.2: A vehicle spring of semi-elliptic type has leaves of 75 mm width and 10 mm thickness and effective length of 900 mm. If the stress is not to exceed 220725 kPa, when the spring is loaded to 4905 N, estimate the number of required leaves and the deflection under the condition. If the spring is just flat under the load, what is the initial radius? Take $E = 196.2 \times 10^6$ kPa.

Given data

Semi-elliptic type leaf spring

$b = 75$ mm

$t = 10$ mm

$L = 900$ mm

$$\sigma_b = 220725 \text{ kpa}$$

1 pa = 1 N/m²

$$= 220725 \times 10^3 \text{ pa}$$
$$= 220725 \times 10^3 \text{ N/m}^2$$
$$= 220.725 \text{ N/mm}^2$$
$$P = 4905 \text{ N}$$
$$E = 196.2\times10^6 \text{ kPa} = 196.2\times10^3 \text{ N/mm}^2$$

Solution

We know that,
For the semi-elliptical spring,

$$\text{Bending stress } \sigma_b = \frac{3\,P\,L}{n\,b\,t^2}$$

$$220.725 = \frac{3\times4905\times900}{n\times75\times10^2}$$

$$n = 8$$

$$\text{Deflection of the spring } y = \frac{2.85\,P\,L^3}{E\,n\,b\,t^3}$$

$$= \frac{2.85\times4905\times900^3}{196.2\times10^3\times8\times75\times10^3}$$

$$= 86.56 \text{ mm}$$

$$\text{Camber (Initial radius) } R = \frac{L^2}{8Z} + \frac{Z}{2}$$

$$= \frac{900^2}{8\times86.56} + \frac{86.56}{2}$$

$$= 1212.98 \text{ mm}$$

Problem 10.3: A truck spring has 10 leaves and is supported at a span length of 100 cm, with a central band of 80 mm wide. A load of 6 kN is applied at the center of spring whose permissible stress is 300 N/mm². The spring has a ratio of total depth to width of about 2.5. Determine the width, thickness, deflection and length of all leaves. Assume inside diameter of eye is 20 mm.

Given data

$n = 10$

$2L_1 = 100 \text{ cm} = 1000 \text{ mm}$

$l = 80 \text{ mm}$

$2P = 6 \text{ kN} = 6000 \text{ N}$

$\sigma_b = 300 \text{ N/mm}^2$

$\frac{nt}{b} = 2.5$

Solution

(1) Determination of width and thickness of the leaf

Assume $n_e = 2$ if not given

$n = n_e + n_g$

$10 = 2 + n_g$

$\therefore \quad n_g = 8$

Usually, the full length leaves are stressed more than the graduated leaves and hence design is based on stress induced in full length leaves.

$$\sigma_{be} = \frac{18\,P\,L}{b\,t^2\,(3n_e + 2n_g)}$$

$$\sigma_{be} = \frac{18 \times 3000 \times 460}{4t \times t^2\,(3 \times 2 + 2 \times 8)}$$

$$\left[\begin{aligned} \frac{nt}{b} &= 2.5 \\ \frac{10t}{b} &= 2.5 \\ b &= 4t \end{aligned}\right]$$

$t = 9.79 \text{ mm}$

$\therefore \quad b = 4t = 4 \times 9.79 = 39.16 \text{ mm}$

(2) Determination of deflection of leaf spring

$$y = \frac{12\,P\,L^3}{E\,b\,t^3\,(3n_e + 2n_g)}$$

Assume, $E = 2 \times 10^5 \text{ N/mm}^2$

$$y = \frac{12 \times 3000 \times 460^3}{2 \times 10^5 \times 39.16 \times 9.79^3\,(3 \times 2 + 2 \times 8)}$$

$y = 21.56 \text{ mm}$

(3) Determination of length of all the leaves

Length of smallest leaf $L_1 = \frac{2L}{n-1} \times 1 + \ell$

$$= \frac{2 \times 460}{10-1} \times 1 + 80$$

$$= \frac{920}{9} \times 1 + 80$$

$$= 182.22 \text{ mm}$$

Length of 2nd leaf $L_2 = \frac{920}{9} \times 2 + 80 = 284.44$ mm

Length of 3rd leaf $L_3 = \frac{920}{9} \times 3 + 80 = 386.66$ mm

Length of 4th leaf $L_4 = \frac{920}{9} \times 4 + 80 = 488.88$ mm

Length of 5th leaf $L_5 = \frac{920}{9} \times 5 + 80 = 591.11$ mm

Length of 6th leaf $L_6 = \frac{920}{9} \times 6 + 80 = 693.33$ mm

Length of 7th leaf $L_7 = \frac{920}{9} \times 7 + 80 = 795.55$ mm

Length of 8th leaf $L_8 = \frac{920}{9} \times 8 + 80 = 897.77$ mm

Length of 9th leaf $L_9 = \frac{920}{9} \times 2 + 80 = 1000$ mm

Length of master leaf $L_{10} = 2L_1 + \pi\,(d+t)2$

$$= 1000 + \pi\,(20+9.79)\times 2$$

$$= 1187.17 \text{ mm}$$

Problem 10.4: A locomotive semi-eliptical laminated spring has an overall length of 1 m and sustains a load of 70 kN at its centre. The spring has 3 full length leaves and 15 graduated leaves with a central band of 100 mm width. All the leaves are to be stressed to 400 MPa, when fully loaded. The ratio of the total spring depth to that of width is 2. $E=210$ kN/mm^2. Determine:

1. Thickness and width of the leaves
2. Initial gap that should be provided between the full length and graduated leaves before the band load is applied.
3. Load exerted on the band after the spring is assembled.

Given data

$2L_1 = 1m = 1000$ mm

$2P = 70$ kN $= 70000$ N $\Rightarrow P = 35000$ N

$n_e = 3$

$n_g = 15$

$\ell = 100$ mm

$\sigma_b = 400$ Mpa $= 400$ N/mm^2

$\frac{nt}{b} = 2$

$E = 210$ kN/mm^2 $= 210\times10^3$ N/mm^2

Solution

(1) Thickness and width of the leaves

Usually, the full length leaves are stressed more than the graduated leaves and hence design is based on stress induced in full length leaves.

$$\sigma_{be} = \frac{18\,P\,L}{b\,t^2\,(3n_e + 2n_g)}$$

$$\sigma_{be} = \frac{18\times3000\times460}{9t\times t^2\,(3\times3+2\times15)}$$

$$t = 12.1 \text{ mm}$$

$$\left[\begin{array}{l} \frac{n\,t}{b} = 2 \\ \frac{18t}{b} = 2 \\ b = 9t \end{array}\right] \quad \begin{aligned} 2L &= 2L_1 - \ell \\ &= 1000\text{–}100 \\ &= 900 \text{ mm} \\ \therefore L &= 450 \text{ mm} \end{aligned}$$

$\therefore \quad b = 49t = 4 \times 9.79 = 39.16$ mm

(2) Initial gap

$$C = \frac{2PL^3}{Enbt^3} = \frac{2\times35000\times450^3}{210\times10^3\times18\times108.9\times12.1^3} = 8.74\text{mm}$$

(3) Load exerted on the band

$$P_b = \frac{2Pn_en_g}{n\,(3n_e + 2n_g)} = \frac{2\times35000\times3\times15}{18\,(3\times3+2\times15)} = 4487.17 \text{ N}$$

Problem 10.5: A truck has 12 number of leaves, two of which are full length leaves. The spring supports are 1.05 m apart and the central band is 85 mm wide. The central load is to be 5.4 kN with a permissible stress of 280 Mpa. Determine the thickness and the width of the steel spring leaves. The ratio of total depth to width of the spring is 3. Also determine the deflection of the spring.

Given data

$n = 12$

$n_e = 2$

$2L_1 = 1.05\ m = 1050\ mm$

$\ell = 85\ mm$

$2P = 5.4\ kN = 5400\ N \quad \therefore P = 2700\ N$

$\sigma_b = 280\ Mpa = 280\times10^6\ pa = 280\times10^6\ N/m^2 = 280\ N/mm^2$

$\frac{nt}{b}=3$

1 pa = 1 N/m²

Solution

(a) Determination of thickness and width of the spring leaves

We know that,

Effective length of the spring

$$2L = 2L_1 - \ell, \text{ When band is used}$$

$$= 1050 - 85$$

$$= 965\ mm$$

$$\therefore \quad L = 482.5\ mm$$

Number of graduated leaves

$$n = n_e + n_g$$

$$12 = 2 + n_g$$

$$n_g = 10$$

Usually, the full length leaves are stressed more than the graduated leaves and hence design is based on stress induced in full length leaves.

$$\sigma_{be} = \frac{18\,P\,L}{b\,t^2\,(3n_e + 2n_g)}$$

$$\sigma_{be} = \frac{18\times2700\times482.5}{4t\times t^2\,(3\times2+2\times10)}$$

$$\left[\frac{nt}{b}=3, \quad \frac{12t}{b}=3, \quad b=4t\right]$$

$$t = 9.3\ mm$$

$$b = 4t = 4\times9.3 = 37.2\ mm$$

(b) Determination of deflection of the spring

$$y = \frac{12\,P\,L^3}{E\,b\,t^3\,(3n_e + 2n_g)}$$

$$y = \frac{12 \times 2700 \times 482.5^3}{0.21 \times 10^6 \times 37.2 \times 9.3^3\ (3 \times 2 + 2 \times 10)}$$

$$y = 22.28 \text{ mm}$$

Problem 10.6: A semi-elliptical laminated automobile spring is made of 50 mm wide and 3 mm thick plates. The length between the supports is 65 cm and the width of the band is 6 cm. The spring has two full length leaves and five graduated leaves. If the spring carries a central load of 160 kg, find

(a) maximum stress in full length and graduated leaves for an initial condition of no stress in the leaves.

(b) maximum stress if the initial stress is provided to cause equal stress when loaded,

(c) the deflection in parts (a) and (b).

Given data

Semi-elliptical leaf spring

$b = 50$ mm

$t = 3$ mm

$2L_1 = 65$ cm $= 650$ mm

$\ell = 6$ cm $= 60$ mm

$n_e = 2$

$n_g = 5$

$2P = 160$ kg $= 1600$ N $\quad \therefore P = 800$ N

Solution

(a) Bending stress induced in graduated leaves

$$\sigma_{bg} = \frac{12\,P\,L}{b\,t^2\ (3n_e + 2n_g)}$$

$$2L = 2L_1 - \ell$$

$$= 650 - 60$$

$$= 590 \text{ mm}$$

$$\therefore \quad L = 295 \text{ mm}$$

$$= \frac{12 \times 800 \times 295}{50 \times 3^2\ (3 \times 2 + 2 \times 5)}$$

$$= 393.3 \text{ N/mm}^2$$

Bending stress induced in full length leaves

$$\sigma_{be} = \frac{18\,P\,L}{b\,t^2\,(3n_e + 2n_g)}$$

$$= \frac{18 \times 800 \times 295}{50 \times 3^2\,(3 \times 2 + 2 \times 5)}$$

$$= 589.9\ \text{N/mm}^2$$

(b) Final stress $\sigma_b = \dfrac{6\,P\,L}{n\,b\,t^2}$

$$= \frac{6 \times 800 \times 295}{7 \times 50 \times 3^2} = 449.5\ \text{N/mm}^2$$

(c) Deflection which is common to both full length and graduated leaves

$$y = \frac{12\,P\,L^3}{E\,b\,t^3\,(3n_e + 2n_g)}$$

$$= \frac{12 \times 800 \times 295^3}{0.21 \times 10^6 \times 50 \times 3^3\,(3 \times 2 + 2 \times 5)}$$

$$= 54.33\ \text{mm}$$

The deflection in the spring due to the applied load is same as without initial stress.

DESIGN OF COIL SPRINGS

Let,

P Axial load on the spring

d Diameter of the spring wire

D Mean coil diameter

n Number of active coils

G Modulus of rigidity

= 73575×10^6 N/m² for automobile springs

Deflection of the spring can be expressed as

$$y = \frac{8\,n\,P\,D^3}{G\,d^4} \qquad \text{........ (1)}$$

Allowable shear stress in the material of the spring can be calculated as

$$\tau_s = \frac{Gdy}{\Pi D^2 n} \quad \text{......... (2)}$$

Energy stored per unit volume

$$= \frac{\tau_s}{4G} \quad J/m^3 \quad \text{......... (3)}$$

Problem 10.7: A typical coil suspension spring has 10 effective coils of a mean diameter 125 mm and made out of wires of diameter 15 mm. The spring is designed to carry a maximum static load of 3531.6 N. Calculate the shear stress and the deflection under the above loading.

If a maximum shear stress of 637650 kPa is allowable in the material, then what is the possible clearance in the spring? Take the value of G=73575×10³ kPa.

Given data

n = 10

D = 125 mm

d = 15 mm

P = 3531.6 N

τ_{ma} = 637650 kPa = 637.65 N/mm²

G = 73575×10³ kPa = 73575 N/mm²

Solution

(a) Deflection of the spring

$$y = \frac{8nPD^3}{Gd^4}$$

$$= \frac{8 \times 10 \times 3531.6 \times 125^3}{73575 \times 15^4}$$

$$= 148.15 \text{ mm}$$

(b) Shear stress induced

$$\tau_s = \frac{Gdy}{\Pi D^2 n}$$

$$= \frac{73575 \times 15 \times 148.15}{\Pi \times 125^2 \times 10}$$

$$= 333.1 \text{ N/mm}^2$$

(c) Possible clearance in the spring

$$\tau_{max} = \frac{G\,d\,y_{max}}{\Pi D^2 n}$$

$$637.65 = \frac{73575 \times 15 \times y_{max}}{\Pi \times 125^2 \times 10}$$

$$y_{max} = 283.61 \text{ mm}$$

∴ Possible clearance in the spring is

$$y_{max} - y = 283.61 - 148.15 = 135.46 \text{ mm}$$

DESIGN OF TORSION BAR SPRINGS

Let,

L	Length of the effective part of torsion bar
W	Dynamic load acting on the lever arm of the torsion bar = 2 × Static load
l	Length of the arm of torsion bar
M_t	Moment or torque applied on the torsion bar
d	Diameter of the bar
θ	Angular deflection of the bar
τ_s	Torsional stress developed in the bar
G	Modulus of rigidity = 73575 Mpa for spring steel

(a) Angular deflection of the bar

$$\theta_{max} = \frac{5760 M_t L}{\pi^2 d^4 G} \text{ degrees}$$

(b) Diameter of the torsion bar can be calculated from

$$\tau_{max} = \frac{16 M_t}{\pi d^3}$$

(c) Length of the torsion bar

$$L = \frac{\pi G \theta_{max} d}{360 \tau_{max}}$$

Note: θ_{max} should be in degrees

(d) Linear spring rate or load rate

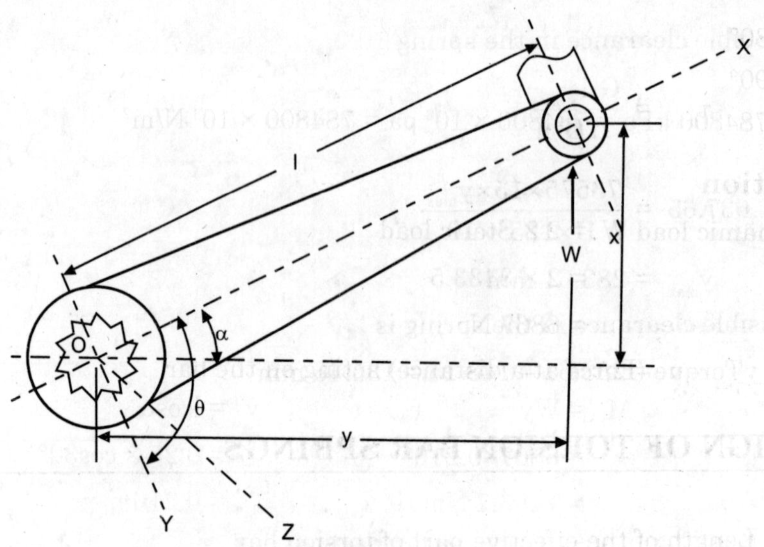

A torsion-bar arm extending upward is shown diagrammatically.

Let,

α Upward inclination of the arm with horizontal
   = Zero, when the arm is horizontal

x Linear deflection from the horizontal = lsinα

y Effective arm length = lcosα

Linear spring rate or load rate

$$r = \frac{\pi d^4 G + 32LWx}{32L(\ell^2 - x^2)}, \quad N/m$$

Problem 10.8: A torsion bar suspension is to be designed to support a maximum static load of 3433.5 N at the end of a leaver arm 250 mm long. The deflection of the lever above the horizontal is to be 30° with a total angle of deflection of 90°. Assuming a state allowable stress of 784800 kPa, calculate

i. the diameter of the bar
ii. the effective length, and
iii. the load rate

Given data

Torsion bar spring

Static load = 3433.5 N

l = 250 mm = 0.25 m

a = 30°

q = 90°

t = 784800 kPa = 784800 × 10^3 pa = 784800 × 10^3 N/m^2

Solution

Dynamic load W = 2 × Static load

= 2 × 3433.5

= 6867 N

Torque (Force at a distance) acting on the bar

M_t = Wy y = lcosα

= 6867×0.216 = 0.25 × cos30°

= 1486.7 Nm = 0.216 m

(i) Diameter of the torsion bar

It can be calculated from,

$$\tau_{max} = \frac{16M_t}{\pi d^3}$$

$$784800 \times 10^3 = \frac{16 \times 1486.7}{\pi d^3}$$

d = 0.0212 m

= 21.2 mm

(ii) Effective length of the torsion bar

$$L = \frac{\pi G \theta_{max} d}{360 \tau_{max}}$$

Take, G = Modulus of rigidity = 73575 Mpa for spring steel

$$L = \frac{\pi \times 73575 \times 10^6 \times 90 \times 0.0212}{360 \times 784800 \times 10^3}$$

= 1.56 m

(iii) Load rate

$$r = \frac{\pi d^4 G + 32LWx}{32L(\ell^2 - x^2)}$$

x = lsina = 0.25 × sin30°= 0.125 m

$$= \frac{\pi \times 0.0212^4 \times 73575 \times 10^6 + 32 \times 1.56 \times 6867 \times 0.125}{32 \times 1.56(0.25^2 - 0.125^2)}$$

= 38264.9 N/m

Brake System

Brakes are the devices for slowing and stopping the vehicle. The braking operation is reverse of the acceleration to the vehicle. In the acceleration the heat energy of the fuel is converted into the kinetic energy of the vehicle, whereas in the braking operation the kinetic energy of the vehicle is converted into heat, which is dissipated into the atmosphere.

PURPOSE

Brakes are designed to:

1. decelerate (slow down) the vehicle.
2. stop the vehicle, and
3. allow the vehicle to be parked on a slope.

They are therefore extremely essential equipment on an automobile to ensure safety driving.

REQUIREMENTS OF A BRAKE

1. Must be strong enough to stop the vehicle during emergency within the desired shortest distance.
2. The effectiveness of the brakes should remain constant (anti-fade characteristics).
3. Should be safe in operation, effective, progressive and consistent in response to brake pedal.

FACTORS AFFECTING THE BRAKES

The factors which affect the ability of the brakes of a vehicle are as given under.

1. Radius of the brake drum. The equation, $Ft = \frac{F_b \times R_b}{R_t}$ shows that a higher brake drum radius increases the retarding force produced at the ground.

2. Radius of tyre. From the above equation it is clear that a bigger tyre radius would decrease the retarding force at the ground.
3. Pressure on the brake shoe. From the above equation it is also clear that the amount of pressure applied at the brake lining increases the braking effect directly.
4. Area of the brake lining. The area of the brake lining also increases the braking effect directly.
5. Coefficients of friction. Both the coefficients of friction, between drum and lining as well as between tyre and ground, increase the braking effect directly. The too high coefficients must be avoided which may cause locking of the wheel.

PRINCIPLE

A moving vehicle cannot stop immediately when the engine is disconnected from the power train, due to inertia (the tendency of a moving object to continue moving). This inertia must be reduced in order to bring the vehicle to a halt.

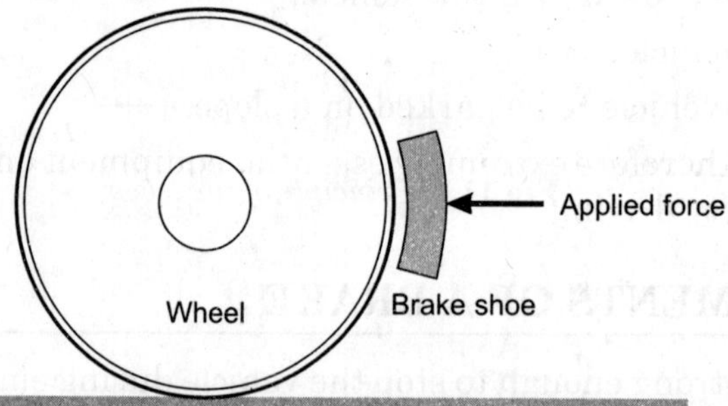

Fig.11.1: *Principle of braking*

The engine converts heat (thermal energy) into kinetic energy (energy of motion) to move the vehicle. In contrast, the brakes change this kinetic energy back into heat energy to stop the vehicle. Generally, brakes work by causing a fixed object to be pressed against a rotating object. The braking effect is obtained from the friction that is generated between the two objects. The heat formed at this time is dissipated to the atmosphere.

BRAKING FORCE

Brake pedal operation is based on the principle of the lever and converting a small pedal force into large force acting on the master cylinder.

Based on Pascal's law, the hydraulic force generated in the master cylinder is transmitted via pipe line to individual wheel cylinders. It acts on brake linings and disc pads to generate a braking force.

Operating force

The principle of the lever is applied by the brake pedal as follows.

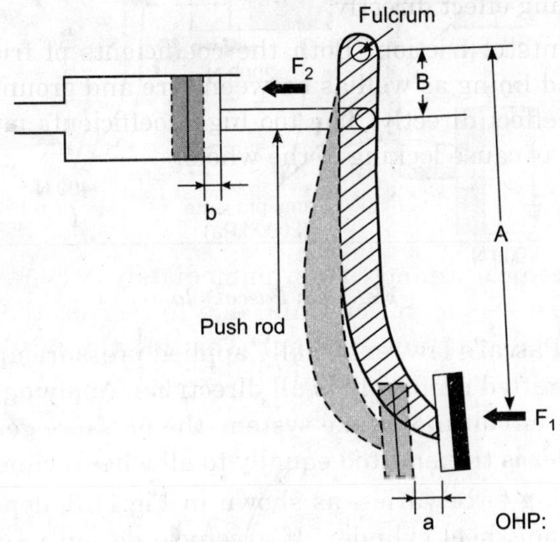

Fig.11.2: *Principle of the lever*

$$F_1 \times A = F_2 \times B$$

$$\therefore \quad F_2 = F_1 \times \frac{A}{B}$$

Where, F_1 : pedal force

F_2 : push rod output force

A_1 : distance from center to brake pedal to fulcrum

B : distance from push rod to fulcrum

Amount of movement:

$$\frac{b}{a} = \frac{B}{A}$$

$$a = b \times \frac{A}{B} \quad \text{(or)} \quad b = a \times \frac{B}{A}$$

a : amount of movement of pedal edge

b : amount of movement of push rod

Pascal's law

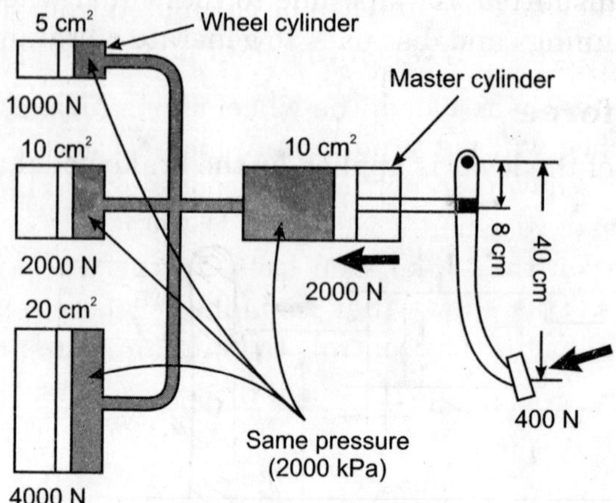

Fig.11.3: *Pascal's law*

According to Pascal's law, externally applied pressure upon a confined fluid is transmitted uniformly is all directions. Applying this principle to a hydraulic circuit in a brake system, the pressure generated in the master cylinder is transmitted equally to all wheel cylinders.

The braking force varies, as shown in Fig.11.3, depending on the diameter of the wheel cylinder. If a vehicle design requires a larger braking force at the front wheels, for example, the designer will specify larger cylinders for the front.

CO-EFFICIENT OF FRICTION AND TYRE ADHESION

Frictional force opposes the motion of the vehicle. Consequently it consumes power and produces heat. Frictional force occurs between the sliding tyre and the road surface when wheel rotation is locked by brakes. The ability of a vehicle to stop depends on the coefficient of friction between the contacting surfaces. Maximum usable coefficient of friction occurs between the tyre and road surface. The amount of energy that can be absorbed by the brakes depends upon the coefficient of friction of the brake materials, brake diameter, brake surface area, shoe geometry, and the pressure used to actuate the brake. Stopping a car suddenly means very high friction, resulting in high brake temperature.

The strength of the force pressing a shoe against a drum governs the resistance to rotation of a road wheel. During this operation the road surface has to drive the wheel around. The limit of this driving force is reached when the resistance of the brake equals the maximum frictional force that is produced between the tyre and road. The latter is called the adhesive force and can be calculated from the expression:

Adhesive force = Load on wheel × Co-efficient of friction

When the limit is reached, the wheel starts to skid, so extra force on the brake shoe will not produce any increase in the rate at which the vehicle slows down, no matter how good the braking system. This situation is apparent when a vehicle is braked on a slippery surface. Slight pressure on the brake soon locks up the wheel and very poor braking results. This shows that the adhesion between the tyre and road is the main factor that controls the minimum stopping distance.

Road adhesion is affected by:

- type of road surface.
- condition of surface, e.g. wet, dry, icy, greasy, etc.
- design of tyre tread, composition of tread material and depth of road.

STOPPING DISTANCE

Stopping distance is the distance travelled by the vehicle from the time the driver foot off the accelerator pedal for braking to the time the automobile comes to a full stop.

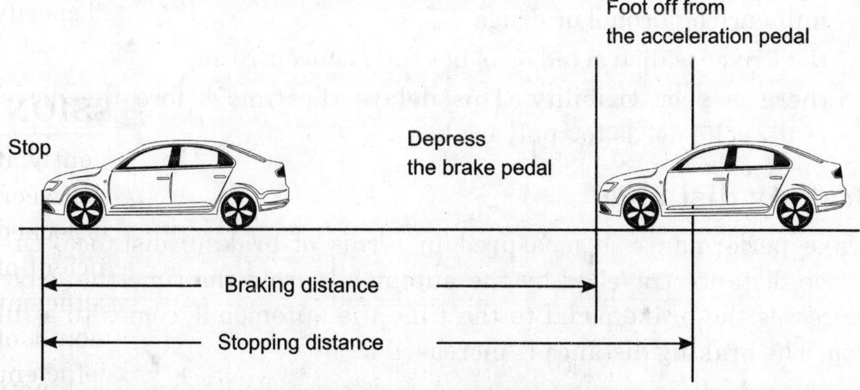

Fig.11.4: *Brake performance*

Kinetic energy is the force that keeps the vehicle moving. Kinetic energy is dissipated as heat by the brakes during application of brakes. The kinetic energy of a vehicle during braking is given by,

$$KE = \frac{1}{2} mv^2$$

Where, KE = kinetic energy, J

m = mass, kg

v = velocity, m/s

When a vehicle is moving at a given velocity then,

Kinetic energy = Work stored

= Retarding force × Stopping distance

$$\therefore \text{Stopping distance (S)} = \frac{\text{Kinetic energy}}{\text{Retarding force}} = \frac{\frac{1}{2}mv^2}{F}$$

Stopping distance is extremely important for emergency braking. The stopping distance of a vehicle depends on

- the thinking distance
- the braking distance

Stopping distance = Thinking distance + Braking distance

Thinking distance

The distance travelled by the vehicle from the point the driver realises he needs to brake to when he applies the brakes.

The thinking distance is increased if

- the vehicle is moving faster. The driver is ill, tired or under the influence of alcohol or drugs
- the driver is distracted or is not fully concentrating
- there is poor visibility. This delays the time before the driver realizes he needs to apply brakes.

Braking distance

Brake performance is measured in terms of braking distance, that is, the distance travelled by the automobile from the time the driver depresses the brake pedal to the time the automobile comes to a full stop. The braking distance is increased if

- the vehicle is moving faster. A faster vehicle has greater kinetic energy which increases the braking distance.
- there are poor weather/road conditions. The friction between the tyres and the road is decreased. A wet or greasy road surface reduces the amount of friction between the tyres and road, which increases the braking distance.
- The vehicle is in poor condition, e.g. brakes and tyres are worn out, tyres are not inflated properly.

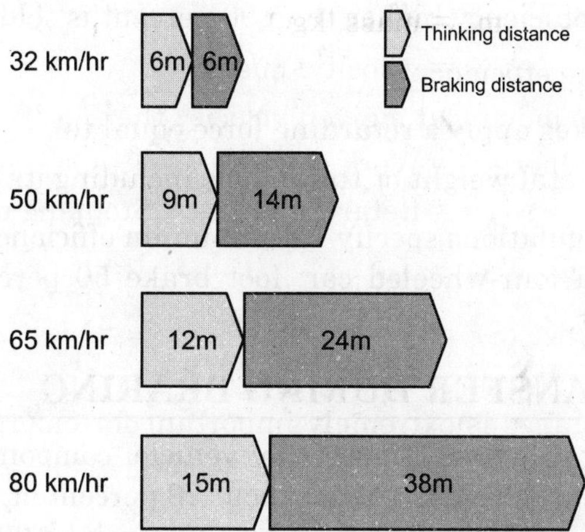

Fig.11.5: *Effect of vehicle speed on thinking distance and braking distance*

BRAKING EFFICIENCY

The braking efficiency of a vehicle is defined as the ratio of braking force produced (retarding force) to the total weight of the vehicle. Thus,

$$\eta_{brake} = \frac{\text{Braking force}}{\text{Weight of vehicle}} \times 100$$

Gravitational acceleration "*g*" can be used as the standard for brake efficiency. If a vehicle on a level road was decelerated by the brakes at the rate of 9.81 m/s^2, the braking efficiency is said to be 100 percent. 100 percent brake efficiency = deceleration at the rate of *g*

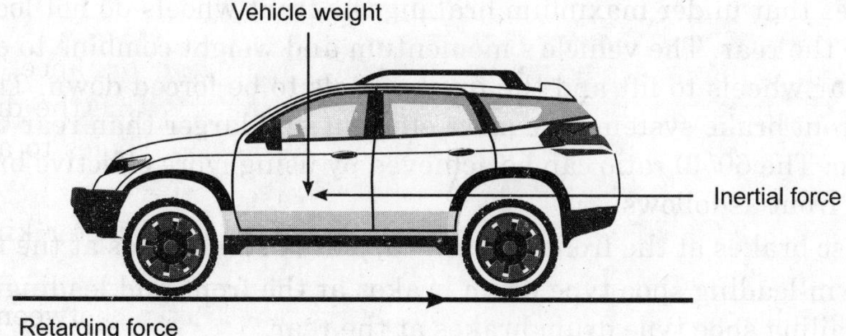

Fig.11.6: *100 percent brake efficiency*

To produce an efficiency of 100 percent, the brakes must give, at the road surface, a retarding force equal to the total weight of the vehicle as shown in Fig.11.6. This would require a coefficient of friction between the tyre and road of 1.0. Since this friction value is far above the average, an efficiency higher than 90 percent is seldom achieved.

100 percent brake efficiency

$$= \begin{Bmatrix} \text{Brakes apply a retarding force equal to} \\ \text{the total weight of the vehicle including its load} \end{Bmatrix}$$

Statutory regulations specify the minimum efficiency for each type of vehicle, e.g. a four-wheeled car: foot brake 50 percent, and hand brake 25 percent.

WEIGHT TRANSFER DURING BRAKING

During braking, the weight of the vehicle components transfers forward. On a rear-wheel drive car, about 70 percent of weight ends up on the front. On a front-wheel drive car, as much as 90 percent of the car weight ends up on the front during heavy braking. Due to weight transfer, the ratio of braking effect at the front and at the rear wheels should be minimum of 60% at front and 40% at rear.

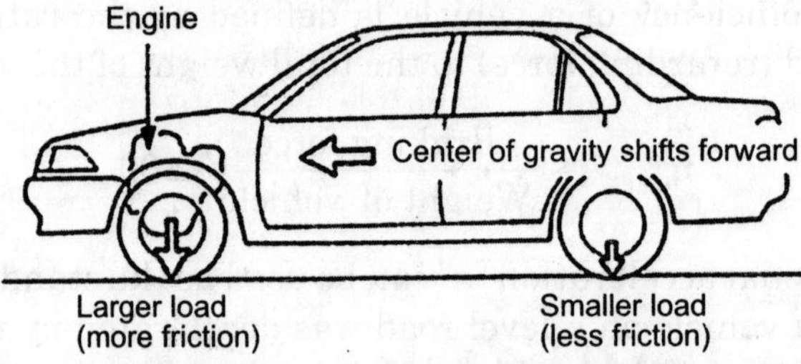

Fig.11.7: *Weight transfer during braking*

This 60/40 apportionment is chosen for safety reasons because it ensures that under maximum braking the front wheels do not lock up before the rear. The vehicle's momentum and weight combine to cause the rear wheels to lift and the front wheels to be forced down. This is why front brake systems are more efficient and larger than rear brake system. The 60/40 ratio can be achieved by using more effective brakes at the front as follows:

- disc brakes at the front and low capacity drum brakes at the rear.
- twin-leading shoe type drum brakes at the front and leading-and-trailing shoe type drum brakes at the rear.
- larger area wheel cylinders at the front.
- larger diameter drums or discs at the front.

As disc brakes are more efficient than drum brakes, disc brakes are mostly used at the front and drum brakes are used at the rear. This is the most common arrangement in present day cars.

BRAKING TORQUE

The braking torque is the twisting action caused by the drum or disc on the shoes or caliper anchors during the application of brakes. The amount of torque is determined by the effective axle height and stopping force between the tyre and road surface.

Brake torque on the front wheels is absorbed by the knuckle and suspension control arm. In rear, it is absorbed by the axle housing and the leaf spring or control arm. Braking torque during an emergency stop is much higher than accelerating torque at full throttle. Brake supporting and anchoring members must, therefore, have sufficient strength to withstand these high braking loads.

BRAKE-FADE

Excessive heating of the brakes following prolonged or repeated braking can result in fade. Heat causes temporary changes in the friction properties of the material used for brake pads and linings and the brakes become less efficient as they get hotter. Normal efficiency generally returns when they cool again.

If one brake goes on harder than the others, it may fade much sooner and cause uneven braking, which may lead to a skid. Brake pads and linings also wear away faster at higher temperatures.

ENGINE BRAKING

Engine braking is a method of braking a vehicle without using the foot brake. This method uses the rotational resistance of the engine to help reduce the vehicle's speed.

For example, on a steep downhill slope, when the accelerator pedal is released with the transmission engaged, the drive wheels continue to turn, forcing the engine (i.e. crankshaft) to turn, even though no fuel is being supplied to the cylinders. However, the crankshaft resists turning due to the resistance of the air being compressed in the cylinders, friction between moving parts, etc. The drive wheels therefore turn more slowly, causing the vehicles to slow down.

Furthermore, the lower the gear, the faster the crankshaft turns at the same vehicle speed. And since the crankshaft's resistance to rotation is higher at higher engine speeds (when no fuel is being supplied), the engine's braking force increases with the lower gear.

TYPES OF BRAKES

Brakes that are used on motor vehicles can be grouped into several types depending on their purpose.

1. The service brakes, operated by a foot pedal, which slow or stop the vehicle.
2. The parking brakes, operated by a foot pedal or hand lever, which hold the vehicle stationary when applied.
3. Auxiliary brakes are used in combination with the ordinary service brakes by diesel trucks and other heavy-duty vehicles.

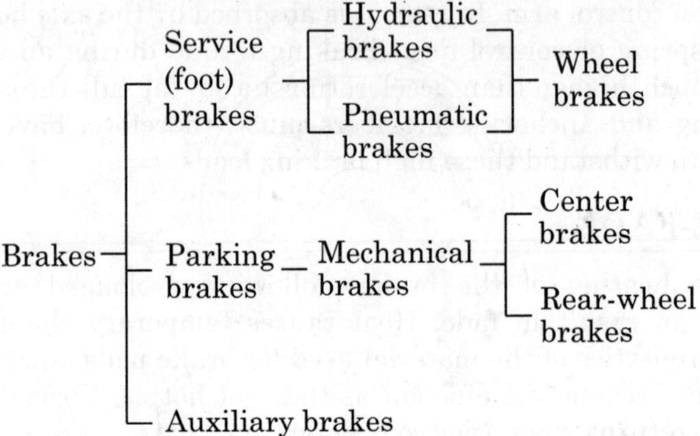

SERVICE BRAKES

Service brakes are also known as foot brakes or foundation brakes. Different types of service brake system are used in different vehicles, depending on the application:

- Mechanical brakes
- Hydraulic brakes
- Air brakes

Most automotive service brakes are hydraulic brakes. Dual-circuit hydraulic brake systems are used in passenger cars and small commercial vehicles up to 7.5t. Combined disc/drum brakes or just disc brakes are usually used.

The brake master cylinder is often driven by a brake booster. Advantages of hydraulic brake systems over air brake systems are their higher efficiency and their smaller dimensions.

The service or foundation brakes on many medium and heavy-duty trucks and buses are operated by air pressure (pneumatic). These are air brakes.

All these braking systems depend on friction between moving parts and stationary parts for their stopping force.

There are two main types of service brake: drum brake and disc brake.

DRUM BRAKE

In a drum brake, braking power is obtained by causing non-rotating shoes to be pushed (by a cam in mechanical brakes or hydraulic pressure in hydraulic brakes or compressed air in air brakes) against the inner surface of a drum that rotates together with the wheel. When the pressure is released, return springs pull the shoes back to their rest position.

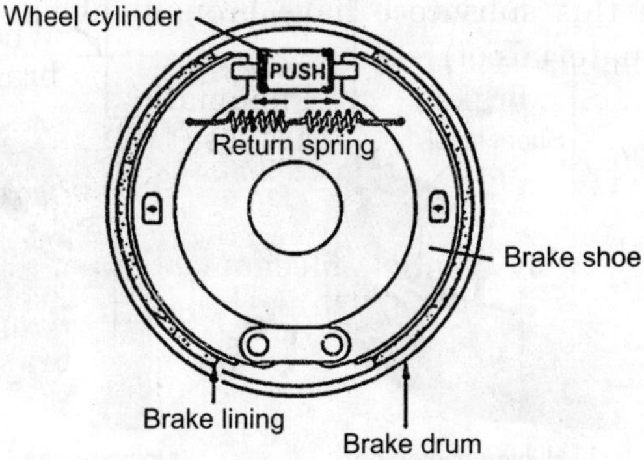

Fig.11.8: *Components of a drum brake*

The drum brake consists mainly of the following components:

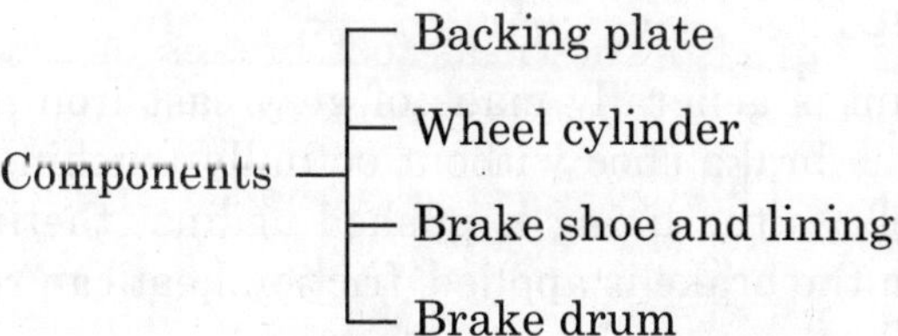

Backing plate

The backing plate is a pressed steel plate bolted to the rear axle housing or rear axle carrier. Since the brake shoes are fitted to the backing plate, all of the braking force acts on the backing plate.

Wheel cylinder

The wheel cylinder is used to expand the brake shoes against the brake drum. One or two wheel cylinders are used for each wheel. Some systems have two pistons (operated by hydraulic pressure) to operate two shoes, one at each side of the wheel cylinder, while other systems have only one piston to operate only one shoe.

Brake shoe and brake lining

The brake shoe is semi-circular in shape. Brake shoes are usually made of steel plates. The brake lining is attached to the shoe by rivets (on large vehicles) or by adhesives (on small vehicles). The material used for lining must have the following qualities:

- good frictional properties
- wear resistance
- heat resistance

Linings were once largely made of asbestos, but the health risks associated with this substance have brought about its replacement with synthetic materials.

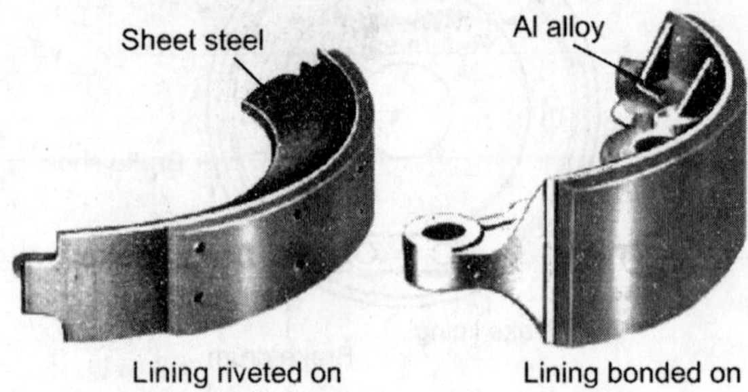

Fig.11.9: *Brake shoe and brake lining*

Brake drum

The brake drum is generally made of gray cast iron. It is positioned very close to the brake shoe without actually touching it and rotates with the wheel. As the lining is pushed against the inner surface of the drum when the brake is applied, friction heat can reach as high as 200°C to 300°C.

SELF-ENERGIZING ACTION

In the drum type brake, braking power is obtained by causing non-rotating shoes to be pushed against the inner surface of a drum that rotates together with the wheel. There are two types of brake shoe: the leading (or primary) shoe and the trailing (or secondary) shoe. The shoe facing the front of the vehicle is called the leading shoe and the shoe facing the rear of the vehicle is called the trailing shoe.

The action of the drum, in which it tries to force the leading shoe to rotate with it, is called self-energizing or self-servo action. Self-energizing action creates a greater braking force results from a relatively small pedal effort. Although leading shoes provide more braking power, there is a disadvantage to this in that they wear faster than trailing shoe.

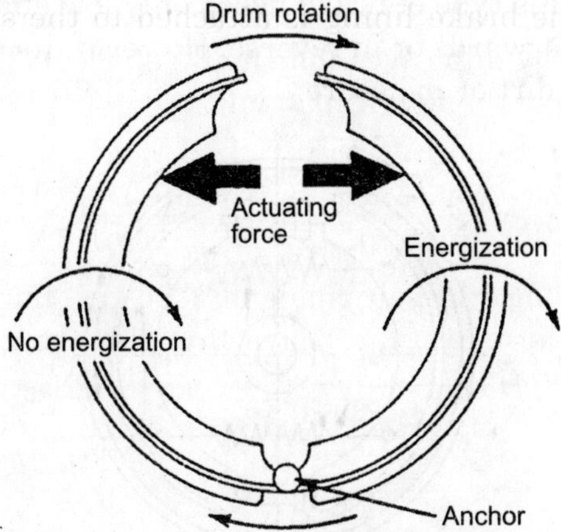

Fig.11.10: *Self-energizing action*

TYPES OF DRUM BRAKES

LEADING-AND-TRAILING SHOE TYPE

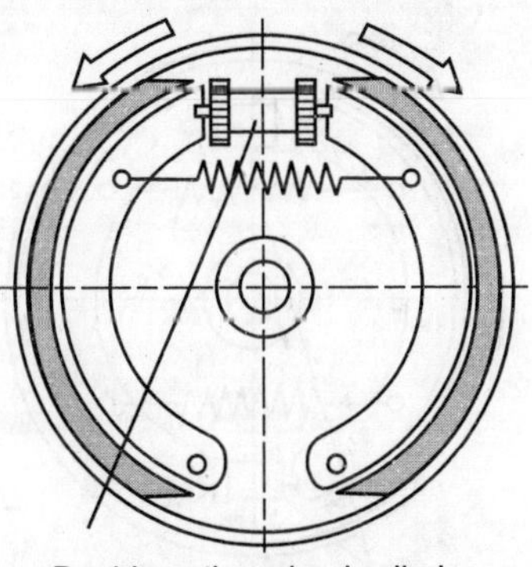

Fig.11.11: *Leading-and-trailing shoe drum brake*

In this design, both shoes pivot about the same point and a single expander unit is employed to force the other ends of the shoes in an outward direction with the shoes being applied in opposite directions.

The leading shoe wears faster than the trailing shoe because of the self-energizing action. This type is used in the rear brakes of passenger cars and small commercial vehicles.

DUO-SERVO BRAKE

On the Duo-servo brake, the self-servo action of the leading shoe is utilised to increase the contact pressure of the second leading shoe. The support bearing is therefore movable, and the shoes are located at the wheel cylinder. This type of brake needs only low effort to apply and is equally efficient forwards or in reverse. However, its action can be very badly affected by dirt or moisture.

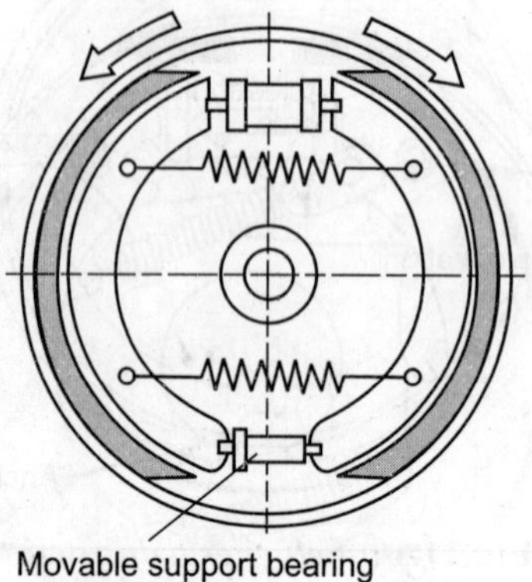

Fig.11.12: *Duo-servo brake*

TWIN-LEADING SHOE TYPE

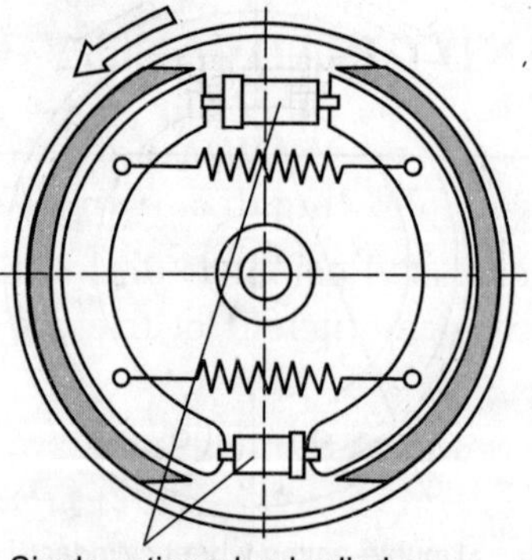

Fig.11.13: *Twin-leading shoe drum brake*

This type is used in front brakes of passenger cars and commercial vehicles. This design has two wheel cylinders, each having pistons at one end. When the brake is applied when the vehicle is moving forward, both shoes function as leading shoes.

When the drum is rotating in the forward direction, this type exerts a high braking force. However, a disadvantage of this type is that when the drum rotates in the opposite direction (reversing the car), both shoes work as trailing shoes exerting only a small braking force.

DUO-DUPLEX BRAKE

Two leading shoe brakes incorporating two double-acting wheel cylinders are also made. Their braking action is equally good whether the vehicle is moving forwards or in reverse. They are known as Due-duplex brakes.

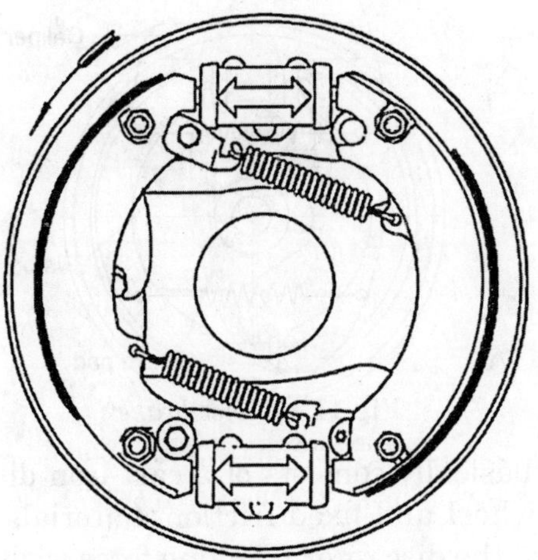

Fig.11.14: *Duo-duplex brake*

ADVANTAGES AND DISADVANTAGES OF DRUM BRAKES

The drum brake has the following disadvantages:

1. Brake fading when hot. Poor heat dissipation may cause brake to fade and also reduce co-efficient of friction between friction lining and brake shoe.
2. Brake inspection requires removal of parts.
3. Self-servo effect.
4. Poor water recovery. Because it has an enclosed design, when the brake cavity gets wet due to rain, floods, or driving over a puddle, the water cannot be expelled right away. The collected water can reduce the frictional properties of the braking system, and this can be dangerous because the vehicle has a decreased stopping ability.

However, the drum brake has many advantages:

1. Self-applying action
2. Simple hand brake arrangement
3. The drum brake is housed inside the dished area of the road wheel and is thus protected against dirt to a large extent. It is also easy to install a mechanically-operated parking brake.
4. Production and purchase costs are cheaper.

DISC BRAKE

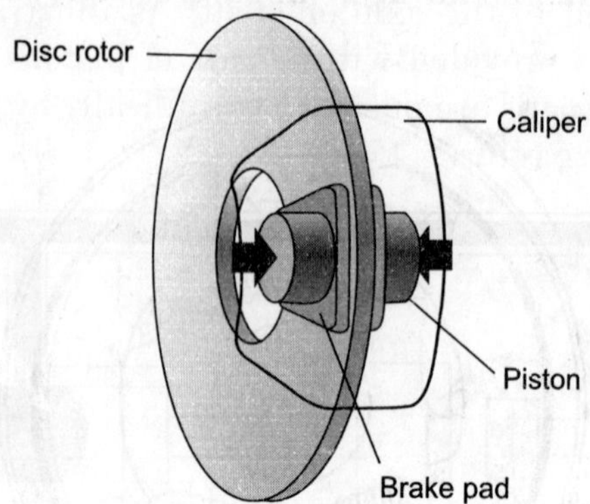

Fig.11.15: *Disc brake*

A disc brake basically consists of a cast-iron disc (disc rotor) that rotates with the wheel and fixed friction materials (brake pads) that are pushed against the disc rotor. Braking force is generated by friction between the disc and the disc pads.

COMPONENTS OF DISC BRAKES

A disc brake basically consists of a cast-iron disc (disc rotor), brake pads and caliper.

DISC ROTOR

Generally, the disc rotor is made of gray cast iron and is either solid or ventilated. The ventilated type disc rotor consists of a pair of hollow discs to ensure good cooling, both to prevent fading and to ensure longer pad life.

BRAKE PAD

A disc pad is usually a baked mixture of metallic fibers and resin containing a small amount of metal powder. This type is called the semi-metallic disc pad. A slit is provided on the rotor side of the pad to indicate the pad thickness (allowable limit) so that the pad wear can be checked easily.

CALIPER

The caliper, also called the cylinder body, holds the pistons and is provided with channels through which the brake fluid is supplied to the cylinders.

Calipers are grouped into the following types by the way in which they are installed: fixed caliper type and floating caliper type.

Fixed caliper type

The caliper is fixed to the axle or strut. As illustrated in Fig.11.18, the fixed caliper is provided with a pair of pistons. Braking force is obtained when the pads are pushed hydraulically by the piston against both ends of the disc rotor.

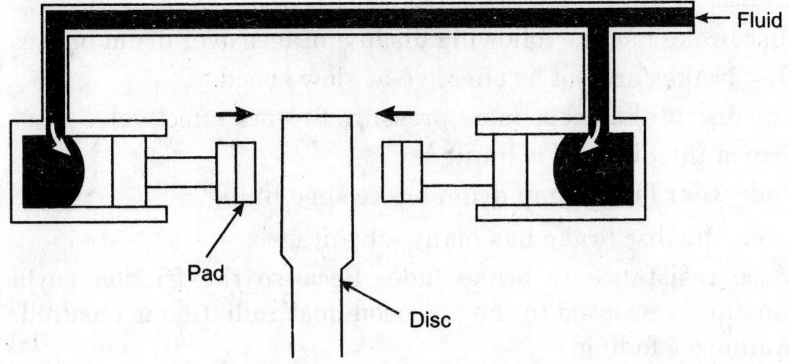

Fig.11.16: *Fixed caliper type*

The fixed caliper is the most basic design and ensures accurate operation. However, it requires a large number of components. For this reason, the fixed caliper type disc brakes are used only for some high performance models today.

Floating caliper type

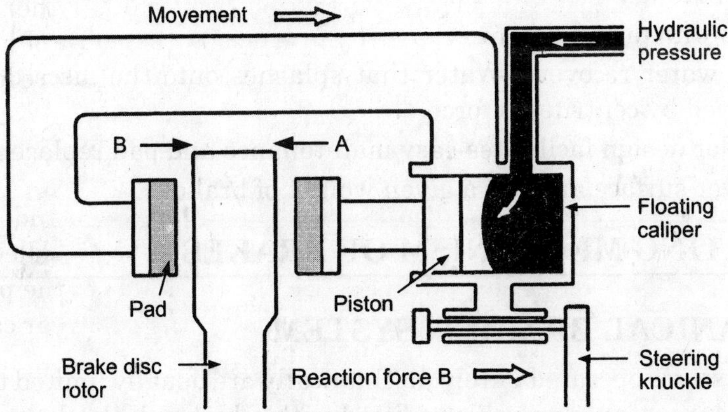

Fig.11.17: *Floating caliper type*

As illustrated in Fig.11.19, the piston is located in one side of the caliper only. Hydraulic pressure from the master cylinder pushes the piston (force A) and thus presses the pads against the disc rotor. At the same time equal hydraulic pressure (reaction force B) acts on the bottom of the cylinder. This causes the caliper to move to the right and presses the pad located opposite the piston against the disc rotor to exert braking force.

ADVANTAGES AND DISADVANTAGES OVER DRUM BRAKES

The disc brake has the following disadvantages over drum brakes:

- Disc brakes are not as effective at slow speed.
- The disc brake needs more pressure to work effectively.
- Size of the disc pad is limited.
- Pads wear faster than drum brake shoe lining.

However, the disc brake has many advantages:

- More resistance to brake fade. Because the friction surface is constantly exposed to the air, good heat radiation is ensured. This minimizes fading.
- Better cooling. Friction surfaces directly exposed to cooling air.
- The disc brake has the same stopping power in forward or reverse.
- Disc brakes have progressive action. The greater the effort applied to the pedal the more effective the brake will be.
- Disc brakes are self-adjusting.
- Disc brake pads can be easily checked for wear.
- High braking efficiency
- High stability
- Water and dirt resistant
- Good water recovery. Water that splashes onto the discs can be removed by centrifugal force.
- Simpler design facilitates easy maintenance and pad replacement.
- Greater surface area for a given weight of brake

OPERATING MECHANISM OF BRAKES

MECHANICAL BRAKING SYSTEM

Brakes systems operated purely mechanically are usually limited today to the parking brakes on vehicles fitted with a hydraulic brake system or the service brakes of lightweight and medium-weight motorcycles and single-axle trailers.

The efficiency of a mechanical brake system is low (in the region of 50%). In cold, wet weather or frost, the actuating linkage may freeze and impair operation of the brakes.

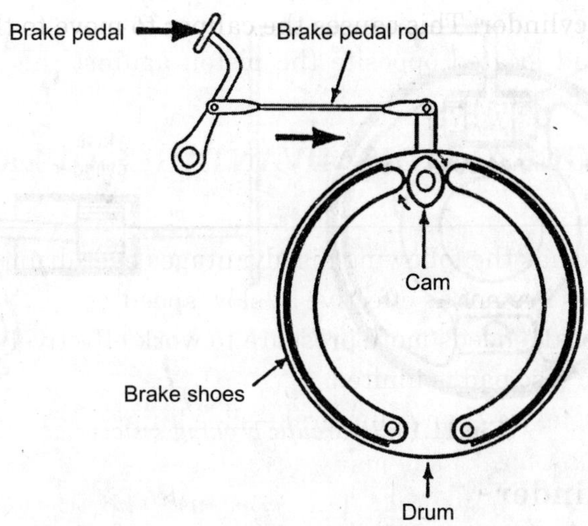

Fig.11.18: *Mechanical braking system*

The brake cables are wires passing over rollers or moving within a metal outer sheath. To reduce friction and prevent icing up, the outer cables are packed with grease. Screw adjusters are provided for accurate brake adjustment.

Brake equalization between the left and right wheels on an axle can be ensured by incorporating a balance beam in the linkage or by means of a pulley.

Lengthy brake rods must be guided halfway along to avoid vibration or buckling. The linkage can be arranged to generate additional force by incorporating levers at the appropriate points.

The rod or linkage is connected to pivots on the brake back plates. In contrast to wire cables, no equalization is possible nor can wear be compensated for by means of screw adjusters.

HYDRAULIC OPERATED BRAKE SYSTEM

Fluid is not compressible. Any effort applied at one place will be transmitted without loss of pressure or movement anywhere in the system. The hydraulic brakes can respond much faster than other types and are simpler in design. Hydraulic brakes also have superior design flexibility. Due to these advantages, hydraulic brakes are most commonly used with passenger cars and light-duty trucks today.

Hydraulic service brakes comprise the following operating mechanisms that transmit the brake applying force and the braking mechanisms that generate the braking force.

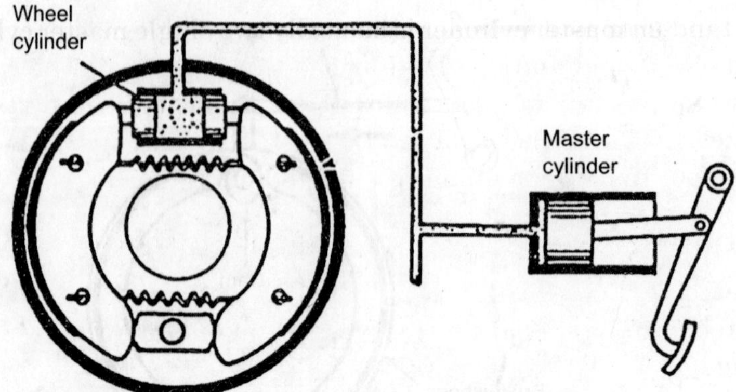

Fig.11.19: *Hydraulic braking system*

Master cylinder

The master cylinder converts the motion of the brake pedal into hydraulic pressure. It consists of the reservoir tank which contains the brake fluid as well as the piston and cylinder which generates the hydraulic pressure.

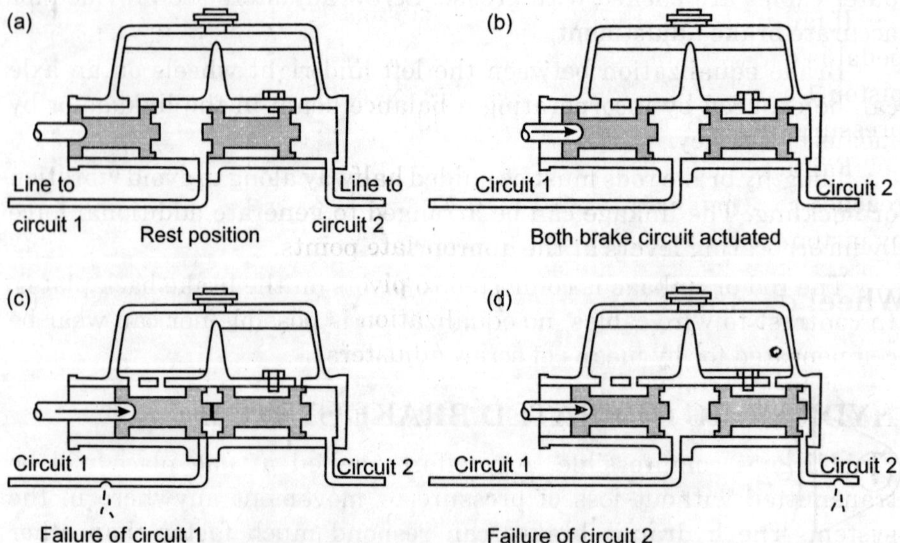

Fig.11.20: *Tandem master cylinder*

There are two types of master cylinder: the single type and tandem type. The tandem type master cylinder is used more commonly than the single type. The tandem master cylinder has separate hydraulic systems for front and rear wheels so that if one of these systems becomes inoperative, the other system can still function properly.

The tandem master cylinder is basically two single master cylinders mounted end-to-end and operated by the same pushrod. The cylinder contains two pistons, dividing it into two chambers. The first piston is operated by the brake pedal and a push rod. The second piston is normally operated by the fluid pressure created by the first.

Operation

Two reservoirs or one divided, supply fluid through ports to each half of the master cylinder. Separate reserves of brake fluid keep one system operating if the other drains away.

When the brake pedal is depressed, the fluid in chamber 1 becomes pressurized and moves fluid into the first system. At the same time, this pressure also moves piston 2 and pressurises the fluid in chamber 2 and the other system to the same pressure. Upon release of the brake pedal, return springs and seals return the wheel brake units to their rest positions and fluid can return to the master cylinder and reservoirs. Inside the master cylinder, return springs for both pistons ensure that they return to their correct positions against the stops.

If failure occurs in system 1, no pressure will be built up when the pedal is depressed. There will be extra pedal travel until the bosses on piston 1 and piston 2 touch. Further movement of the brake pedal will pressurize system 2 in the normal way.

Failure of system 2 also results in extra pedal travel until piston 2 reaches the limit of its travel. From then on chamber 1 is pressurized by piston 1 in the usual way.

Wheel cylinder (Slave cylinder)

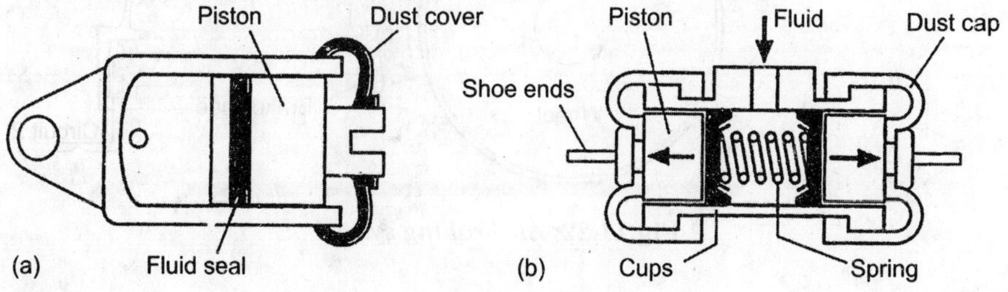

Fig.11.21: *Wheel cylinder*

These are sometimes called expanders or actuators. It is the purpose of the wheel cylinder pistons to force the brake shoes into contact with the brake drum. There are detail differences, but all of them work on the same principle.

The wheel cylinder consists of a cylindrical casting made from cast iron or aluminium alloy. It is rigidly attached to the back plate, but is sometimes free to slide slightly to centralize its position.

There are two basic types: single acting and double acting. Internally both have a machined bore, but in a single-acting cylinder one end is blind and contains only one piston. A double-acting cylinder is open at both ends and contains two pistons separated by a spring. The pistons are free to move inside the cylinder.

Rubber cups or seals are fitted to each piston to prevent fluid from leaking out of the system. Correct fitting is essential to avoid brake failure.

AIR OPERATED (PNEUMATIC) BRAKE SYSTEM

Most heavy commercial vehicles now use air operated brakes. In this system compressed air (around 7 bar) is used to actuate the brake mechanism. Air pressure from the compressor is transmitted through pipes equally in all directions, in exactly the same way as fluid pressure in a hydraulic system. The actuation of each brake is similar to the simple cam brake through diaphragm in the brake chamber.

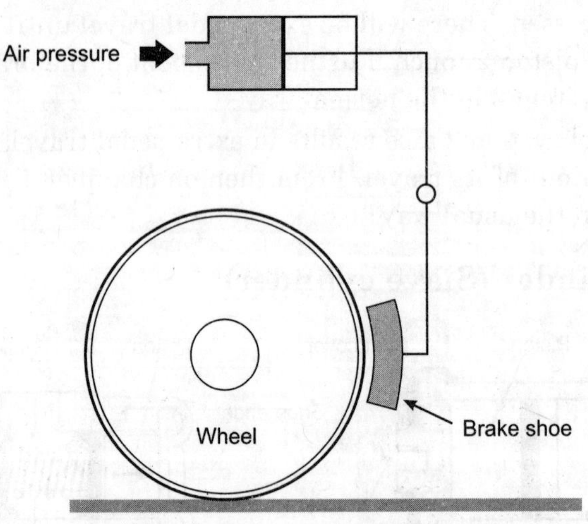

Fig.11.22: *Air braking system*

Air compressor

The compressor is a piston type driven by the engine or in some cases, the gearbox. It supplies air under pressure to a reservoir tank for storage. A governor regulates the pressure in the reservoir tank and when the operating pressure is exceeded, it is vented to atmosphere. The reservoir has to be strong enough to withstand pressures of around

7 bar. The brake valve, which is operated by the driver's foot pedal, allows the compressed air to travel from the reservoir to the diaphragm in the brake chamber at each wheel. The valve is designed to permit controlled application of the brakes.

The brake chambers are fitted with a movable diaphragm, which operates a rod connected to the brake shoe operating mechanism. The admission of air pressure to the diaphragm in the brake chamber moves the diaphragm and linkage to apply the brake. A quick-release valve is used to speed up the release of air pressure to the atmosphere once the brakes have been released. A relay valve speeds up the application and release of the rear brakes.

Increasing safety legislation has required manufacturers to provide more sophisticated air brake systems. The system described would fail if any air pipe fractured. More recent systems contain dual or triple brake lines.

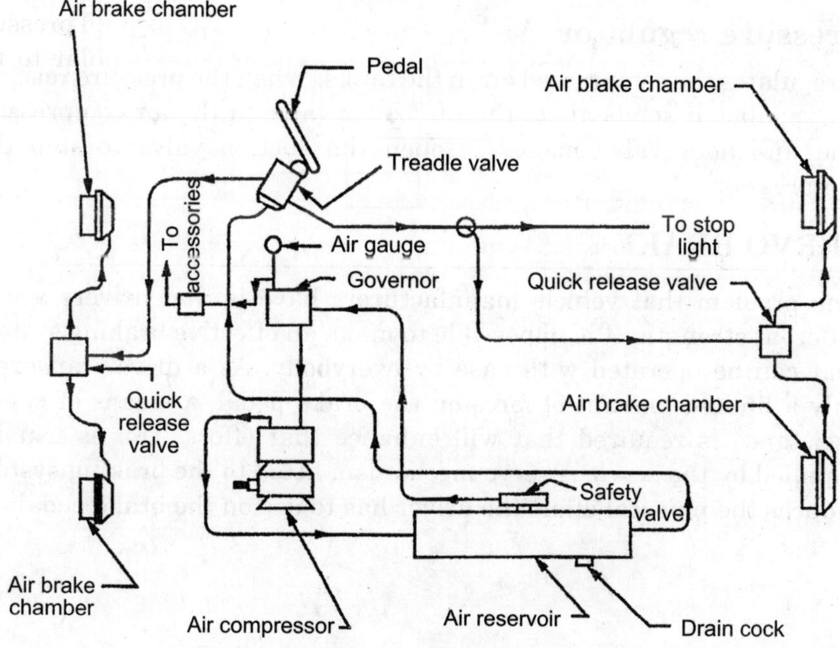

Fig.11.23: *Layout of Air braking system*

Air tank

There are 3 kinds of tanks, wet tank, front tank and rear tank. Wet tank is to remove moisture from compressed air. Front and rear tank accumulate the compressed air for each front and rear system.

Brake valve

The brake pedal operates this valve. It sends compressed air to relay valve according to the depressing condition of the pedal. To control the front and the rear system simultaneously, there are two valves inside of it.

Relay valve

Each of the front and the rear system has this valve. It works according to the compressed air from brake valve and controls the compressed air from the tank to the brake chamber. When the brake pedal is released, the compressed air in the brake chambers is exhausted from this valve.

Brake chamber

It receives the compressed air from relay valve and makes to rotate the brake cam in the drum brake.

Pressure regulator

It regulates the compressed air in the tanks. When the pressure reaches the regular, it sends air to the un-loader valve in the air compressor. The un-loader valve makes to open the suction valve to stop the compression.

SERVO BRAKE

One problem that vehicle manufacturers have is that drivers are of different strength. It is impossible to make an effective braking system that can be operated with ease by everybody. As a driver can exert only a limited amount of force on the brake pedal, a means of power assistance is required that will enhance that effort. This is usually supplied by the servo. A servo mechanism fitted to the braking system reduces the physical effort the driver has to use on the brake pedal.

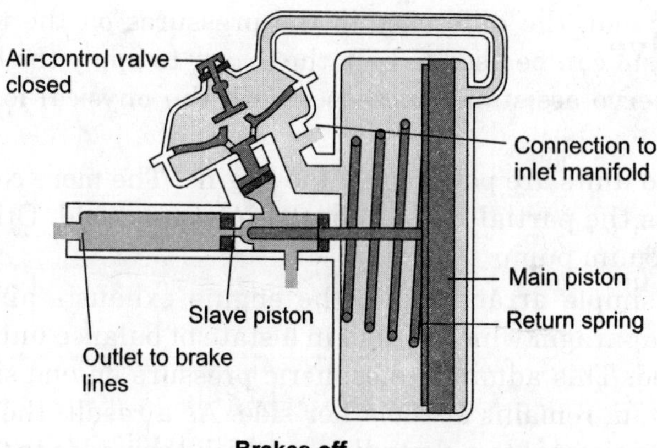

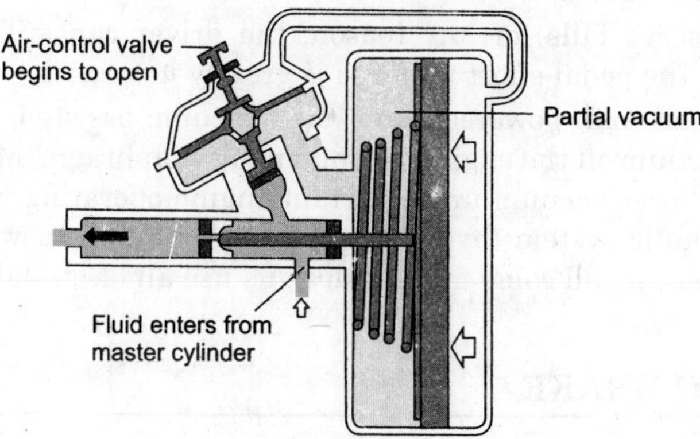

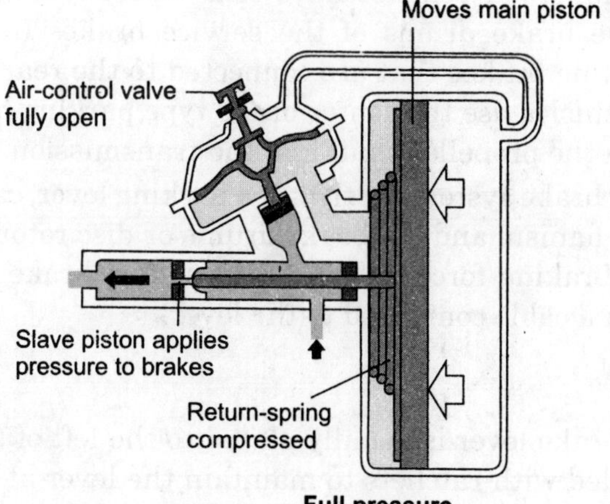

Fig.11.24: *Servo brake*

The servo cylinder contains a diaphragm. When air is exhausted from one end of the cylinder and atmospheric pressure is admitted to the other end, the difference in the pressures on the two sides of the diaphragm can be used to help the driver to apply the brakes. The amount of servo assistance increases with the physical force used on the pedal.

All servo units are powered by the engine. The more common type of servo uses the partial vacuum in the inlet manifold. Others have a separate vacuum pump.

In one simple arrangement, the engine exhausts air from both sides of a diaphragm which is held in a state of balance until the brake pedal is used. This admits atmospheric pressure on one side, while a partial vacuum remains at the other side. As a result, the diaphragm moves, applying force to a slave piston, which then adds to the pressure applied by the driver.

If the servo fails for any reason, the driver can still apply the brakes, but the pedal effort required is greatly increased.

Most gasoline powered cars use vacuum-assisted brakes by drawing vacuum off the engine or a pump to a diaphragm. Most diesels, because of lower vacuum under certain engine operating parameters, use a hydraulic system (hydro-boost) powered by the power steering hydraulic pump and some modern designs use an electrically powered system.

PARKING BRAKE

Parking brakes are mainly used for parking the vehicle. Passenger cars and small commercial vehicles have rear wheel type parking brakes that share the brake drums of the service brakes (foot brakes), or exclusive parking brakes that are connected to the rear wheels. Large commercial vehicles use the center brake type parking brakes that are fitted between the propeller shaft and the transmission.

A parking brake system consists of a parking lever, cable or rod type operating mechanism and the brake drums or disc rotor and shoe that generate the braking force. Pulling the parking brake lever operates the brakes via a cable connected to the lever.

Lever

The parking brake lever is usually placed to the left of the driver. The lever is provided with ratchets to maintain the lever at the position to which it was set. Pressing the button at the end of the lever trips a pawl to allow free movement of the lever.

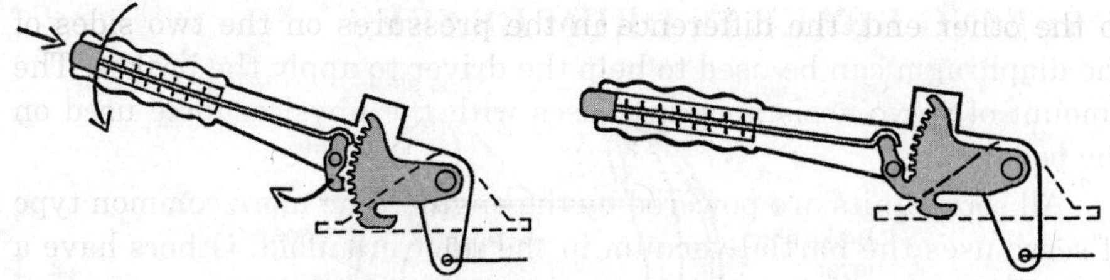

Fig.11.25: *Parking brake lever*

Intermediate lever, rod and cable

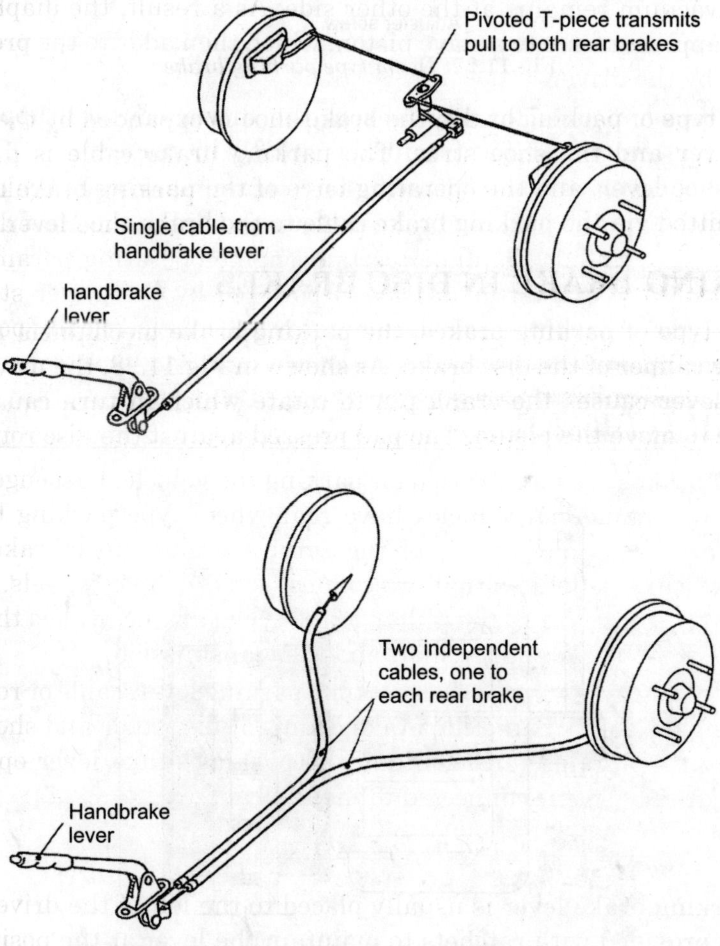

Fig.11.26: *Parking brake cable arrangement*

The parking brake cable transmits the lever movement to the brake drum sub-assembly. The intermediate lever is provided to increase the operating force.

PARKING BRAKE IN DRUM BRAKES

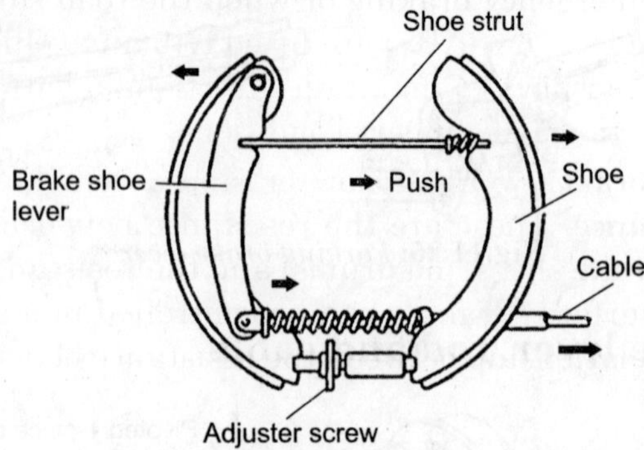

Fig.11.27: *Drum type parking brake*

In this type of parking brake, the brake shoe is expanded by the brake shoe lever and the shoe strut. The parking brake cable is fitted to brake shoe lever, and the operating force of the parking brake lever is transmitted via the parking brake cable to the brake shoe lever.

PARKING BRAKE IN DISC BRAKES

In this type of parking brakes, the parking brake mechanism is built into the caliper of the disc brake. As shown in Fig.11.28, the movement of the lever causes the crank pin to rotate which in turn causes the spindle to move the piston. The pad pressed against the disc rotor as a result.

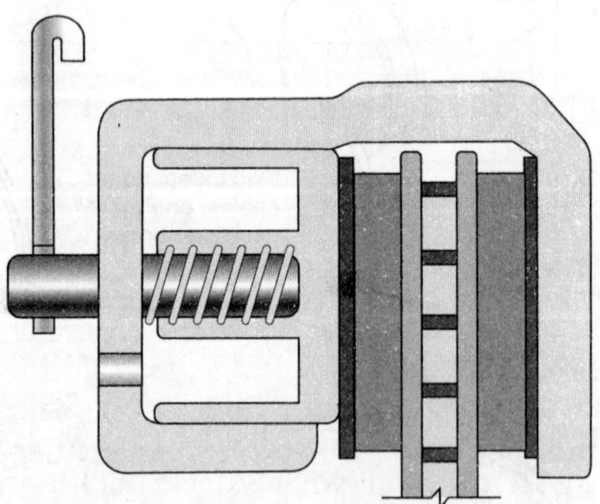

Fig.11.28: *Disc type parking brake*

ANTI-LOCK BRAKE SYSTEM (ABS)

Friction between the tyre and the road is essential for all aspects of vehicle control. The brakes may lock up and control of the vehicle may be lost under emergency braking or when the road surface is slippery. Anti-lock brake systems (ABS) are fitted to many vehicles to overcome this danger. Such devices maintain directional control, and decrease the stopping distance in slippery conditions.

Ordinary brakes work to slow or stop a vehicle by utilizing two types of resistance. These are the resistance between the brake pads and discs (or brake shoes and drums) and the resistance between tyres and the road surface. Braking can be controlled in a stable manner if the following relationship between the resistance between the road and tyres exists:

Brake system resistance < Resistance between tyre and road

However, if this relationship is reversed, the tyres will lock up and the vehicle will begin skidding:

- Brake system resistance > Resistance between tyre and road

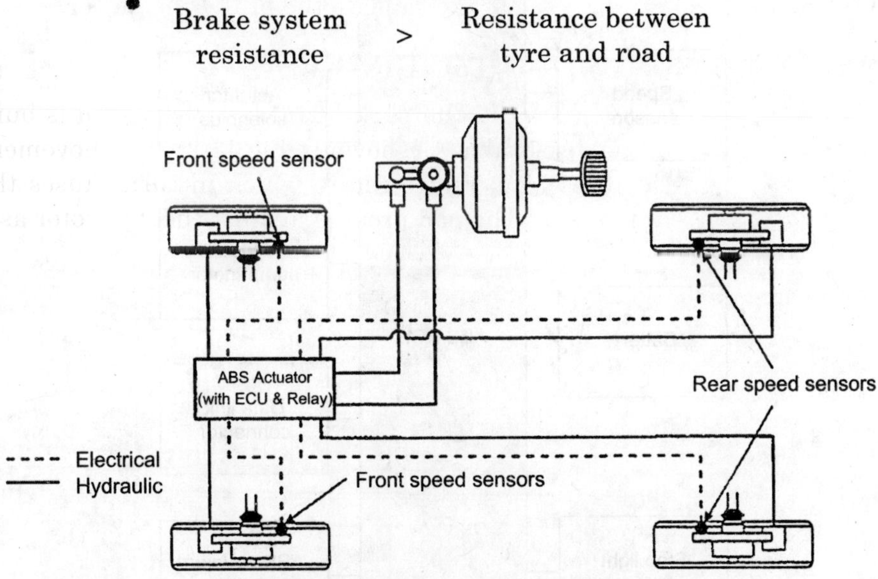

Fig.11.29: *ABS braking system*

As a result, if the front wheels lock up, it will become impossible to steer the car. If the rear wheels lock up, the differences in the coefficients of friction (m) between the left and right sides on the road surface will cause a tail spin to occur.

The ABS controls the hydraulic pressure acting on the wheel cylinders to prevent the wheels from locking up when the brakes are applied on a slippery surface or during panic braking. It also maintains directional stability during braking so that the vehicle can be steered.

Components and operation

The heart of the ABS system is an electronic control unit (ECU) which is provided with inputs from sensors and gives outputs to the controls. The sensors are located at each wheel and send pulse generated signals to the ECU. They indicate whether each wheel is revolving and how fast it is turning. The ECU then controls the output of a hydraulic pump assembly. Based on the input signals from the wheel sensors, the ABS system's ECU knows the speed of each wheel. When the brakes are applied, the wheels and the vehicle start to slow down. The ECU monitors the reducing pulses from each of the wheel sensors. If a wheel seems about to lock up, the ECU reduces the pressure at that wheel until the tyre grips again.

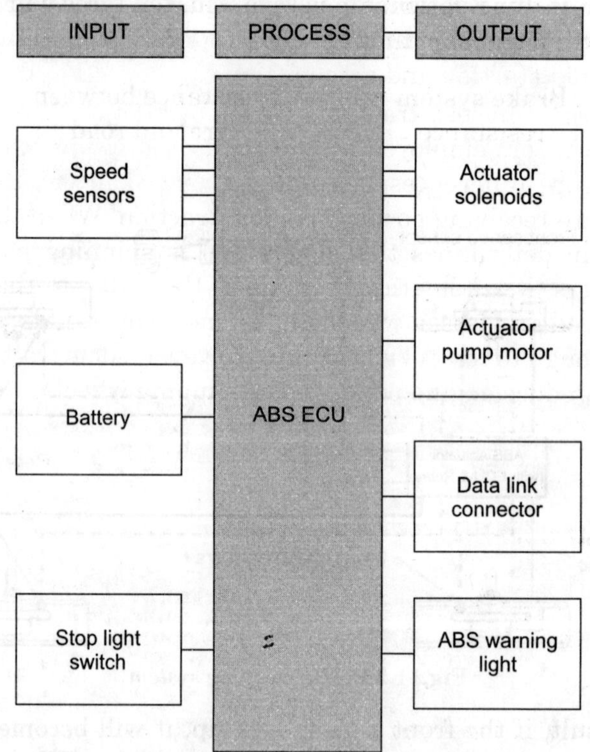

Fig.11.30: *ABS control system*

TRACTION CONTROL SYSTEM (TCS)

Traction Control System (TCS) also known as vehicle stability assist system. On roads with a low coefficient of friction (m), such as snowy, icy or wet roads, the drive wheels spin if the vehicle is started or accelerated quickly, resulting in loss of torque and possibly causing vehicle skidding.

The maximum torque that can be transmitted to the wheels is determined by the coefficient of friction generated between the road and the tyres. If the attempt is made to transmit torque exceeding that level to the wheels, it will become easy for the wheels to spin.

Maintaining the torque at a level appropriate to the coefficient of friction at such times is sometimes difficult for the driver. In most cases, when attempting to start the vehicle quickly, the driver presses the accelerator pedal too hard and the wheels spin, causing a loss of traction and of torque.

The traction control system lowers the engine's torque, regardless of the driver's intentions, when the wheels start to spin, while at the same time controlling the brake system, thereby reducing the torque transmitted to the road surface to appropriate levels. In this way, stable and rapid starting and acceleration can be achieved.

In modern vehicles, traction-control systems utilize the same wheel-speed sensors employed by the antilock braking system. These sensors measure differences in rotational speed to determine if the wheels that are receiving power have lost traction. When the traction-control system determines that one wheel is spinning more quickly than the others, it automatically "pumps" the brake to that wheel to reduce its speed and lessen wheel slip. In most cases, individual wheel braking is enough to control wheel slip. However, some traction-control systems also reduce engine power to the slipping wheels.

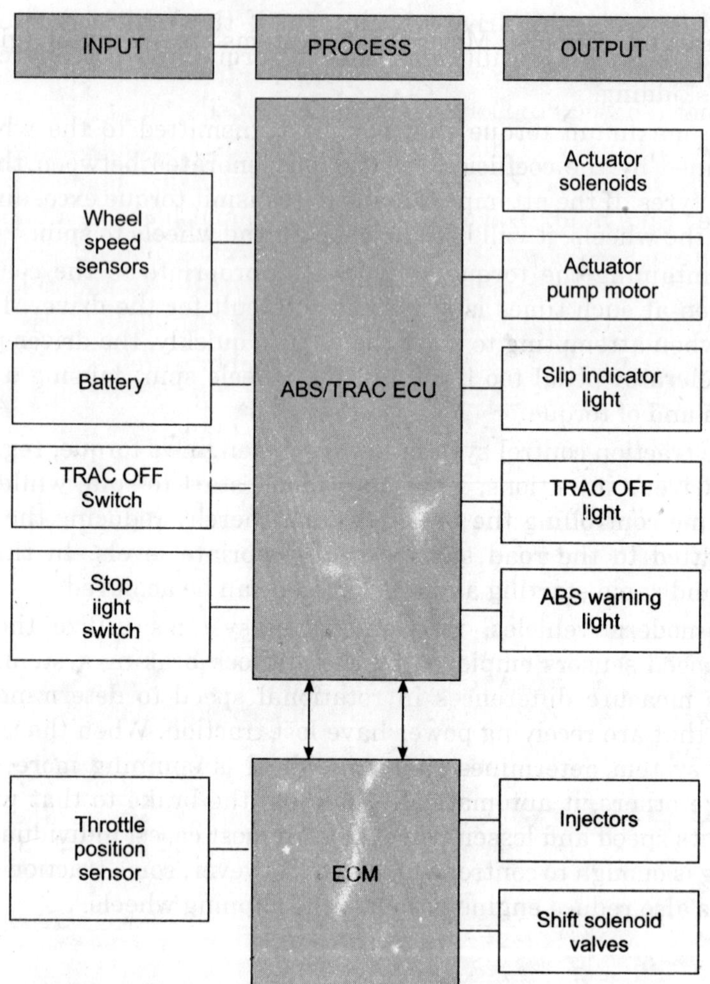

Fig.11.31: *Traction control system*

Traction control does not have the ability to increase traction. It just attempts to prevent a vehicle's wheels from spinning. For drivers who routinely drive in snowy and icy conditions, traction control, antilock brakes, and snow tires are must-have safety features.

Traction control is basically an anti-lock braking system in reverse. Anti-lock brakes prevent the wheels from locking up when the car is decelerating. Traction control keeps the wheels from locking up or skidding when the car is accelerating.

BRAKE FLUIDS

Brake fluid is a chemically inert hydraulic fluid used to transmit force and motion. It also lubricates the pistons in the master cylinder, wheel cylinders and calipers. Most brake systems use a glycol (glycerin-alcohol) fluid with additives to meet the requirements. Some vegetable and mineral based fluids are also recommended.

A wide variety of materials is used within a standard braking system. Several types of metal, rubber and plastics come into contact with the brake fluid. Brake fluid must be compatible with all materials in the brake system and maintain stability under varying conditions, both in temperature and pressures. Because of these conditions, brake fluid must possess the following six characteristics.

1. Viscosity-must be free-flowing at all temperatures.
2. High boiling point-must remain liquid at high operating temperatures without vapourisation.
3. Non-corrosive- must not attack metal, plastic or rubber parts.
4. Water tolerance-must be able to absorb and retain moisture that collects in the system. This characteristic is called hygroscopic. Water causes pitting in the brake system.
5. Lubricating ability-must lubricate pistons and cups to reduce wear and internal friction.
6. Low freezing point-must meet a certain freezing point as established by safety standards.

BRAKE BLEEDING

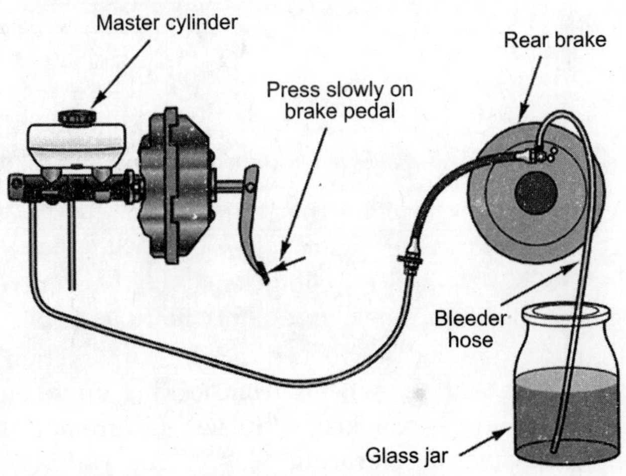

Fig.11.32: *Brake bleeding*

Most brakes should be bled every two or three years to keep your brake system at optimum performance. Small amounts of air can become trapped in the brake line, creating a spongy feeling when you step on the pedal. If large amounts of air enter the brake line, your vehicle can suffer a complete braking failure. Air can enter into your brake system during certain types of servicing or if there is a leak. The main steps in the operation are:

- Ensure that the reservoir is full.
- Attach one end of rubber tube to bleeder valve and immerse other end in fluid contained in jar.
- Open bleeder valve and slowly pump pedal until there are no more bubbles appear. Close bleeder valve as pedal is being depressed.
- Repeat at all wheel cylinders.
- Top up reservoir.

Wheel and Tyre

Vehicles ride on pneumatic tyres filled with pressurised air. Tyres are the only vehicle components that come into direct contact with the road surface. However, it goes without saying that tyres cannot be used independently on a vehicle. They must be mounted on wheels in order to be used. The wheel and tyre correspond to the human foot and shoe. They roll along the road while supporting vehicle weight. The tyre serves to absorb the road shocks and prevent them from entering into body.

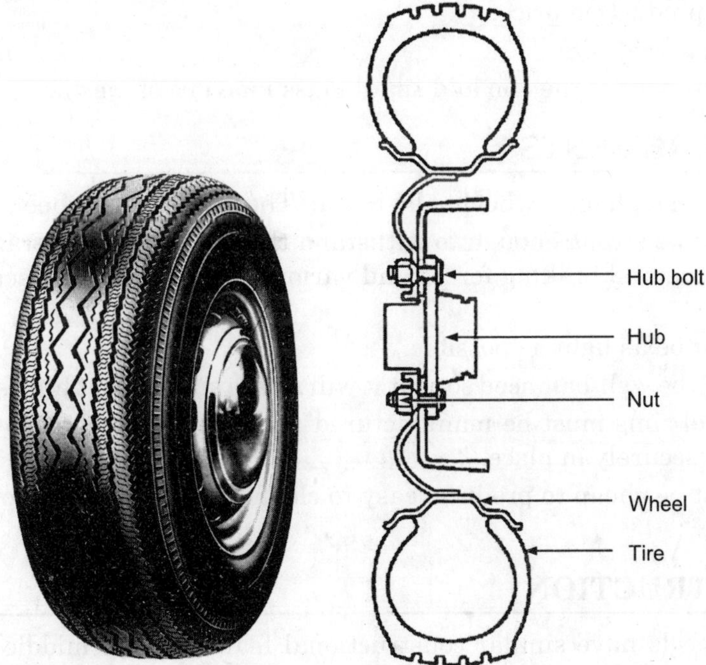

Fig.12.1: *Wheel and tyre*

WHEEL

Tyres are, of course, not installed directly onto the automobile but are mounted on wheels. Vehicle wheels have developed from wooden spoked wheels via cast wheels to the sheet metal disc wheel of today. This is the most commonly used wheel in automobiles at the present time. The wheel must be able to resist and transmit all forces which act between the road and the vehicle. The following essential demands are made on the vehicle:

- adequate rim stability.
- firm fit of the tyre on the rim.
- firm and secure connection with the wheel hub.
- good dissipation of frictional heat.
- adequate space for accommodating the brake system.

The following travelling comfort is demanded:

- vertical and lateral impact must be as small as possible.
- unbalance at circumference must be kept low.
- attractive design.
- simple fitting of tyres to the rim and of wheel to the hub.

Production should be based on the following:

- low production price.
- long service life.
- low weight of the rim and small mass moment of inertia.

REQUIREMENTS

It is not enough for a wheel to be round. The automotive wheel:

1. must be strong enough to withstand the vertical and lateral loads, driving and braking forces and various other forces imposed upon it.
2. must be as light as possible.
3. must be well-balanced so that it will rotate smoothly at high speeds.
4. wheel rims must be manufactured accurately enough to keep the tyre securely in place.
5. must be cheap to produce, easy to clean and simple to remove and refit.

CONSTRUCTION

Most wheels have similar constructional features. The middle of the wheel rim is recessed and is called the well base. This well enables the removal and refitting of a tyre. The outer edge of the wheel rim is called the rim flange and this is normally tapered. The taper ensures that, as the tyre is inflated, the tyre bead is forced into the taper, creating an air-tight seal. The rim shown in Fig.12.1 is also humped, to prevent the tyre from readily dislodging if it is deflated.

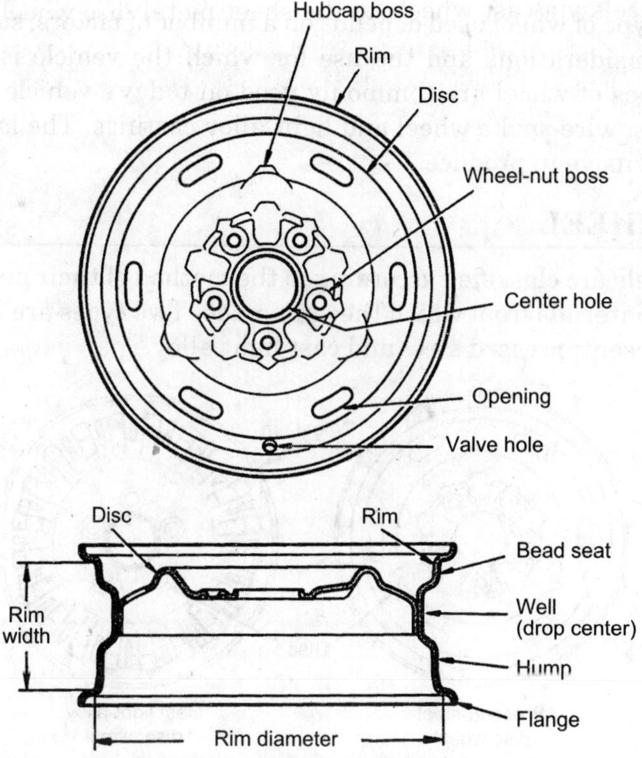

Fig.12.2: *Construction of a wheel*

The wheel (disc) is secured to the threaded studs (hub bolts) on the axle hub by the wheel nuts. The wheel-nut bosses and the ends of the wheel nuts are tapered so that the wheel will automatically center itself on the axle hub during assembly.

TYPES OF WHEELS

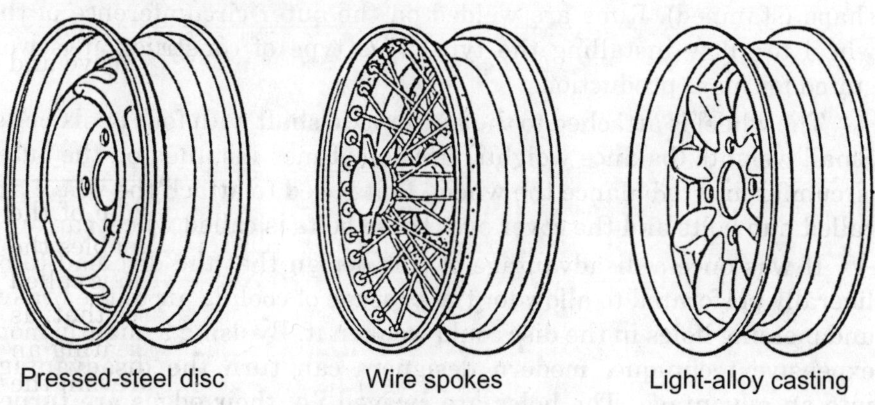

Fig.12.3: *Types of wheels*

The type of wheel used depends on a number of factors, such as cost, design considerations and the use for which the vehicle is intended. Three types of wheel are commonly used on today's vehicles: pressed-steel discs, wire-spoke wheel and light alloy castings. The last two are more expensive to produce.

DISC WHEEL

Disc wheels are classified according to the method of their manufacture and the materials from which they are made. Two types are in common use at present: pressed steel and cast light alloy.

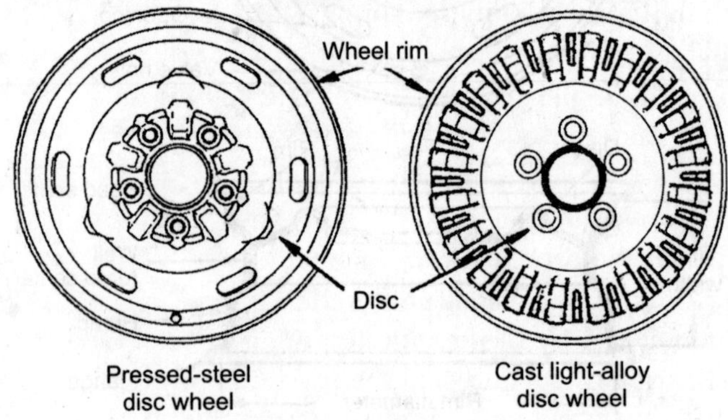

Fig.12.4: *Disc wheels*

Pressed-steel disc wheel

Most automobiles today are fitted with pressed-steel disc wheels. They are light, strong, stiff and resistant to accidental damage. They are also easy to produce in large numbers at low cost.

The disc wheel made of a single steel plate pressed into suitable shape (stamped). Rims are welded on the outer circumference of the wheel to allow installing the tyre. This type of construction is well suited for mass production.

The wheel is attached to the hub or axle shaft with four or six bolts. Small weights (balance weights) are sometimes installed on the outer circumference to balance the wheel. Bolts used to attach the wheel are called hub bolts and the cover over these bolts is called wheel cap.

It was once a disadvantage of this design that the disc had to be liberally perforated to allow for the passage of cooling air to the brakes and piercing holes in the disc could weaken it. By using a slightly more expensive technique, modern designers can turn the disadvantage into an advantage. The holes are swaged i.e. their edges are turned smoothly inwards and this may actually increase the strength of the wheel. Swaging of this kind is now a standard technique on production cars.

Cast light-alloy wheel

During recent years, the lightness of the wire wheel and the ease of cleaning of the pressed steel wheel have been combined in a wheel made of light alloy. Although much more costly, the appeal of attractive styles of aluminium and magnesium alloy wheels has made this type popular for use on the high performance type car. A light alloy wheel is cast to shape and then machined to give the final finish. At present the number of production cars fitted with these wheels as standard is relatively small, but it is increasing.

Because of their light weight compared with steel, aluminium and magnesium alloys permit the use of thicker sections, which promise stiffness and distribute stresses over a wide area.

The wider rim possible with light-alloy wheels is the main reason for their use on some sports cars. They allow wider tyres to be fitted, so improving road-holding, especially on corners.

Light alloys are also good conductors of heat and therefore disperse the heat generated by brakes and tyres more quickly than steel does. Some precautions are necessary with light-alloy wheels. They react badly to salt spray and must be checked regularly for signs of corrosion.

There is also a danger of electrolytic corrosion, which may occur when steel comes into contact with a light alloy. To prevent this, the holes through which retaining or locating studs pass should be greased and if balance weights are used, avoid those which are held in place by steel clips or screws.

WIRE-SPOKE WHEEL

The earliest type of wheel still in production is the wire-spoke variety, a light but very strong wheel now used mostly for motorbikes and sports and racing cars. All loads on the wheel are transmitted from the rim to the hub by spokes made of steel, which are much stronger in tension than in compression.

Spokes individually have little resistance to bending stresses, so they have to be laced in a complex pattern, criss-crossing in three planes. This ensures that all the complex loads fed into a wheel are resolved into tensile loads (that is, pulling rather than pressing or bending loads), evenly distributed among an adequate number of spokes.

Assembling a wire-spoke wheel is a skillful operation. Each spoke is hooked at one end into the hub and its other end is pushed through a hole in the wheel rim, where a tapered nut (the nipple) is screwed down to pull the spoke tight. If any spokes are too loose or too tight, the relatively flimsy rim will distort.

Advantages

- Strong and light.
- Centre lock mounting. Single large nut screwed on to the hub to retain the wheel. It is therefore very easy to change the wheel when required, because only a single nut has to be removed.
- Ensures good circulation of air to the brake system.

Disadvantages

- Wire wheels are expensive because of their complicated construction.
- Difficult to clean.
- The pierced rim of a wire wheel makes it impossible to fit tubeless tyres, as they need airtight rims.

SPOKES ARRANGEMENT

The car wheel is subjected to extreme loads and stresses even in normal road use. It has to withstand the car's weight and forces of acceleration, braking and cornering and it often has to cope with several of these stresses at once. Cornering, for instance, can be combined with acceleration or braking, which subjects the wheel to different stresses at the same time.

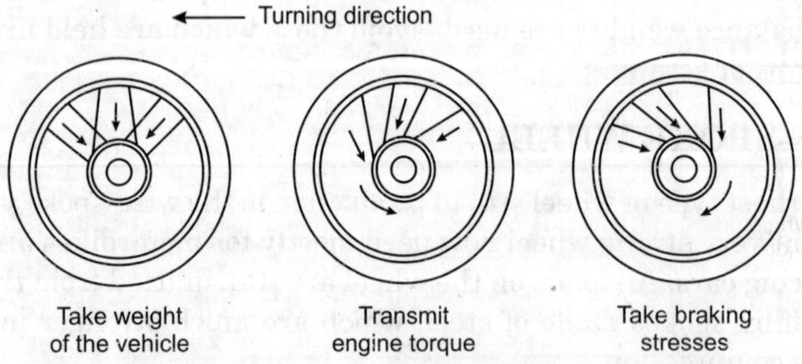

Fig.12.5: *Stresses from different directions*

For steering control the wheels must be of rigid construction. There is no difficulty in obtaining the necessary stiffness with a cast light-alloy wheel, because the short spoke portion can be cast in the form of substantial radial ribs.

With a pressed-steel wheel, the spoke portion is usually of near-conical shape for extra lateral (side-to-side) stiffness.

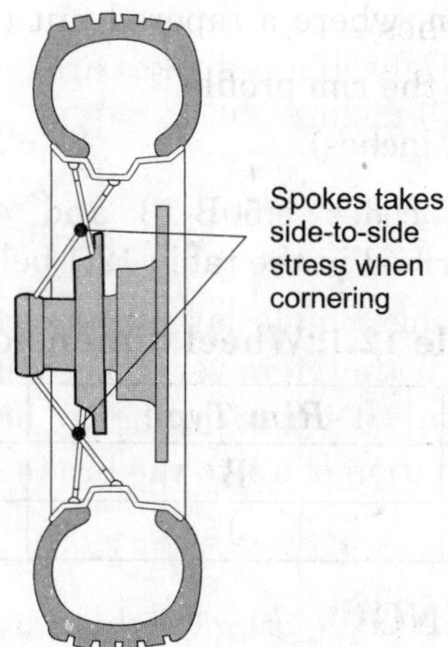

Fig.12.6: *Section showing traingular arrangement of spokes*

In the traditional wire wheel, where the spokes are the only connection between the hub and the rim, this essential side-to-side stiffness is achieved by using a relatively wide hub with two or three sets of spokes arranged at different angles. The spokes are arranged in pairs, inclined so that, with the hub, they form a series of rigid triangles for resisting side-to-side forces when the car is cornering.

Backward and forward weight transfer, when the car is braking or accelerating, is absorbed by the steel spokes acting alternatively in tension. Properly designed disc or alloy wheels cope readily with these forces.

WHEEL DIMENSIONS

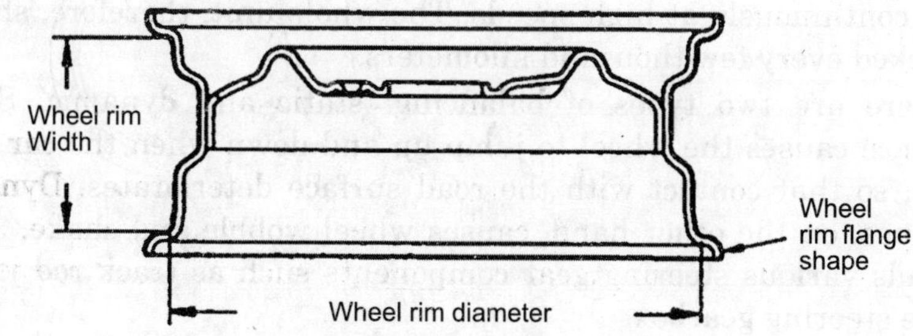

Fig.12.7: *Wheel dimensions*

$$\frac{4\frac{1}{2}}{①} - \frac{J}{②} \times \frac{13}{③}$$

The wheel is denoted by a code number which contains the following in sequence.

1. Width of rim (in inches).
2. A letter identifying the rim profile.
3. Diameter of rim (in inches).

For example, the codes 5.50B-13 and 4½ J-14 respectively represents wheels described in the table 12.1 below:

Table 12.1: Wheel dimensions

Width (W)	*Rim Type*	*Diameter (D)*
5.50	B	13"
4½	J	14"

WHEEL BALANCING

Due to improvements in engine performance, handling and braking performance it is becoming possible for automobiles to operate at higher and higher speeds every year. At high speeds, an unbalanced wheel assembly (wheel plus tyre) can create vibrations which are transmitted to the body through suspension components, causing annoyance to the driver and passengers. A wheel that is not properly balanced may affect steering control even at moderate speeds. Therefore, it is necessary to balance the wheel assemblies properly in order to eliminate such vibrations and this operation is what is called wheel balancing. Wheel balancing involves balancing the weight of the entire wheel assembly i.e. the wheel with the tyre attached.

Wheels, tyres and tubes are usually checked for balance within certain limits before they leave the factory. But the balance of a wheel can be upset by damage and that of a tyre by wear, especially on a car driven continuously at high speeds. The whole unit, therefore, should be checked every few thousand kilometers.

There are two types of balancing: static and dynamic. Static imbalance causes the wheel to jump up and down when the car is in motion, so that contact with the road surface deteriorates. Dynamic imbalance, on the other hand, causes wheel wobble and shake. This overloads various steering gear components such as track rod joints, and the steering gearbox.

STATIC BALANCING

Static balancing can be done, as its name implies, with the wheel almost stationary. Small lead weights are clipped to the wheel rim to counter balance any local heaviness of rim and tyre at the opposite point on the wheel.

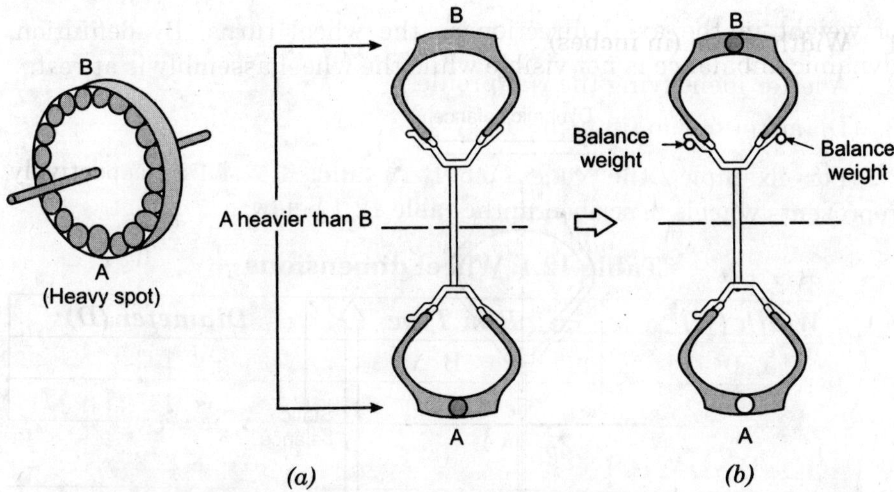

Fig.12.8: *(a) Statically unbalanced wheel and (b) Counter balancing local heaviness (statically balanced)*

From the Fig.12.12a, the portion A at the bottom is clearly heavier than its counterpart on the opposite side of the wheel B. If the weight is not uniformly distributed around the axis as shown, the wheel is said to be statically unbalanced.

Attaching a weight W_2 equal to A's extra weight W_1 to B, which is 180° opposite A (Fig.12.13) and the same distance from the center, will eliminate the vibration problem, because W_2 will counter balance W_1. Since it is impossible to attach weight to the tyre tread itself, two counter balance weights (Fig.12.12b) of equal size are attached to the inside and outside of the disc wheel rim, opposite part A.

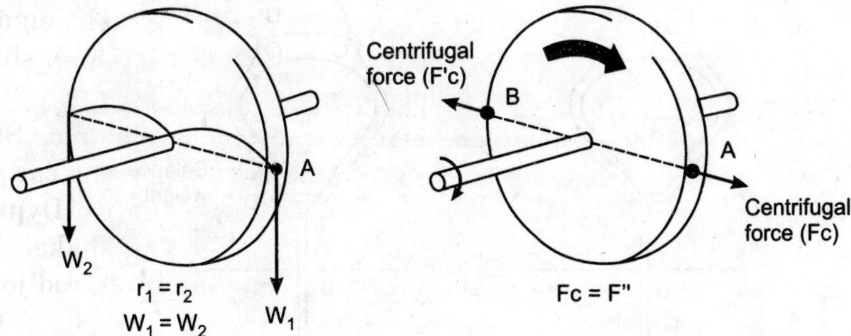

Fig.12.9: *Statically balanced wheel*

DYNAMIC BALANCING

Whereas static balance refers to the balance of weight in the radial direction under static conditions, dynamic balance refers to the balance of weight in the axial direction as the wheel turns. By definition, dynamic unbalance is not visible while the wheel assembly is at rest.

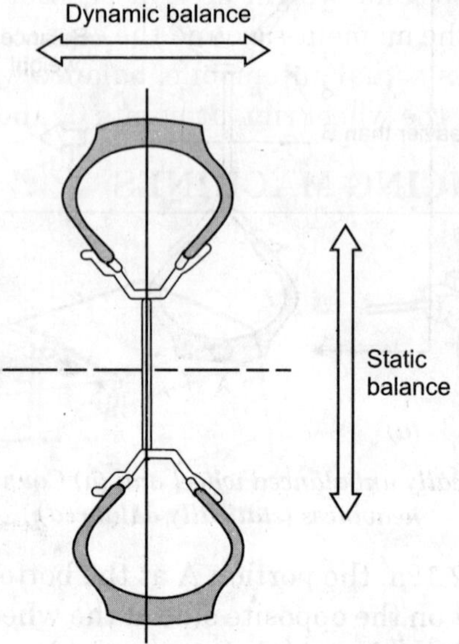

Fig.12.10: *Static and dynamic balance*

A wheel that is in static balance can still need dynamic balancing. Dynamic balancing of a wheel is done when it is rotating. The effect of a wheel being dynamically out of balance is as if it had a concentration of weight at one point on the edge of its rim so that, as the wheel spins, it wobbles rapidly from side to side. The wider the rim the greater the chances of this imbalance being troublesome.

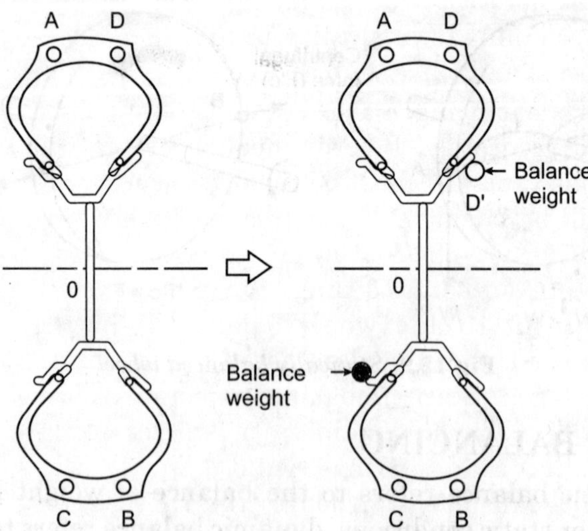

Fig.12.11: *Dynamic balance*

Improper dynamic balance is corrected by attaching two weights to the wheel assembly, one with the same weight as A at position C and another with the same weight as B at position D. Attaching these weights will cancel the moments around the center G_0, thus eliminating the vibrations. On an actual automobile, balance weights of the proper size are attached on the wheel rim, at points C' and D'.

WHEEL-BALANCING MACHINES

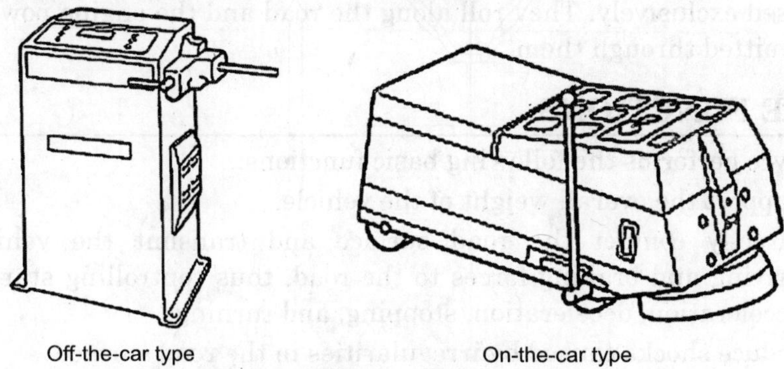

Fig.12.12: *Wheel-balancing machines*

Each wheel has to be statically and dynamically balanced. Both static and dynamic balance can be checked and eliminated on wheel balancing machines which most large garages have.

These machines are normally set up permanently at a suitable point in the workshop and used to balance complete wheels and tyres which have been removed from the vehicle. If wheel balancing on the vehicle is called for, a mobile balancing machine must be used. It is capable of determining total imbalance, that is to say imbalance in the tyre, wheel, brake and wheel hub as a complete assembly.

On modern balancing machines, imbalance down to grams can be determined accurately. Before wheels removed from a vehicle for balancing, stones and dirt must be removed from the tyre tread or else the results will be falsified. The old balance weights, if any, must be removed and the tyre checked to ensure that it does not run out of true. Vertical run out not exceeding 0.8 mm can usually be tolerated without affecting tyre and wheel balance. If more severe vertical run out is detected, it can be eliminated by turning the tyre while in position on the rim. Static imbalance is established by allowing the wheel to swing. If it will remain stationary in any position, it is in static balance. If it always swings round to the same position, suitable balance weights must be attached to the rim. The mobile balancing machine usually performs the static and dynamic balancing work in a single operation and displays a combined value. The machine indicates the point of imbalance as well, so that the required balance weights can be attached correctly.

TYRE

The tyres of the vehicle are intended to moderate the effects of uneven road surfaces, to improve the driving qualities and to make high speeds possible by low ground friction. Today pneumatic types are used exclusively. They roll along the road and the engine power is transmitted through them.

TYRE FUNCTIONS

The tyre performs the following basic functions:

- support the overall weight of the vehicle.
- directly contact the road surface and transmit the vehicle's driving and braking forces to the road, thus controlling starting, acceleration, deceleration, stopping, and turning.
- reduce shocks caused by irregularities in the road surface.

TYRE CHARACTERISTICS

Tyres have four characteristics that are important in understanding their design. These include:

Tyre traction

Traction is defined as a tyre's ability to grip the road and to move or stop the vehicle.

Ride and handling

Tyres are measured by their ride and handling ability. This is an indication of the degree of comfort a tyre delivers to vehicle passengers. It is also a measure of the responsiveness to the driver's steering actions.

Rolling resistance

Rolling resistance is a term used to describe the weights of force required to overcome the resistance of a tyre to rotation. As rolling resistance of a tyre decreases, fuel mileage typically increase.

Noise

All tyres make a certain amount of noise. Tyres can be made 'quiet' by scrambling or changing the size, length, and shape of the tread elements. Scrambling prevents the sound frequency buildup that would develop if all of the patterns of the tread were spaced evenly.

TYRE CONSTRUCTION

The following illustrates the basic construction of a tyre.

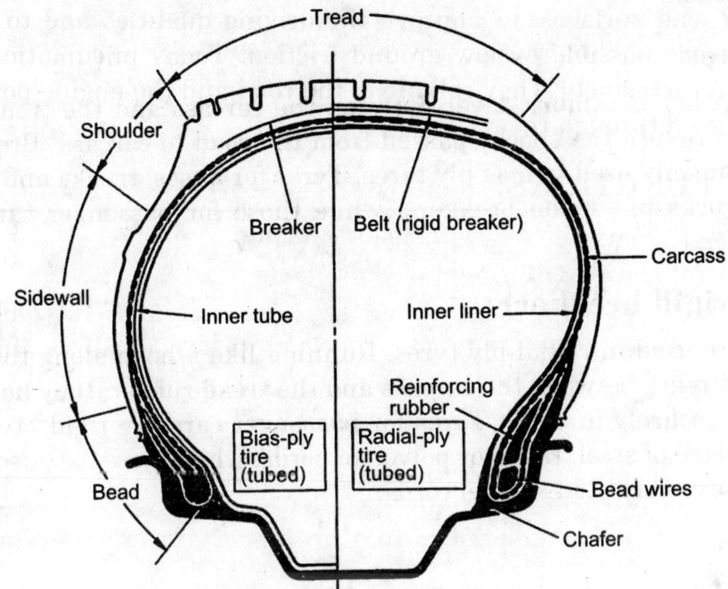

Fig.12.13: *Construction of a tyre*

Carcass (casing)

The carcass is the framework of the tyre. It must be rigid enough to hold the high pressure air and flexible enough to absorb load changes and impact. It consists of plies (layers) of tyre cords (sheets of parallel strands of a strong material) bonded together with rubber. The cords in bus and truck tyres are generally made of nylon or steel, while passenger car tyres use polyester or nylon. Tyres are generally classified, according to the direction of the cords, into radial-ply and bias-ply types.

Tread

The tread is an external rubber layer protecting the carcass from wear and external damage caused by the road surface. It is the part that directly contacts the road and transmits the vehicles driving and braking forces to the road. The tread pattern consists of grooves moulded into the tread surface, and is designed to help the tyre more effectively transmit these forces to the road.

Side wall

The side walls are rubber layer that cover the sides of the tyre and protect the carcass from external damage. The largest, most flexible tyre constituents, they continuously flex under the loads applied during travel. They bear the manufacturer's name, tyre size, and other information.

Breaker

The breaker is a fabric layer between the carcass and the tread and helps to reduce the shocks passed from the road to carcass. Breakers are commonly used in bias ply tyres. Tyres for buses, trucks and light-duty trucks use nylon breakers, while those for passenger cars use polyester.

Belt (rigid breaker)

Belts are used in radial-ply tyres. Running like a hoop along the tyre circumference between the carcass and the tread rubber, they hold the carcass securely in place. Tyres for passenger cars use rigid breakers constructed of steel, rayon or polyester cords, while those for buses and trucks are fabricated of steel cords.

Bead

To keep the tyre from being thrown off the rim by the various forces acting on it, the free edges or sides of the layers are wound around strong steel wires called bead wires. The pressurised air inside the tyre forces the beads out against the rims of the wheel and holds them securely in place. The beads are protected from damage caused by chafing against the rim by hard strips of rubber called chafer strips.

TYRE TYPES

There are various methods used to classify tyres, as can be seen below.

Classification by construction

- Classification by the methods of affixing the ply cords which make up the carcass of the tyre
  - Bias-ply tyre (cross-ply tyre)
  - Radial-ply tyre
  - Belted-bias tyre
- Classification by the method used to hold air
  - Tubed tyre

- Tubeless tyre

Classification by air pressure

- High-pressure tyre
- Low-pressure tyre
- Extra low-pressure tyre

Classification by use

- Classification by the type of vehicle on which the tyre is meant to be used
  - Passenger car tyres
  - Light truck tyres
  - Trucks and bus tyres
- Classification by the road surface on which the tyre is meant to be used
  - Snow tyre
  - Spiked tyre
  - All-weather tyre
  - Sand tyre
- Others
  - Compact spare tyre
  - Low-profile tyre

Here we will describe the functions and features of tyres with emphasis on the classification of the tyres by the construction and air pressure.

BIAS-PLY AND RADIAL-PLY TYRES

Tyres are classified, according to the construction of the carcass, into bias-ply and radial-ply tyres.

BIAS-PLY TYRES

The bias-ply tyre carcass is made up of alternating layers of ply cords bonded together, and lying at an angle of 30° - 40° relative to the tyre's circumferential centerline. This arrangement supports loads applied both along the tyre circumference and across the diameter.

However, when the tyre is subjected to a vertical load from the road surface, the ply cords tend to 'squirm' or deform as shown in the illustration 12.18b below.

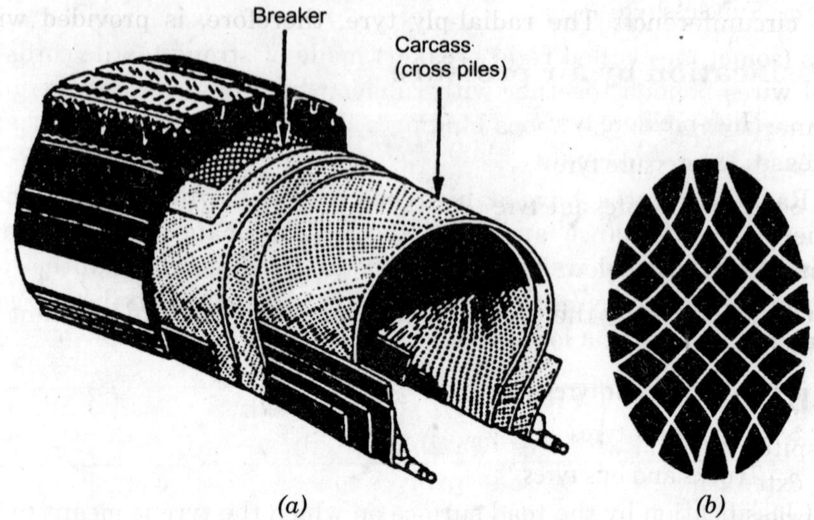

Fig.12.14: *(a) Sectional bias-ply tyre and (b) Bias-ply tyre squirm*

Bias-ply tyre tread tends to 'squirm' as it meets the road, while radial-ply tyre tread tends to remain apart. Bias-ply tyres give a softer ride, but are inferior to the radial-ply tyres in terms of cornering performance and wear resistance.

RADIAL-PLY TYRES

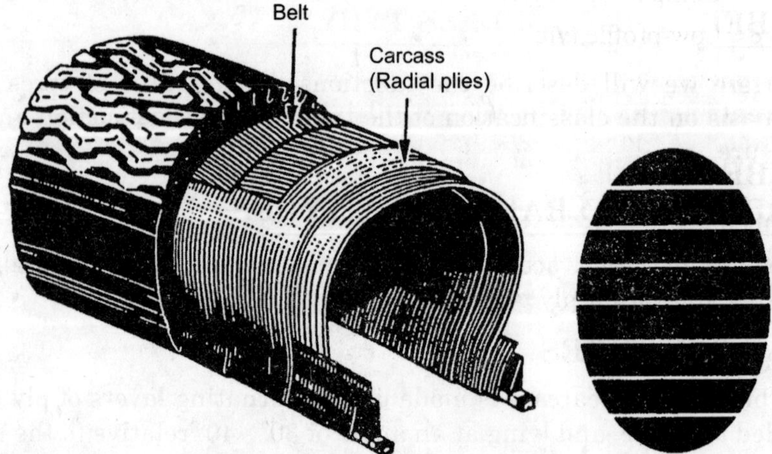

Fig.12.15: *Sectional radial-ply tyre*

The radial-ply tyre carcass consists of layers of cords bonded together with rubber perpendicular to the tyre circumference. This construction gives the tyre great flexibility in the radial direction. Nevertheless, such a carcass alone cannot withstand the full load applied along the tyre circumference. The radial-ply tyre, therefore, is provided with belts (sometimes called rigid breaker) made of strong textile cords or steel wires bonded together with rubber which bind the carcass in a manner similar to the hoops binding a barrel. This arrangement also increases the tread rigidity.

Radial-ply tyres have a high belt rigidity, which means their cornering performance and high speed performance are good and rolling resistance is low. Tyre wear resistance is high also, but they are inferior to bias-ply tyres in terms of riding comfort when the car goes over small bumps at a low speed.

BELTED-BIAS TYRES

In spite of the advantages of radial-ply, it has not been adopted to any extent, mainly because it involves considerable outlay to re-equip factories. Manufacturers compromise by using their existing equipment to make cross-ply casings and moulding semi-rigid breakers under the tread, to produce a belted-bias design that incorporates some features of both the cross-ply and the radial-ply tyres.

In addition, this form of construction places the shoulders of the tyre right at the edges of the breaker belt, which is an ideal position for independent flexing of the tread and the walls. The casing is usually made of polyester cord and the breaker of glass fibre.

TUBED AND TUBELESS TYRES

The pneumatic tyres used in automobiles are of two types: the conventional tubed type and the tubeless type.

TUBED TYRE

A tubed tyre has a separate rubber inner tube to retain pressurised air inside the tyre. The tyre valve or air valve, which protrudes through a hole in the wheel rim, is bonded to the inner tube. If punctured, a tubed tyre deflates quickly.

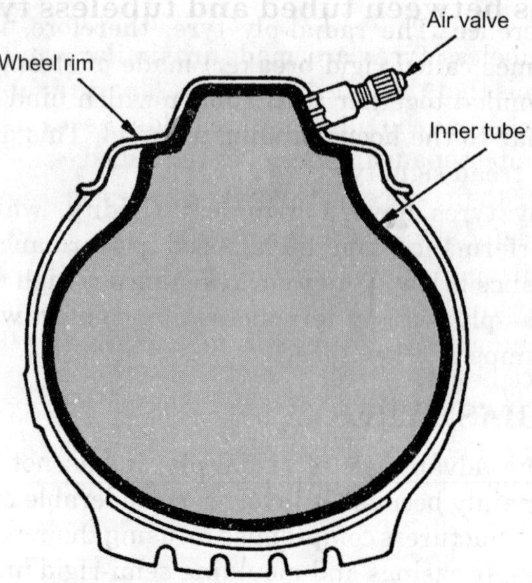

Fig.12.16: *Tubed tyre*

TUBELESS TYRE

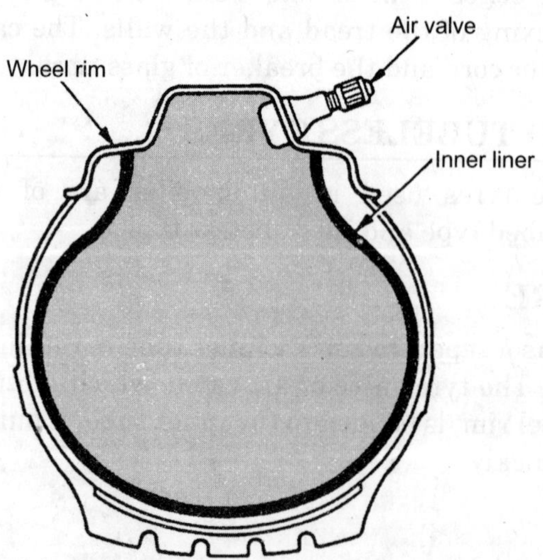

Fig.12.17: Tubeless tyre

The tubeless tyre, as the name indicates, has no inner tube. The pressurized air is held in by an inner liner, a specially formulated thick rubber lining with high sealing effectiveness. Since the tubeless tyre does away with the inner tube, the air valve is secured directly to the wheel rim.

Differences between tubed and tubeless tyres

Tubed and tubeless tyres are made in similar ways. The difference between them is that the tubed tyre has a separate rubber inner tube to hold the air, whereas the tubeless tyre holds the air itself. The tubeless tyre has a number of advantages over the tubed:

- It is easier to fit, since no tube has to be inserted.
- Lower weight, since there is no tube.
- Greater driving safety, since a sudden tyre burst can be almost completely ruled out. When punctured it deflates slowly, because the soft lining has a self-sealing effect. Slow loss of pressure can be soon noticed by the driver.
- Reduced heat build-up, since there is no friction between outer cover and inner tube. Furthermore, the air inside the tyre keeps cooler because it is in direct contact with the wheel rim.
- A temporary puncture repair can be made without removing the wheel, by stopping the hole with a special rubber plug.
- Tubed tyres are still necessary for wire wheels, because of the difficulty of making an air seal at the spoke ends.

TYRE MATERIALS

For many years, tyres were made entirely of natural rubber, but now-a-days most tyre manufacturers have given it up almost completely, in favour of a variety of synthetic rubbers. These, used in various blends, produce a wide variety of tyres for different cars, conditions and purposes. Natural rubber is used in some blends and also remains the best material for impregnating and coating the carcass fabrics.

The synthetic most widely used in tyres is styrene butadiene rubber (SBR). It has much less bounce than natural rubber, which means a softer ride, and when used in tyre treads SBR maintains closer contact with the road, providing a good grip, especially in wet weather. It also has excellent resistance to abrasion. Another synthetic used in tyres is poly butadiene (PB). This is hard-wearing and less sensitive to temperature than other synthetics, but too much of it in a blend makes tyres scream in the dry and slip in the wet. PB is usually mixed in small quantities with SBR, with natural rubber or with both.

Other synthetics are used in small quantities. But whatever the blend, it must have other additives, including oil, carbon black and sulphur. Oil improves road holding at the expense of rapid tyre wear. Carbon black adds abrasion resistance, and sulphur acts as a vulcanizing agent.

The number of plies and the relative angle between the cord and the tyre bead governs the tyre characteristics. In the past cotton was the main cord material, but modern tyres use manmade fibres such as rayon and nylon. These materials are much stronger and offer greater resistance to the heat built up by the flexing of the tyre. Heavy vehicles and some cars require an even stronger material, so steel cords are needed in these cases.

The bead consists of a number of hoops of steel wire which are responsible for retaining the tyre on the rim.

TYRE SPECIFICATION CODING SYSTEM

The tyre sidewall usually bears the codes representing the tyre width, the inside diameter (rim diameter) and the ply rating. A high-speed tyre usually bears an additional code such as H, S, etc., a radial ply tyre, an R. Some coding systems include the aspect ratio.

Ply rating

This indicates the number of plies that are used in the construction of the tyre. The more plies that are used, the stronger the tyre will be. Due to improvements in tyre construction, the actual number of plies used has steadily reduced. For example, a tyre marked '6 ply/4 ply' has the same load carrying capacity as a 6 ply tyre, but is constructed with only 4 plies. Because there is no common standard for the definition of ply rating, a more reliable indicator of a tyre's strength is the load index.

Aspect ratio

This term describes the relationship between a tyre's cross-sectional width and its sectional height. This is usually expressed as a percentage.

Early pneumatic tyres had aspect ratios of 100 percent, which meant that the width of the tyre was approximately the same as its sectional height. Now that vehicles have become faster, many are fitted with tyres that have a low aspect ratio. This means that the tyre's sectional height is smaller than its cross-sectional width. A tyre which is 5 inches wide and 4 inches from tread to rim has an aspect ratio of 80 percent. Some of the advantages of using tyres with a low aspect ratio include the following:

- More of the tyre is in contact with the road surface, which improves the vehicle's road-holding and steering characteristics.
- Because of the tyre's smaller sectional height, the vehicle's centre of gravity is lower.

They do have the disadvantage of a more harsh ride as the sectional height is reduced.

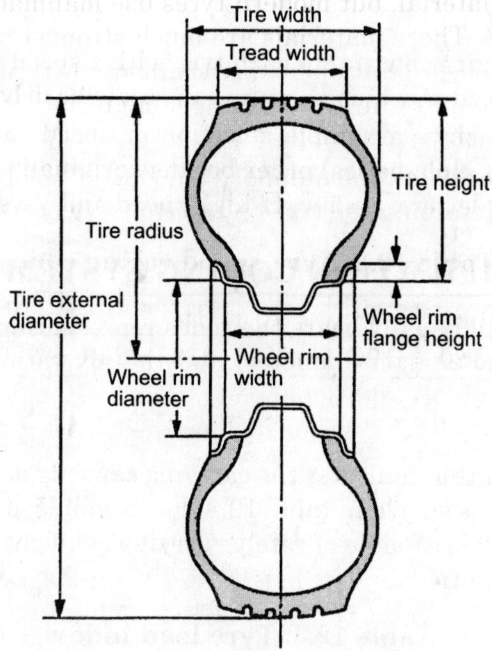

Fig.12.18: *Tyre specification coding system*

- **Bias-ply Tire**

6.45 (1) 14 (3) 4PR (4)

- **Radial-ply Tire**

195 (1) / 70 (5) H (2) R (6) 14 (3)

- **International Standardization Organization (ISO) Tire Coding System**

195 (1) / 70 (5) R (6) 14 (3) 86 (7) H (2)

① Tire width in inches (bias tires) or millimeters (radial tires)

② Maximum permissible speed

③ Wheel rim diameter in inches

④ Maximum load-carrying capacity expressed as a ply rating. (A 4PR tire has a strength equivalent to a tire with 4 layers of cotton cord.)

⑤ Aspect ratio (Tire Height/Tire Width) as a percentage

⑥ Radial-ply tire

⑦ Load-carrying capacity (load index)

Speed rating

Tyre manufacturers now stamp each tyre with a speed code letter. This indicates the speed at which the tyre can safely be driven.

Table 12.2 shows a sample selection of speed rating codes. The letter H is out of alphabetical order because originally there were only three speed ring letters: S-slow, H-high speed and V-very high speed.

Table 12.2: Tyre speed rating codes

Code	S	T	U	H	V	W	Z
Speed (km/h)	180	190	200	210	240	270	270 or more

Load index

This is a number that indicates the carrying capacity of a tyre. A sample load index chart is shown in table 12.3. For example, a tyre with a load index code is 102 is capable of safely carrying a weight of 850 kg under normal road conditions.

Table 12.3: Tyre load index

Code	78	82	86	90	94	98	102
Load (kg)	425	475	530	600	670	750	850

TYRE CONTACT PATCH

The tyre contact patch or tyre contact area refers to the area of the tyre that is in contact with the road surface when it is supporting the vehicle weight.

As the tyre's width increases, the area of the patch in contact with the road surface increase. This in turn increases the frictional force which can be exerted before the tyre ceases to grip when cornering fast or braking. Coulomb's law, which states that the frictional force depends only on the normal force (vertical load) and the friction rating, does not apply unrestrictedly to tyres. If an elastic substance such as tyre rubber is in contact with a rough surface (the road), the area of the friction surfaces in contact with one another also plays a major part. However, the large contact patch has disadvantages as well: high steering effort at parking speeds and poor traction in winter on account of the low specific ground contact pressure.

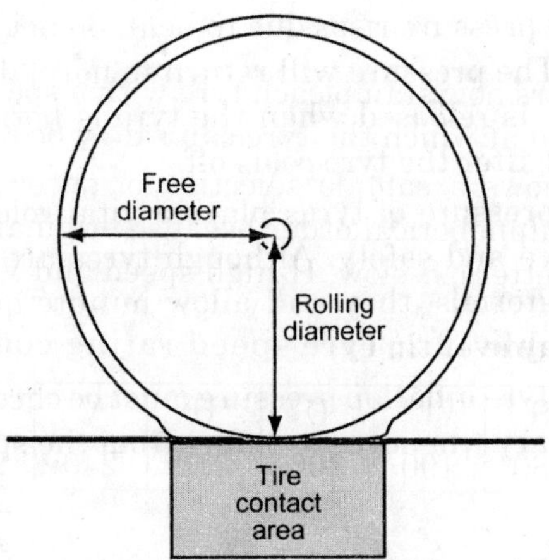

Fig.12.19: *Tyre contact patch*

HYDROPLANING (AQUAPLANING)

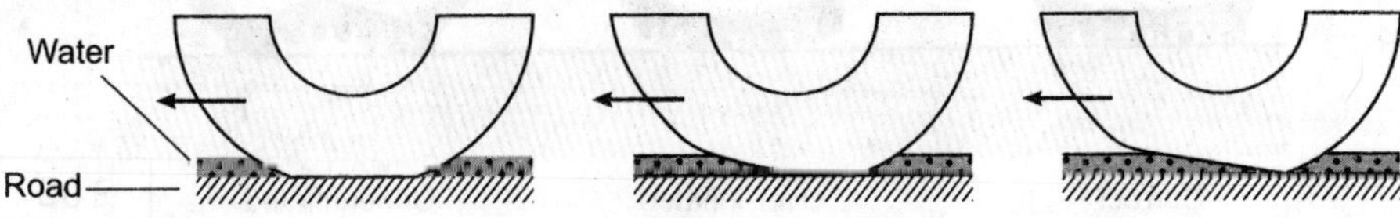

Fig.12.20: *Hydroplanning*

On a dry road the best grip is afforded by a fairly smooth tyre, as this provides the greatest possible contact patch, that part of the tyre is direct touch with the road at any moment. But on a wet road a smooth tyre has hardly any grip and the least trace of water acts as a lubricant. If there is a lot of water on the road, it will form a wedge in front of and underneath a smooth tyre, causing a form of skidding known as hydroplaning or aquaplaning. The tyre, lifted off the road and riding on the water, stops turning and steering control is lost.

TYRE INFLATION PRESSURE

This is governed by the load carried by a given tyre, although manufacturers vary the pressure to modify the steering characteristics. Large-section tyres use a lower pressure for a given load, and so greater comfort is achieved, but the resistance to rolling is increased. The pressure recommended by the manufacturer is applicable to a cold tyre, i.e. it is the pressure before the tyre is used and takes into account the average pressure rise which is caused by the temperature increase during use. Recommended pressure is normally suitable for sustained speeds up to 130 km/h, but a higher pressure is advised if speeds or loads are increased.

If the inflation pressure rises due to heat, do not release the excess air from the tyre. The pressure will return to normal once the tyre has cooled down. If air is released when the tyre is hot, the pressure will drop below normal after the tyre cools off.

The inflation pressure of tyres plays a vital role in terms of total vehicle performance and safety. Although tyres are made up of more or less airtight materials, they still allow minute quantities of air to gradually leak away over time.

Therefore, the tyre inflation pressure must be checked regularly and adjusted as necessary whenever it differs from the specified pressure.

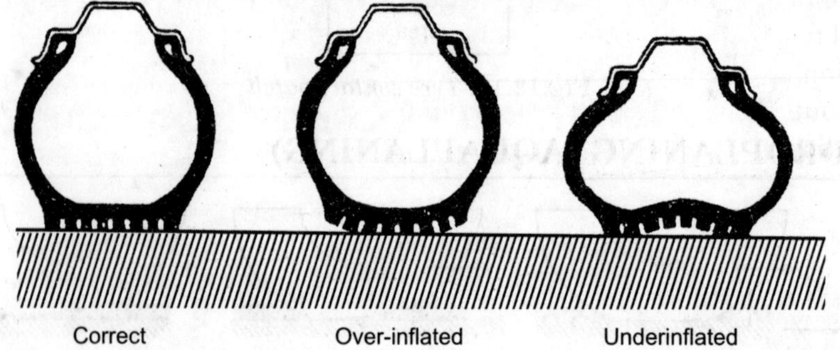

Fig.12.21: *Tyre inflation pressure: correct, over-inflated and under-inflated*

Overinflation

An inflation pressure above the specified level can lead to the following problems:

- The narrower tread width reduces driving stability and braking performance.
- Driving comfort suffers.
- The middle part of the tread wears more quickly.
- The tyre's ply cord layers are extremely tense and prone to damage due to external impacts.

Underinflation

Insufficient inflation can lead to its own special problems:

- Increased friction between the tread and the road wastes energy and consumes more fuel.
- Front tyres become harder to steer.
- The tread shoulders wear more quickly.
- When in motion, the tyre is subjected to greater flexure which sharply increases the tyre's internal temperature. When the

inflation pressure is extremely low and the vehicle is running at high speeds, the tyre may burst, which can be very dangerous.

- At high speeds, the tyre is more likely to generate a standing wave and experience hydroplanning.

TYRE ROTATION

The amount of tread wear depends upon the location of the tyre on the vehicle. For example out of the four tyres, the right rear tyre wears most rapidly, the left rear tyre is second, the right front tyre is third and the left front tyre wears least of all. Therefore to equalize tyre wear and to obtain maximum tyre life, the tyres must be rotated from one wheel to another after every 8000 kilometres service of the vehicle. The rotation of the tyres are shown in the Fig. 12.22. Thereby the wear is distributed more evenly and also changed the direction of tyre rotation periodically.

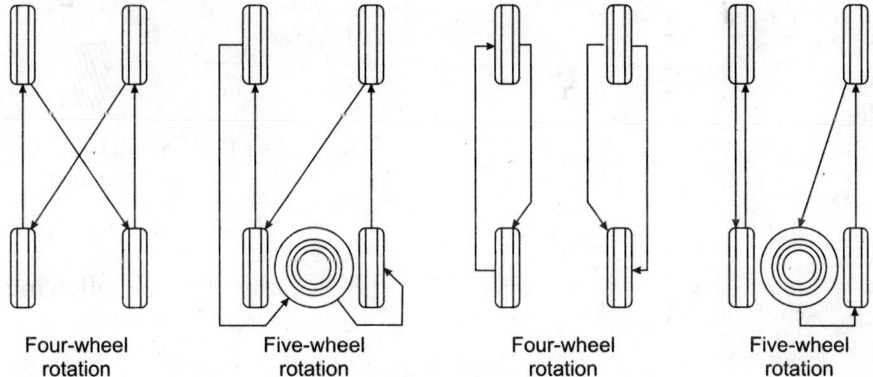

Fig.12.22: *Tyre rotation*

TYRE LIFE

Tyre life, which depends to a very large extent on operating conditions, can be as high as 50,000 km on a passenger car, or 100,000 km on a commercial vehicle. The loads to which the vehicle is subjected must not be exceeded, and the correct tyre pressures must be maintained. Excessive speeds, violent starting and stopping and fast cornering naturally reduce tyre life quite severely. Excessive wear is also caused by incorrect wheel alignment, worn or incorrect shock absorbers, incorrect tyre fitting and poorly balanced wheels and tyres.

CHAPTER 13

Vehicle Electrical System

The electrical system in the present day automobiles can be divided up into the storage battery which serves as the source of electric power, the ignition system which ignites the air-fuel mixture in the cylinders, the charging system that charges the storage battery, the starting system that cranks and starts the engine, and the lighting and signalling system.

The five main headings in the electrical system are:

1. Charging system
2. Starting system
3. Ignition system
4. Lighting system
5. Accessories

CHARGING SYSTEM

The charging system includes the battery, the alternator and the regulator. This system is responsible to keep the battery charged for the regular supply of electric current to the various electrical systems in an automobile. It is not possible for the battery (due to its limited capacity) to supply electric current for long periods. If it is not charged to save it from run down and this

Vehicle Electrical System

The electrical system in the present-day automobiles can be divided up into the storage battery which serves as the source of electric power, the ignition system which ignites the air-fuel mixture drawn in the cylinders, the charging system that charges the storage battery, the starting system that cranks and starts the engine, and the lighting and signaling system.

The five main headings in the electrical system of automobiles are:

1. Charging system
2. Starting system
3. Ignition system
4. Lighting system
5. Accessories

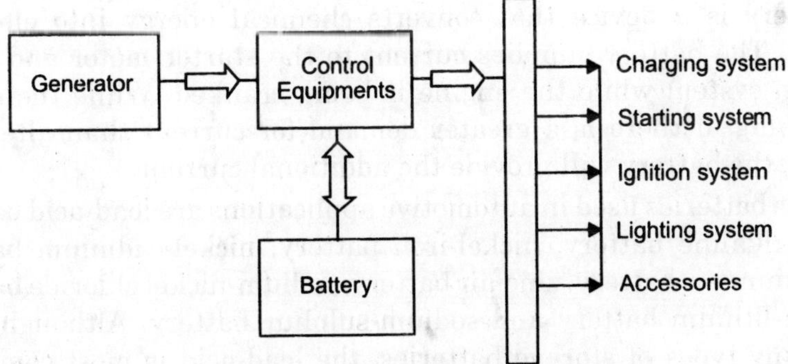

Fig.13.1: *Vehicle electrical system*

CHARGING SYSTEM

The charging system includes the battery, the alternator and the regulator. This system is responsible to keep the battery charged for the regular supply of electric current to the various electrical systems in an automobile. It is not possible for the battery (due to its limited capacity) to supply electric current for long periods. So it is to be kept charged to save it from run down and thus to fulfill the requirements of electric supply to various electrical components. This problem is solved by means of a generator which replaces the current used in the various systems of the automobile during its operation. After the engine has started, the generator charges the battery because it produces a higher voltage than the battery. The regulators control the alternator voltage to prevent overcharging and to maintain the voltage at the proper level.

REQUIREMENTS

The charging system must meet the following criteria (when the engine is running):

1. Supply the current demands made by all loads
2. Supply whatever charging current the battery demands
3. Operate at idle speed
4. Constant voltage under all conditions
5. Efficient power to weight ratio
6. Reliable, quiet, resistance to contamination
7. Low maintenance
8. Provide indication of correct operation

BATTERY

A battery is a device that converts chemical energy into electrical energy. The battery supplies current to the starter motor and to the ignition system while the engine is being cranked. While the engine is running, if there is a greater demand for current than alternator supply, the battery will provide the additional current.

The batteries used in automotive applications are lead-acid battery, nickel-alkaline battery, nickel-iron battery, nickel-cadmium battery, aluminium-air battery, zinc-air battery, sodium-nickel chloride battery, sodium-lithium battery and sodium-sulphur battery. Although there are many types of storage batteries, the lead-acid is most commonly used for automotive service because it has been found to be the most practical and economical type. This battery can be charged and discharged continuously throughout its life.

The main functions of a battery in automobile are as follows:

- It supplies a high value of current to the starter motor and a low value of current to the ignition system when the engine is being started.
- It intermittently supplies current to the electrical units (lights, radio and other electrical accessories) when the total demand

exceeds the power output of the alternator (when the alternator fails to supply the electrical load).

- It controls the voltage of the electrical system (stabilizes over voltage).

PRINCIPLE

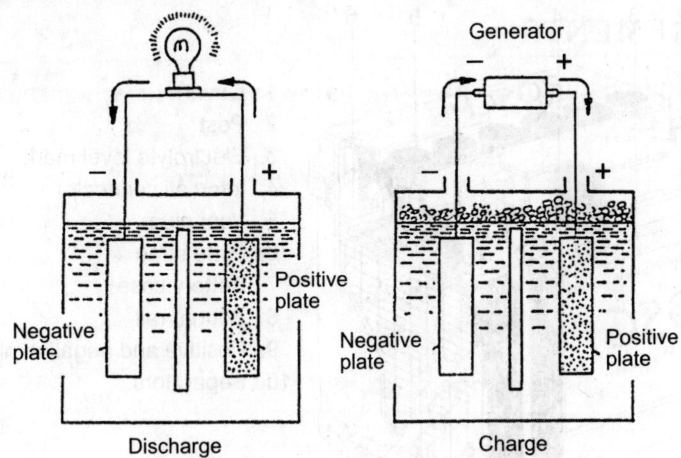

Fig.13.2: *Principle of a battery*

Electrons are moved because of the chemical reaction between the electrolyte and the two kinds of plates. The electrolyte allows electrons to move away from one kind of plate and build up on another. As a result, one kind of plate will have more than the normal number of electrons. The other plate with fewer electrons. In the plate with fewer electrons, the positive protons will predominate giving that plate a positive charge. The plate with more electrons will have a negative charge.

LEAD-ACID BATTERY

It is called lead-acid battery because its active ingredients are the metal lead and sulphuric acid. The difference in voltage between the positive plate (lead peroxide) and the negative plate (spongy lead) is 2.13 V when a sulphuric acid solution of 1.265 – 1.280 specific gravity is added, and the temperature is 68°F - 80°F. Lead peroxide and spongy lead are used in the automotive battery today because no other materials have a greater voltage difference. Other materials have been used and compared with the following results:

Lead peroxide & spongy lead	:	2.13 V
Silver - Zinc	:	1.40 V
Nickel – Cadmium	:	1.20 V

CONSTRUCTION

All Automotive storage batteries are similar in construction and operation but their difference lies in the number of cells, the arrangement of the cells, the number of plates in each cell, the size and thickness of the plates and the type of separators used.

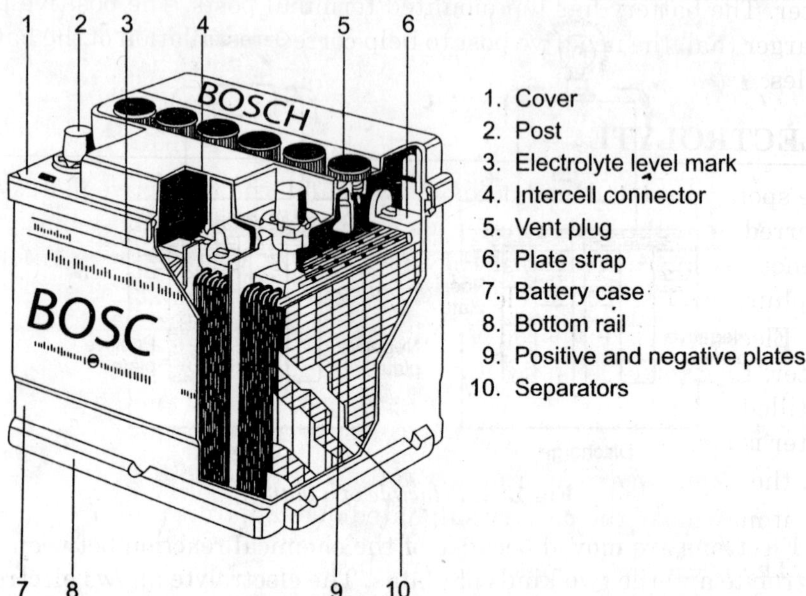

Fig.13.3: *Construction of a battery*

A battery has **cells** basically consists of plates of two different materials immersed in an electrolytic solution that reacts chemically with them to produce an electrical pressure (emf) between the plates. In the storage battery each cell includes an **element** composed of interleaved positive and negative plates, **separators** to keep the plates a part. The positive plates which are brown in colour, have lead peroxide while the negative plates which are grey in colour, have pure lead paste. Single positive and negative plates of the size used in automobile will not create enough reaction to start the engine. So that a number of plates are connected together by their lug ends to form a **plate group**. All the like plates in each cell are connected by a lead strap, soldered to a lead lug. There is always one more negative plate than the positive one in each cell as the outside plate in each cell is a negative plate. The **electrolyte** is a solution of sulphuric acid in water. The voltage depends up on the plate materials and the electrolyte. A fully charged cell will have about 2.1V. The required battery size is made by connecting cells in series. **Cell connectors** are used to connect the positive terminal of one cell to the negative terminal of the next cell. The battery case is a one-piece moulded box with open at the top, made from hard rubber or plastic material, which is shock and acid resistant and mechanically strong. Cell cover is also made of hard rubber. It has holes for terminal posts and for vent plug. **Vent plugs** are provided in the cell cover facilitate the checking of electrolyte and pouring distilled water. The battery has top mounted terminal posts. The positive post is larger than the negative post to help correct installation of the batter cables.

ELECTROLYTE

The spongy lead and lead peroxide, which fill the respective plates, are referred to as the active materials of the battery. But these materials cannot become active until they are covered by a water solution of sulphuric acid called the electrolyte.

Electrolyte is prepared by adding concentrated sulphuric acid to water. Electrolyte is both conductive and reactive. The purpose of the distilled water is to give freedom for the ions to move freely. Ordinary water is not used in an electrolyte because:

- the impurities in an ordinary water may short circuit the cells
- it may make the battery sulphated due to hardness

The electrolyte reacts chemically to combine with the active materials on the plates to form new compound and release electrons at the same time.

The amount of sulphate in the electrolyte is determined by its specific gravity. Specific gravity is the ratio of the weight of a given volume of a liquid divided by the weight of an equal volume of water. Pure water is the basis for this measurement and is given a specific gravity of 1.000 at 80°F. Pure sulphuric acid has a specific gravity of 1.835. The correct concentration of water and sulphuric acid is usually contains about *38% sulphuric acid and 62% water by weight* (fully charged condition). This corresponds to a specific gravity of about 1.280 at 80°F. The following table shows the specific gravity of batteries used in different climates.

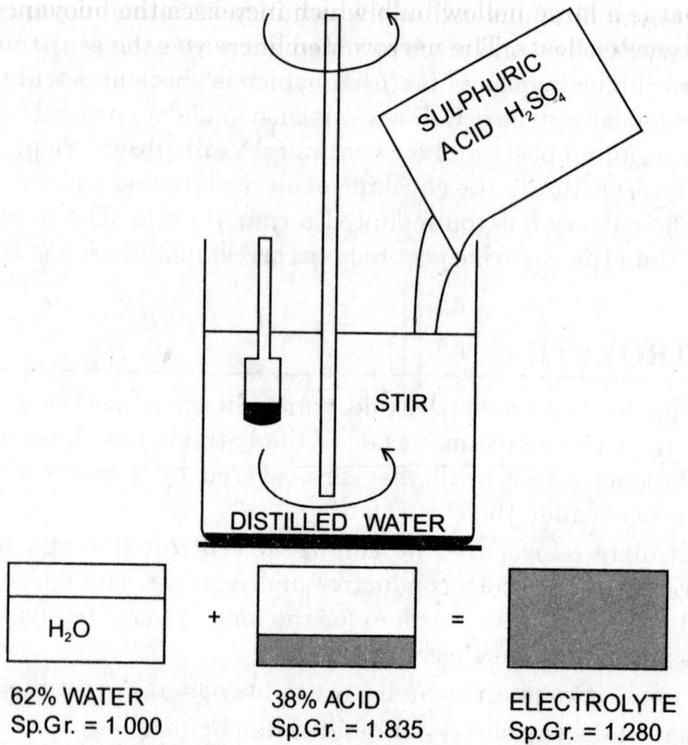

Fig.13.4: *Preparation of electrolyte*

State of charge	Specific gravity		
	In cold climate	In temperate climate	In tropical climate
Fully charged	1.280	1.260	1.225
75% charged	1.260	1.215	1.180
50% charged	1.220	1.170	1.135
25% charged	1.190	1.120	1.090
Discharged	1.110	1.070	1.045

HYDROMETER

A hydrometer is an instrument for measuring the specific gravity of liquids. It generally consists of a small cylindrical glass tube with a lead-weight at its bottom. The hydrometer floats in an upright position. The top part of the hydrometer extends above the liquid surface and the divisions on it indicate directly the specific gravity of the liquid. The glass float is a large hollow bulb which increases the buoyancy so that the hydrometer floats. The narrow stem increases the sensitivity of the hydrometer. The bottom of the hydrometer is made heavier by loading it with lead shots so that it floats vertically.

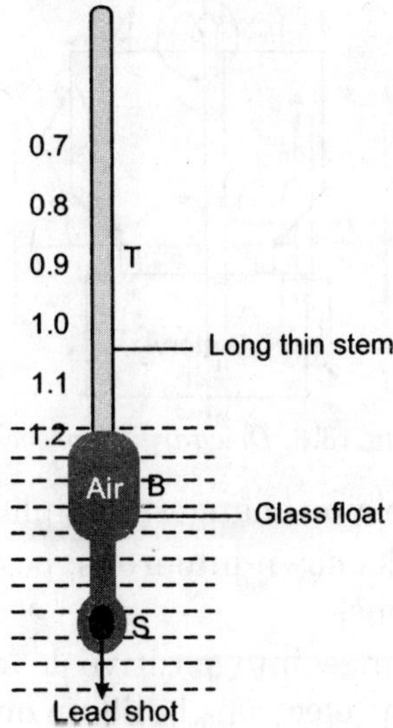

Fig.13.5: *Hydrometer*

Principle

It works on the principle of flotation. Float sinks more in a lighter liquid so as to displace more volume of the lighter liquid whose weight is equal to the weight of the float. Hence, it will sink less in a denser liquid so that it has to displace less volume of the denser liquid whose weight will be equal to the float. Thus, the lowest density is marked on the top of the stem of a hydrometer and the maximum density is marked at the bottom of the graduated stem.

ELECTROCHEMICAL ACTION

Discharge process

The term discharge means to use a battery to provide electric energy. If the two terminals of a lead storage cell are connected to each other through an electric load, electrons flow from the negative electrode through the load to the positive electrode because of the difference in potential. When a battery is being discharged, electrons (given the symbol e^-) leave the negative terminal. The battery produces electrons by means of chemical reactions.

(1) Chemical reaction when discharging

PbO_2	+ $2H_2SO_4$	+ Pb	→ $PbSO_4$	+ $2H_2O$	+ $PbSO_4$
(+plate)	(electrolyte)	(–plate)	(+plate)	(water)	(–plate)

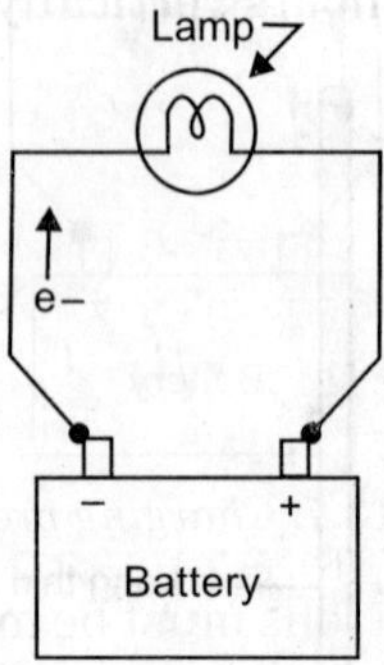

Fig.13.6: *Discharging process*

In the reactions the following changes take place:

- Sulphuric acid breaks down into irons: positive hydrogen ions and negative sulphate ions.
- Lead molecules change into positive lead ions and in doing so release electrons. The electrons build up on the spongy lead plates, which become the negative plates of the battery.
- Lead peroxide also breaks down into positive lead ions and oxygen. The oxygen reacts with the positive hydrogen irons to form water. When water is formed, the positive charges of the hydrogen ions draw electrons away from the lead peroxide plates. Because electrons are removed from the lead peroxide plates, they become positively charged and become the positive plates of the battery.
- The positive lead ions react with the sulphate ions to form lead sulphate.

As the battery discharges, the plate materials and the sulphuric acid are decreasing quantity. The amount of water is increasing. If the discharge reactions were allowed to continue, the battery would become completely discharged. Much of the active material in the positive and negative plates would become lead sulphate.

Charging process

(1) Chemical reaction when charging

$PbSO_4$	+ $2H_2O$	– $PbSO_4$	→ $PbSO_2$	+ $2H_2SO_4$	+ Pb
(+plate)	(water)	(–plate)	(+plate)	(electrolyte)	(–plate)

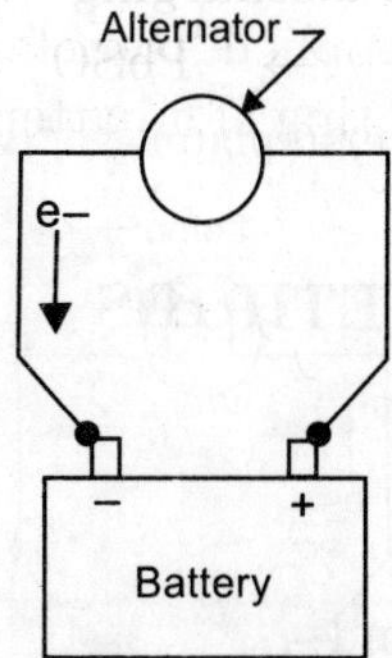

Fig.13.7: *Charging process*

To charge a battery, electrons must be made to flow into the negative terminal. This reverses the chemical reactions. In the automobile, the alternator charges the battery by forcing a flow of electrons into the negative terminal of the battery.

In the charging reactions, the following changes occur:

- Lead sulphate on the plates reacts with the water in the electrolyte to form sulphuric acid.
- Lead sulphate on the plates is also converted into spongy lead on the negative plates and lead peroxide on the positive plates.
- Water is used up.

As the battery charges, the plate materials are recovered. At the same time, the electrolyte is increasing in acid content and decreasing in water content.

BATTERY RATING AND CAPACITY

Energy stored in a battery expressed in terms of period during which it gives the rated current before dropping to the specified final voltage (1.75 Volts per cell) is called its capacity. In simple terms, the rating of a particular battery is determined by how much current it can produce and how long it can sustain this current.

Factors affect battery rating

The rate at which a battery can produce current is determined by the speed of the chemical reaction. This in turn is determined by a number of factors:

- Surface area of the plates and volume of active plate material
- Temperature
- Strength and amount of electrolyte
- Current demanded

The actual current supplied therefore determines the overall capacity of a battery. The rating of a battery has to specify the current output and the time.

BATTERY RATING METHODS

The common ratings are
(i) Ampere-hour rate
(ii) Watt-hour capacity,
(iii) Reserve capacity rate and
(iii) Cold rate.

Ampere-hour capacity

$$\text{Ampere-hour capacity} = \text{Rate of discharge in amperes} \times \text{Number of hours required for discharge}$$

This is now seldom used but describes how much current the battery is able to supply for either 10 or 20 hours. A battery rated at 40 Ah, based on a 10 hour rate, should supply a steady current of 4 A for 10 hours before the final voltage of 1.75 Volts per cell is reached. The 20-hour rate is the most common. This means the amount of current that a battery can deliver for 20 hours before the cell voltage drops to 1.75 Volts. For example, if a battery is capable of delivering 5 amperes for 20 hours before the cell voltage falls to 1.75 Volts, it is rated as 100 ampere-hour.

Watt-hour capacity

A battery may also be rated by watts, which represent the available cranking power. Wattage ratings for most 12-Volt batteries range from 2000 to 4000 Watts.

Watt-hour rating is determined by calculating the ampere-hour rating of the battery times the battery voltage at 0ºF (-17.8ºC).

$$\text{Watt-hour capacity} = \text{Ampere-hour capacity} \times \text{Average voltage during discharge period}$$

Reserve capacity

The reserve capacity rating is the time in minutes a vehicle can be driven after the charging system fails. This is roughly equivalent to the conditions after the alternator fails while the vehicle is being driven at night with the head lights on. The battery alone must supply current to the headlights and the ignition system. The assumed battery load is a constant discharge current of 25A. It is the length of time in minute that a fully charged battery at 80°F (26.7°C) can deliver 25 amperes without the cell voltage dropping below 1.75 Volts. This shows for how long the battery can deliver the electrical operating load, if the alternator fails to supply current.

Cold rating

Large engines require that a starter motor deliver a large amount of cranking torque. Smaller engines require less torque. The greater the torque supplied by a starter motor, the more current it draws from the battery. As a rough guide, a starter motor requires approximately 1A of current for each cubic inch (in^3) of engine displacement. Therefore, a 350 in^3 engine would require a battery with a cold cranking rating of approximately 350 amperes. This rating indicates the power a battery can supply for engine cranking under low temperature conditions. It is the number of amperes that a battery delivers for 30 seconds at 0°F (-17.8°) without the cell voltage falling below 1.2 Volts. The ampere values may be listed as 305, 315, 350 or 440.

Cranking power available from a fully charged battery based on temperature is:

Temperature	Cranking power (percent)
80°F (26.7°C)	100
32°F (0°C)	65
0°F (-17.8°C)	40

BATTERY EFFICIENCY

This is the ability of a battery to supply current. It depends upon temperature and rate of discharge. At low temperature, sulphuric acid cannot work so actively on plates and thus chemical action is highly reduced. So the battery cannot supply high current for long time. At high discharge rates, chemical action takes place only on the surface of the plates due to short time. But during slower rates of discharge, there is sufficient time for the chemical activities to penetrate deep below the plate's surfaces and act in the active material.

When the battery is discharged slowly, it is far more efficient than when it is discharged rapidly. High rates of discharge do not produce as many ampere-hours as low rates of discharge. Hence lower the discharge rate, higher the battery efficiency and higher the discharge rate, lower the efficiency.

The efficiency of a battery can be determined either as the quantity (ampere-hour efficiency) or the energy efficiency (power efficiency).

$$\text{Quantity efficiency} = \frac{\text{Ampere v hour output}}{\text{Ampere v hour input}} \times 100\ \%$$

$$\text{Energy efficiency} = \frac{\text{Watt v hour output}}{\text{Watt v hour input}} \times 100\ \%$$

Battery efficiency is affected by temperature. At 0°C, efficiency falls by 65%. Batteries in warm, tropical climates are more efficient, but may have reduced life spans. Batteries in cold climates have increased operating lives, but are less efficient.

The following table shows the approximate battery efficiencies in relation to battery temperatures:

Battery temperature	Battery efficiency, %
80°F (26.7°C)	100
32°F (0°C)	65
0°F (-17.8°C)	50

BATTERY CHARGING

Charging a battery means causing electrons to flow into the negative terminal. On the car, this job is performed by the alternator. Usually the batteries do not require charge from the external source provided the charging system of the vehicle is working efficiently. In the case of less charging or fault in the system, it becomes necessary to recharge the battery by means of battery chargers. Since most commercially available electricity is alternating current (AC) the charger consists of a transformer to reduce the voltage to a usable value and a rectifier to change the alternating current to direct current. The battery charger has a control switch and an ammeter so the charging rate can be adjusted.

The battery chargers are categorised as slow charger and fast charger.

SLOW CHARGING

In this method, battery is removed from the vehicle and connected to a slow charger to recharge it. Slow chargers are usually constant current machines which use a rectifier to convert AC to DC and a rheostat to adjust current flow to the battery. Slow battery chargers operate at low rate and take considerably longer time to charge a battery. Slow charging permits the battery to be completely charged. The charging rate is one ampere for each positive plate in a cell. For example, a battery having four positive plates in a cell should be charged at a rate of 4 amps. Batteries of different size can be charged at the same time. But the charging rate is determined by the smallest battery. Batteries are safely charged at a high rate if the temperature is below 113°F. The length of time required to recharge a good battery to fully charged condition is 12 to 16 hours at normal rates.

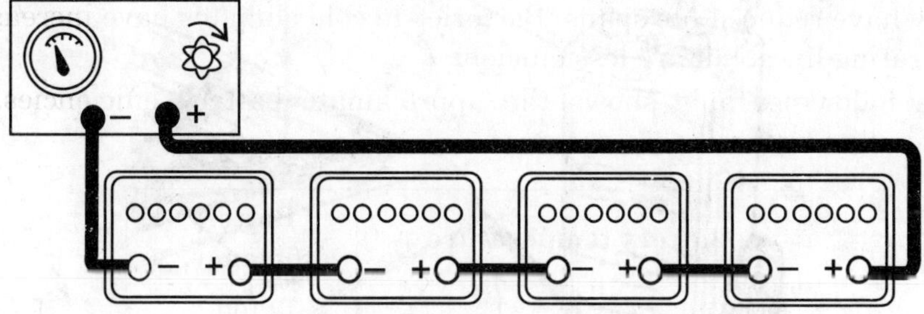

Fig.13.8: *Slow charger*

During slow charging the maximum charging current should be less than 1/10th of the battery capacity. For example, the battery capacity is 40 Ah, and then the charging current is 1/10th of 40 Ah.
i.e. 40/10 = 4 amps or less

Slow charging time can be calculated by the following formula:

$$\text{Charging time (h)} = \frac{\text{State of discharge capacity (Ah)}}{\text{Charging current (A)}} \times (1.2 \text{ to } 1.5)$$

Note: If more than one battery are to be charged, ***connect the batteries in series*** up to the capacity of the charger.

FAST CHARGING

This method does not require battery to be removed from the vehicle, as most fast chargers are portable. Its purpose is to recharge the battery sufficiently to allow the starting motor to start the engine. This is done in a short period of time, usually about 30 to 45 minutes. The battery is brought to a fair state of charge before it reaches at harmful high temperature. Fast charger is usually constant voltage machines with a high capacity so that common charging rates of 80-100 amperes for

6 Volt battery and 50-60 amperes for 12 Volt battery. 30 to 45 minutes of charging time will normally charge the battery so that the starting motor will crank the engine and the battery will not be over heated. Fast chargers do not fully charge a battery in the time allotted for charging. The remaining recharge is done by the alternator when the engine is operating. Some fast chargers have controls for finishing the charging cycle at low rate so the battery can be brought up to a full charge.

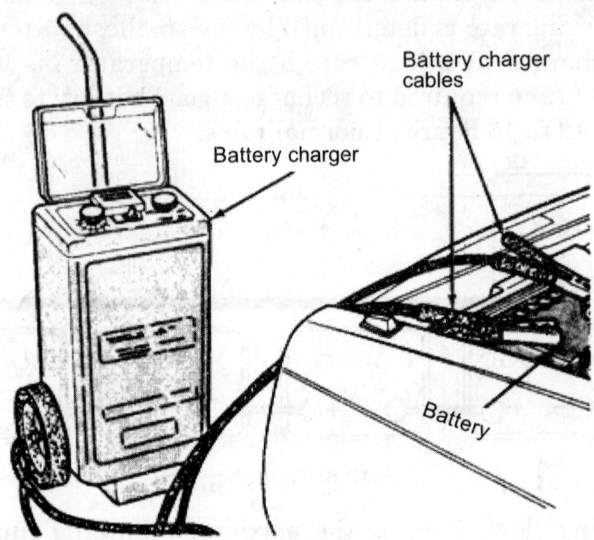

Fig.13.9: *Fast charger*

Determination of the allowable charging amperage and time is very essential if the battery chargers have no test device. Determine the state of the battery from the specific gravity, using the graph, then calculate the correct charging amperage using the following formula. The charging time for quick charging is usually from half to one hour.

$$\text{Correct charging amperage (A)} = \frac{\text{State of discharge capacity (Ah)}}{1 + \text{Charging time}}$$

Note: If more than one battery are to be charged, *connect the batteries in parallel* up to the capacity of the charger.

MAINTENANCE FREE BATTERY

Maintenance free batteries are sealed and never need water. They contain a lifetime supply of water, which is put in during manufacturing. The reason that the batteries do not require water during their lifetime is that the plate grids are made of a special lead alloy and calcium, strontium or cadmium, which greatly reduces gassing. Because there is less of this, there is less water loss.

Sealed batteries have a much longer storage life than non-sealed batteries. This is because antimony used in conventional batteries is not used in maintenance free batteries because it increases gassing. The calcium or strontium provides the needed strength in the grids and in addition greatly reduces gassing and self-discharge. Because the gassing rate is so low, these batteries can be sealed. However, one or more small vent openings are provided to release any pressure formed by gassing. Also, these batteries are materials of a higher purity. The maintenance free batteries are also called "low water loss" batteries.

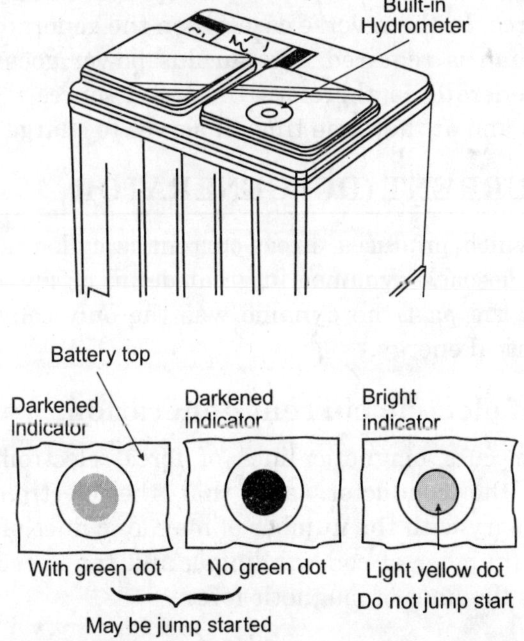

Fig.13.10: *Maintenance free battery*

Many maintenance free batteries have a charge indicator (built-in-hydrometer) built into the cover. The indicator shows the state of charge of the battery (actually only one cell). If the dot that is at the bottom of the indicator is green, then the battery is charged (more than 65% charged). If the dot is black, the battery state of charge is low. A yellow or clear eye means the electrolyte is low and the battery must be replaced.

Advantages of maintenance free batteries

1. Generally do not use water during normal usage.

2. Reduce the corroding of battery cables.
3. Longer shelf life (three times longer than a lead-antimony battery).
4. Resists overcharging.

GENERATOR

The generator is driven by the V-belt which gets drive from the crankshaft. It converts mechanical energy obtained from the engine into electrical energy. Generators can be DC (dynamo) or AC (alternators).

All power required while the automobile is in operation is supplied by the generator, but when this is not enough, the battery also serves as the power source. In the reverse case, when the generator is producing more power than is required, the surplus power goes to charge the battery. The generator is, therefore, the power source when the vehicle is in operation and at the same time, it serves to charge the battery.

DIRECT-CURRENT (DC) GENERATOR

A generator which produces direct current is called a dynamo. This book will not describe dynamos in great detail as few cars now have them. But in the past the dynamo was the only convenient way to generate electrical energy.

Principle of electric current generation

If a conductor cuts magnetic lines of force, electromotive force is generated in the conductor, and that the electromagnetic force produced will vary with the number of magnetic lines of force, number of conductors (number of coil windings), and the speed at which the conductor cuts the lines of magnetic force.

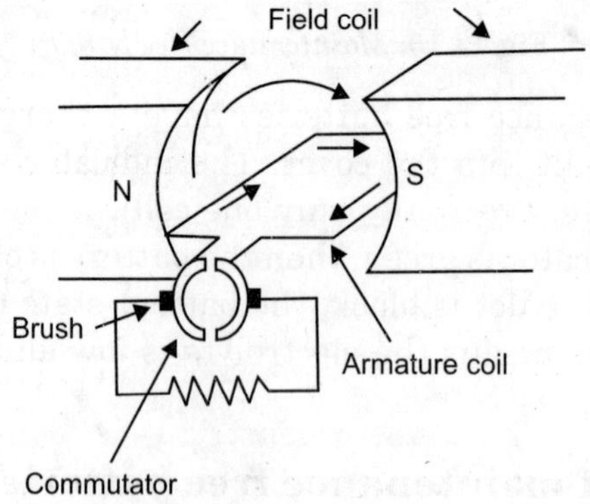

Fig.13.11: *Principle of generator*

As shown in the Fig.13.11, if a conductor is turned clockwise between the N and S poles of a magnet, electromotive force will be generated in the arrow direction in accordance with Fleming's right hand rule.

The generator can be roughly divided up into the field coils that make the magnets to produce the magnetic lines of force, the armature coils that cut the magnetic lines of force, and the commutator that draws out the armature electromotive force.

When the armature coil rotates, current flows in the arrow direction through the armature coil. The induced current flows through the commutator and brush at one side to the load such as the battery and lights, and returns through the brush and commutator at the other side to the armature coil. A part of the current induced in the armature coil flows through the field coils to strengthen the flux.

As the armature coil turns, the position of the conductors reverse, so that current now flows in the opposite direction through the coil; but through the commutator sections and brushes, the current flows in one direction to the outside (direct current).

Generators used in automobiles have a large number of armature coils so that the voltage is generated continuously.

REGULATOR

The regulator is comprised of three relays: a voltage regulator to control the generated voltage, a current regulator to control the output current, and a cut-out relay to prevent reverse flow of current from the battery.

Voltage regulator

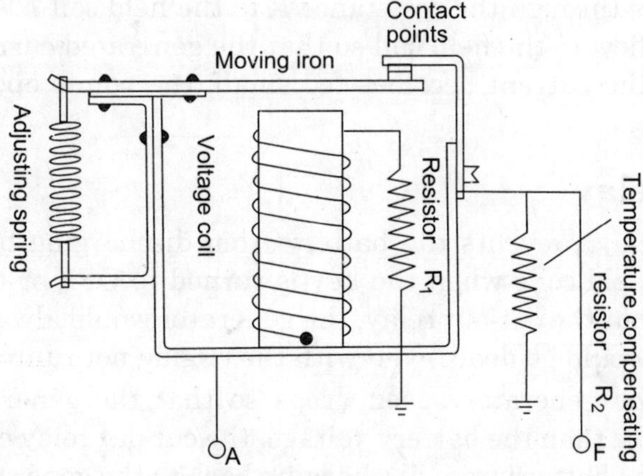

Fig.13.12: *Voltage regulator*

The purpose of this relay is to maintain the generated voltage at a constant value. When the voltage rises, a resistance is inserted in series with the field coil circuit to limit current flow through the field coils and lower the generated voltage. When the voltage drops, the resistance is removed from the field coil circuit to allow the generated voltage to build up again. The repetition of this action maintains the voltage at a constant value.

Current regulator

The purpose of the current regulator (current limiter) is to prevent excessive current from the generator. Its construction and operation is very similar to those of the voltage regulator.

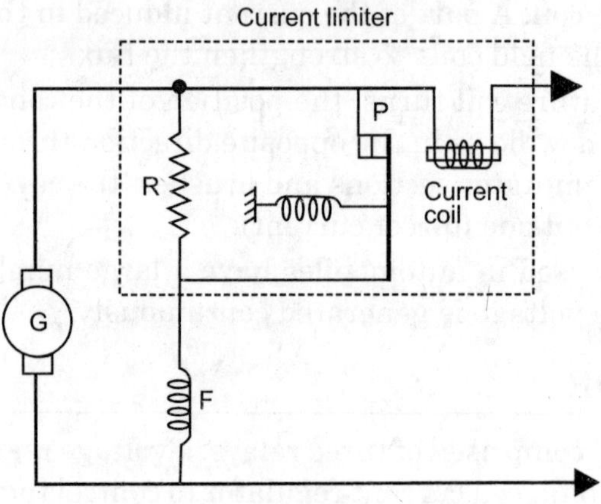

Fig.13.13: *Current regulator*

The contact points are normally closed, but when the generated current rises over a specified value, the points P open so that the current flows through the resistance R to the field coil F. This reduces the current flow to the field coil so that the generated current becomes less. When the current becomes too small, the points open to restore normal state.

Cut-out relay

A cut-out relay prevents the battery from discharging back into the generator's field coils when the key is turned to OFF or the engine is idling. Without the cut-out relay, the generator would always be on and the battery would go dead, even with the engine not running.

When the generator speed drops so that the generated voltage becomes lower than the battery voltage, the cut-out relay contacts open to prevent the battery from discharging back to the generator. The cut-out relay is merely a form of automatic switch.

The construction of the cut-out relay is as shown in Fig.13.14. When the generator is not rotating, the contact points are held open by the tension of the spring, but when the generator starts rotating and its generated voltage reaches the charging voltage, the relay core becomes strongly magnetized through the voltage and current coils, and attracts the iron piece. This causes the contact points to close and permit the generator current to flow to the battery for charging. When the speed drops, the generated voltage also drops so that the core magnetisation weakens, to allow the spring to open the contact points. Thus, the charging circuit is opened to prevent reverse flow of current.

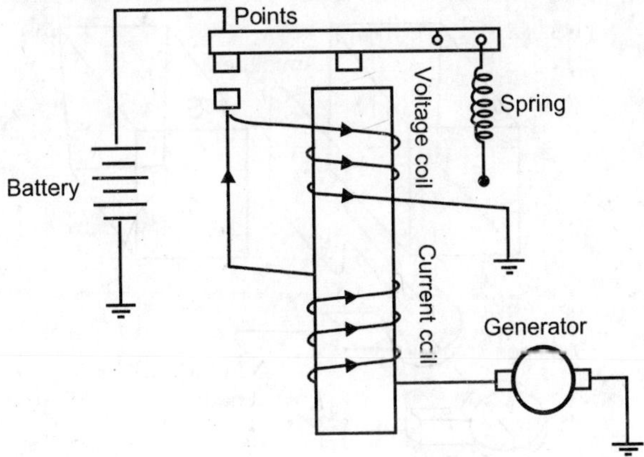

Fig.13.14: *Cut-out relay*

ALTERNATOR

With the traffic growing heavier in recent years, the necessity for making frequent stops has increased engine idling time. The increasing numbers of accessories also consume a larger amount of power. Due to these causes, the output capacity of the direct-current generator is beginning to be over strained and tends to leave the battery in a discharged state. The alternator can overcome this problem. This alternator is compact and lightweight. It also provides better serviceability.

The alternator is the heart of the vehicle's electrical system when the engine is running. It uses electromagnetism to change some of the engine's mechanical energy into electrical energy for powering the vehicle's loads and for charging the battery.

Two different types of alternators are used on vehicles. A conventional alternator and separate voltage regulator were used on all models prior to 1986. A new compact, high-speed alternator with a built-in IC regulator is now used on most models. Both types of alternators are rated according to current output. Typical ratings range from 40 to 80 amps.

PRINCIPLE OF ALTERNATOR

An alternator is designed to change mechanical energy into electrical energy. The working principle of an alternator is the electromagnetic induction (EMI). When the rotor spins, its strong magnetic field cuts across the stator windings. This induces current in the stator windings. If stator windings are connected to a load (for example Bulb), the load would operate.

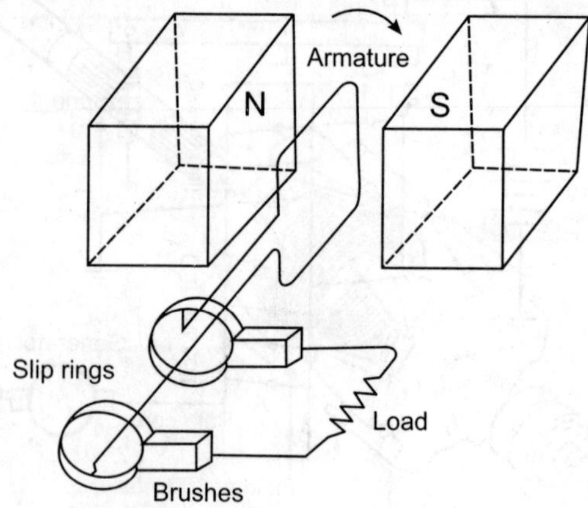

Fig.13.15: *Principle of alternator*

CONSTRUCTION

The alternator is comprised primarily of the stator, rotor, bracket, rectifier (silicon diodes) and brushes.

Stator

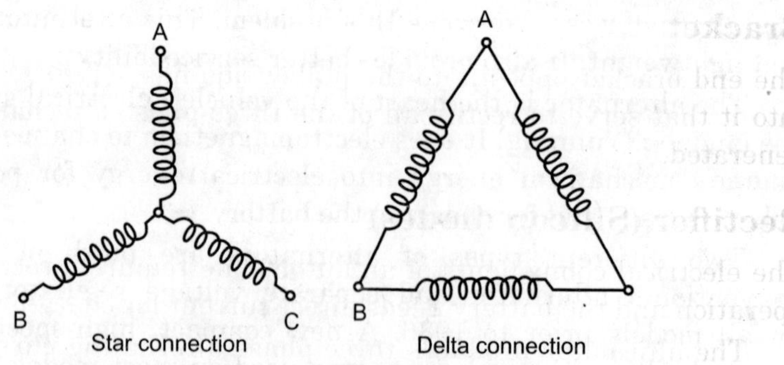

Fig.13.16: *Stator coil connection method*

The stator has three sets of armature coils wound on it to induce a three-phase current. As shown in Fig.13.16, the coils in the stator can be connected in either star or delta form. The star connection is generally used in automobile alternators. The generated wave forms phased 120° apart.

Rotor

In rotor, the number of poles is equal to the number of fingers and range from a few poles to more than a dozen poles.

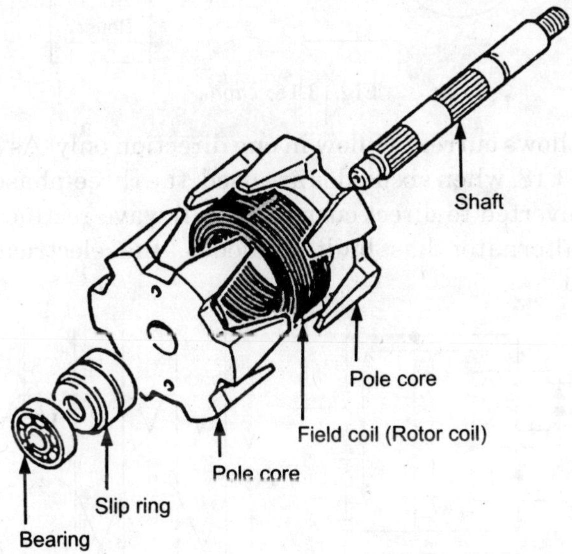

Fig.13.17: *Rotor*

When the current flows through the coils, all fingers at one side of the disk become N poles and all those at the other side become S poles, resulting in the N and S pole to be arranged alternately.

Bracket

The end bracket opposite to the pulley side has six diodes assembled into it that serve to rectify all of the three-phase alternating current generated.

Rectifier (Silicon diodes)

The electrical components of an automobile require direct current for operation and the battery needs direct current for charging.

The alternator produces three-phase alternating current but the charging system of the automobile cannot use this electricity unless it is converted to direct current. Converting alternating current to direct current is called rectification. Rectification can be accomplished in several ways but the automobile alternator uses a simple but effective diode.

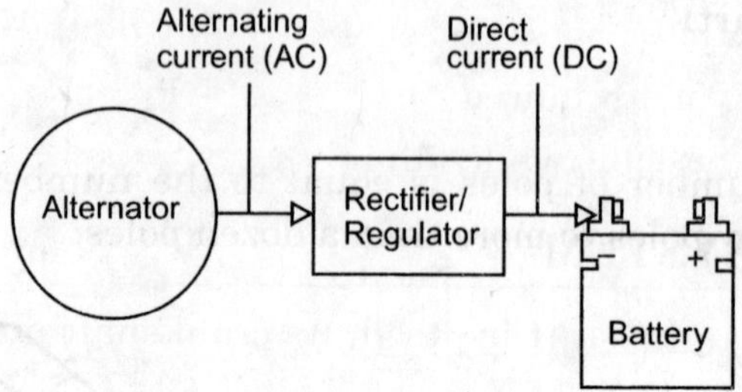

Fig.13.18: *Diode*

A diode allows current to flow in one direction only. As shown in the illustration 13.18, when six diodes are used, the three-phase alternating current is converted to direct current by full wave rectification. As the automobile alternator has built-in diodes, the electricity output is direct current.

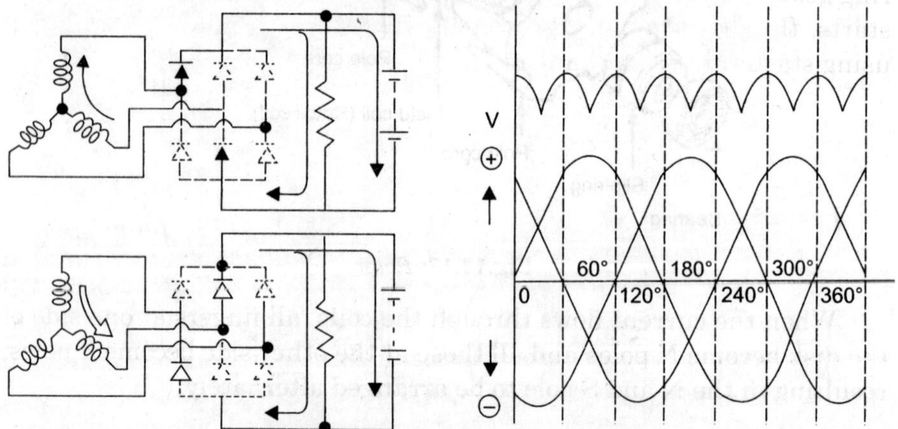

Fig.13.19: *Current rectification*

Thus, we can see that although the current from each coil to the diodes is constantly changing direction in the three lead wires, the direction of the current from the diodes does not change and it forms a circuit with unvarying polarity.

Brushes

The brushes serve to pass a current of 2 to 3 amperes to the rotor coils. As the main current does not flow through these brushes, the brushes have much longer life and hardly require replacement.

ADVANTAGES OF ALTERNATOR

1. Rotor has high strength against speed
2. Good rectifying characteristics
3. Current regulator not required, depending on the design
4. Cut-out relay not required
5. Charging is possible while idling

STARTING SYSTEM

As the engine cannot start by itself, its crankshaft must be turned by some outside force in order to make it start.

Starting methods can be classified into the manual system where the engine is cranked by hand and the self-starting method, where the engine is cranked by an electric motor to start the engine. All automobiles use the self-starting method at present.

The starting motor is installed on the cylinder block. Closing the switch causes the starter motor drive pinion to engage with the flywheel ring gear which allows the motor to crank the engine. When the engine starts, the drive pinion is automatically released from the ring gear using starter drive mechanism.

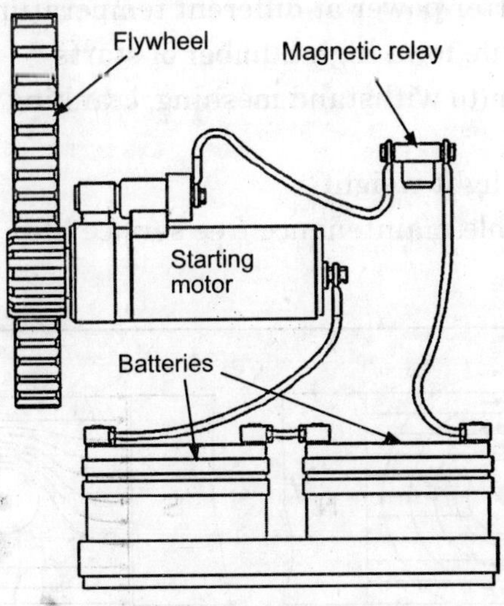

Fig.13.20: *Starting system*

STARTING REQUIREMENTS

An internal combustion engine requires the following criteria in order to start and continue running.

1. Combustible mixture
2. Effective compression stroke
3. Suitable source of ignition
4. Sufficient cranking torque
5. Minimum starting temperature
6. Appropriate starter pinion-ring gear ratio
7. Rated voltage of the starting system
8. Length of the cable

STARTER MOTOR

The starter motor is a small, but powerful, electric motor designed to operate under great electrical loads and to produce great amounts of torque for short periods. The starter motor has a small gear, called a pinion gear, mounted on the end of the armature shaft. The pinion gear engages a large ring gear mounted on the outside edge of the flywheel.

REQUIREMENTS OF STARTER MOTOR

The requirements of starter motor are:

1. Continuous readiness for starting
2. Sufficient starter power at different temperatures
3. Long service life for a high number of starts
4. Robust design (to withstand meshing, cranking, high temperature, etc)
5. Compact and light weight
6. Longest possible maintenance free service life

PRINCIPLE

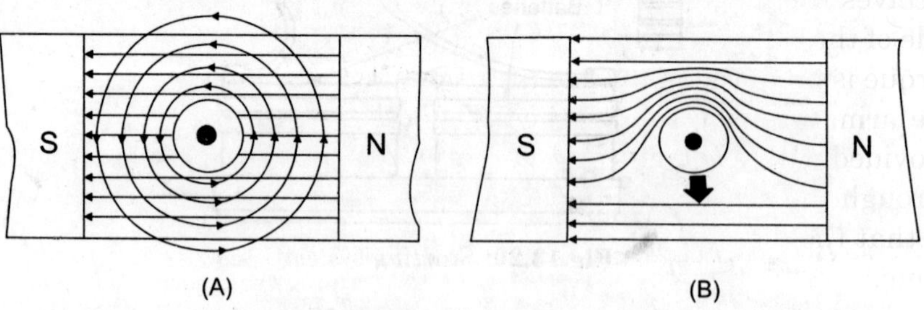

Fig.13.21: *Principle of a starter motor*

A current passing through a conductor produces concentric lines of force around the conductor, the strength of which is proportional to the current. The direction of the lines of force can be determined by the right-hand screw rule. Assume the conductor to be placed between magnetic N and S poles as shown in Fig.13.21A. In this case, the lines of force in opposed direction will counteract each other, while those in the same direction will add up to make up bent lines of force as shown in Fig.13.21B.

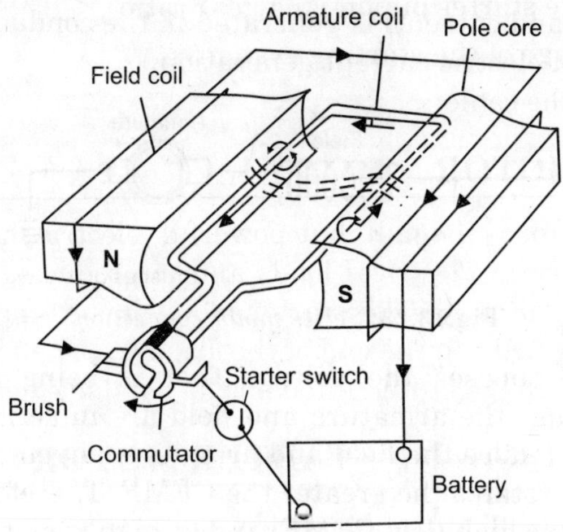

Fig.13.22: *Starter motor*

This pattern develops because the lines of force below the conductor run in opposite direction and counteract each other, while those above the conductor add up and become twice as strong. The dense and bent lines of force act like rubber bands and produce a force tending to push the conductor downward. The starting motor utilises this force to produce torque.

The current from the battery flows through the field coils and armature coils and returns to the battery. The S pole side of the coil receives the magnetic force in a downward direction while the N pole side of the coil receives the magnetic force in a upward direction so that torque is produced in the armature coil in a clockwise direction to cause the armature to rotate in the arrow direction. Since the armature is provided with a commutator and brushes, the direction of current flow through the armature is reversed when the armature coil rotates 180° so that the direction of torque produced in the armature is always the same.

TYPES OF STARTER MOTOR

The four main types of starter motor are referred to as series, shunt, compound, and permanent magnet excitation.

Series motor

In a series motor, the field and armature windings are connected in series. This means that all the armature current must pass through the field winding. When the armature of a motor is rotating, a voltage called a counter electromotive force (CEMF) is induced in the armature. This occurs because whenever a conductor moves through a magnetic field, a voltage (and current) is generated in the conductor. This is the basic principle of electric current generation.

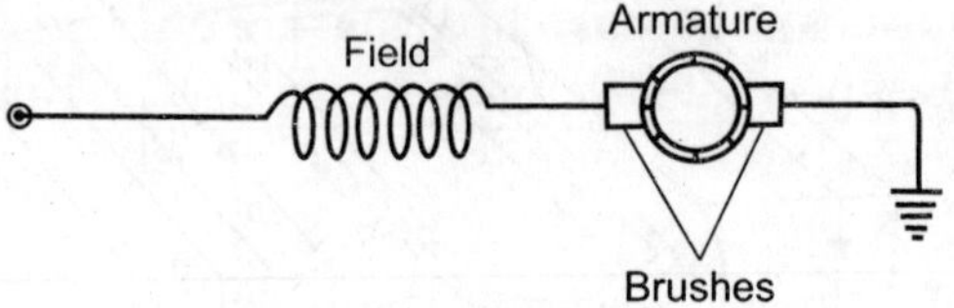

Fig.13.23: *Series motor connection*

The CEMF opposes the voltage that is being applied to the armature. Because the armature and field are in series, the effect of the CEMF is to reduce the field and armature current. The faster the motor armature rotates the greater the CEMF. Therefore, at low rpm, the field current is high (low CEMF) and at high rpm, the field current is low (high CEMF). The torque produced by a motor is directly related to its magnetic-field strength. A series motor can produce a high initial torque at low rpm (Fig.13.24) when its field current and magnetic-field strength are greatest due to low resistance and no CEMF. At higher speeds, the torque output of the motor is reduced. This is an ideal characteristic for a starter motor.

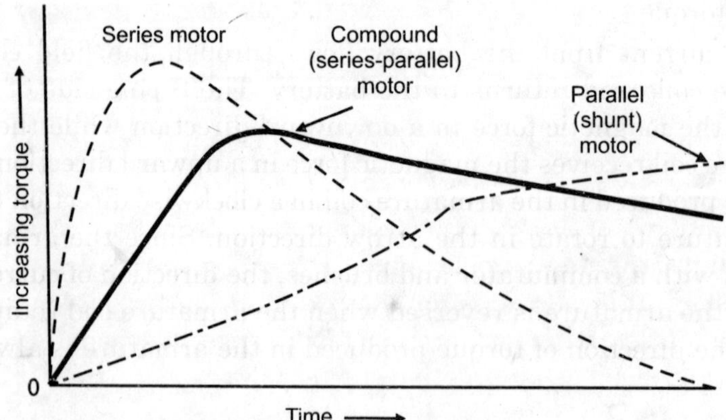

Fig.13.24: *Typical torque and speed relationships of starter motors*

When the starter motor first begins to crank the engine, a large amount of torque is required to overcome the inertia of the engine at rest. Because of the above characteristics series motors are commonly used as automotive starter motors.

Shunt motor

In a shunt motor, the armature and field are connected in parallel. This means the CEMF which is generated in the armature does not affect the field current. The magnetic field strength is constant as motor speed changes. A shunt motor tends to regulate its own speed. If the load on the motor increases, the motor slows down and the CEMF is reduced. A reduced CEMF causes an increase in armature current which, in turn, increases the motor speed. These motors are used if constant speed is more important than starting torque.

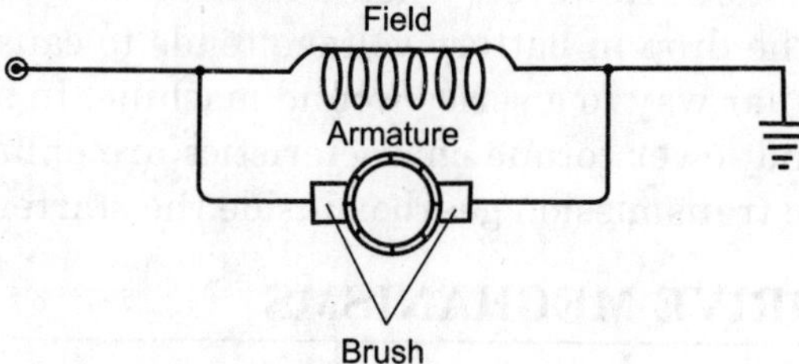

Fig.13.25: *Shunt motor connection*

Compound motor

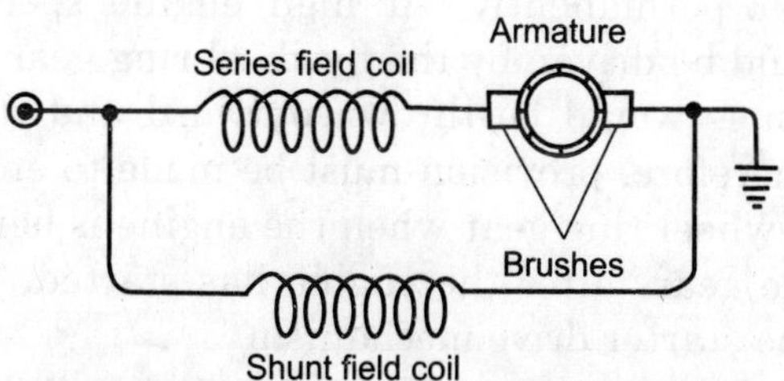

Fig.13.26: *Compound motor connection*

In a compound motor, some of the field windings are in series, and some of the windings are in parallel or shunt. This type of motor combines the characteristics of the series motor and the shunt motor. The compound motor has a high starting torque and a fairly constant speed. Compound motors are also used as automotive starter motors.

Permanent magnet motor

Permanent magnet motors are smaller and simpler compared with the other three discussed. Field excitation, as the name suggests, is by permanent magnet. This excitation will remain constant under all operating conditions.

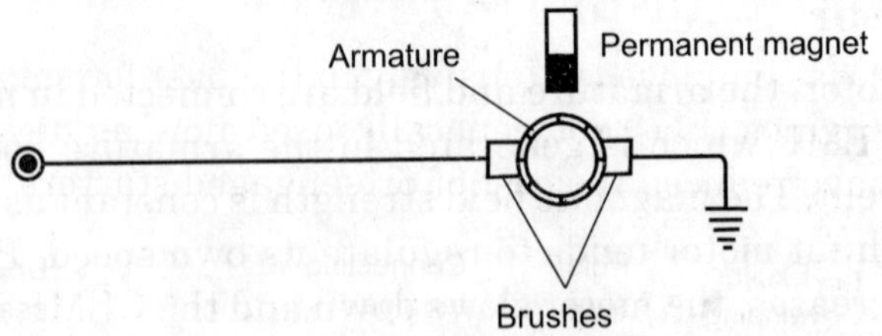

Fig.13.27: *Permanent magnet motor connection*

The characteristics of this type of motor are broadly similar to the shunt wound motors. However, when one of these types is used as a starter motor, the drop in battery voltage tends to cause the motor to behave in a similar way to a series wound machine. In some cases, the higher speed and lower torque characteristics are enhanced by using an intermediate transmission gearbox inside the starter motor.

STARTER DRIVE MECHANISMS

A typical ratio of ring-gear teeth to pinion-gear teeth is 20 to 1. This gear ratio means that the starter motor rotates 20 times faster than the flywheel (and engine crankshaft). Because the starter motor turns 20 times faster than the engine, the pinion and ring gears cannot be meshed together permanently. At high engine speeds, the starter motor shaft would be driven by the flywheel ring gear so fast that the armature windings would be thrown outward and the motor would be damaged. Therefore, provision must be made to engage the pinion gear into the flywheel ring gear when the engine is being cranked and to disengage the gears after the engine has started. This function is performed by the starter drive mechanism.

The common kinds of starter drives used on light vehicles are:

1. Bendix drive (Inertia drive)
2. Improved inertia type
3. Bendix Folo-thru type
4. Pre-engagement drive
5. Sliding armature type (Axial type)
6. Co-axial type
7. Permanent magnet type

Most of these starter drives use a one-way clutch (overrunning clutch) to release the starter drive pinion from the flywheel after the engine has started. The following *types* will now be *discussed* in more detail.

BENDIX DRIVE (INERTIA DRIVE)

This method was employed extensively in the past for small to medium size petrol engines. It is not generally used now as developments in technology have reduced the cost of pre-engaged starters.

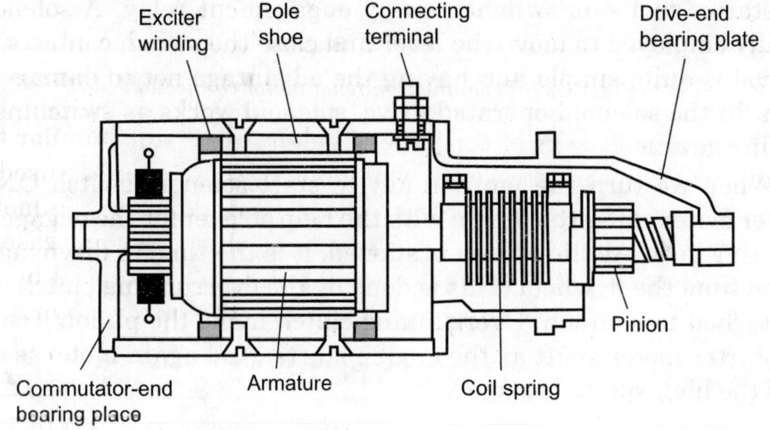

Fig.13.28: *Bendix inertia drive*

Inertia type drive works on the principle of inertia of unbalanced weight. In this type, pinion is mounted on threaded sleeve. When ignition switch is pressed, the starter motor starts rotating and this rotation wound the spring of the drive and also starts rotating the threaded sleeve. Now, the unbalance weight attached to the pinion, restricts the rotation of pinion and this action moves the pinion towards the ring gear. This action causes engagement of pinion with ring gear. The shock at the instant of engagement and disengagement is severe, so a large spring is fitted adjacent to the helix to act as a cushion. Pinion starts rotating ring gear and engine is cranked. The moment an engine starts running at the ideal speed, pinion is driven by the ring gear at high speed and it moves back on the sleeve and gets disengaged from ring gear by high speed inertia effect and starter motor is saved from the terrific centrifugal force.

Disadvantage

The main disadvantage of this drive is lots of moving parts and engagement of pinion is done after starter motor starts. It causes rapid wearing of pinion teeth. To overcome this, pinion must be engaged before starter motor starts rotating. As a result, pre-engage drive is developed.

PRE-ENGAGED DRIVE

The pre-engaged drive is used with both petrol and diesel engines and is now the preferred type.

The name itself suggests the working principle of pre-engage type drive. In this arrangement the drive pinion on the starter motor shaft is pushed axially into mesh with the flywheel ring gear before the starter motor is switched on. When the pinion meshes with flywheel ring gear, the starter motor is switched on by engagement relay. A solenoid is usually employed to move the lever and close the switch contacts. This method is quite simple and having the advantage not to damage gear teeth. In the solenoid operated drive, solenoid works as switching and shifting device.

When we turn the ignition key to start solenoid switch ON, the starter motor shifts the pinion with the help of lever for the engagement with ring gear. As the engine is started, it is the time to disengage the pinion from the flywheel. This is done by the overrunning clutch which is attached to a pinion. Overrunning clutch make the pinion free from the starter motor shaft as the engine starts. And again motor is saved from the high speed damage.

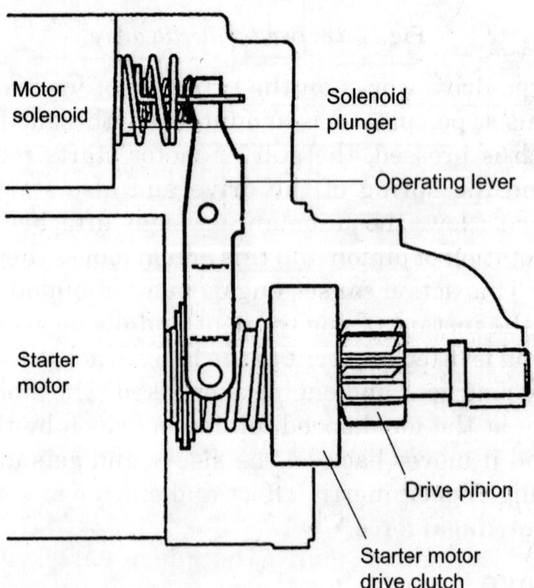

Fig.13.29: *Operation of pre-engagement type drive*

Advantages of pre-engaged drive

As compared with an inertia starter pinion drive, the advantages of the pre-engaged type of drive are as follows.

1. A smaller pinion can be utilised to provide a higher reduction ratio, so that cranking torque is increased to improve cold starting capability.
2. The pinion drive is not so highly stressed, because less energy has to be absorbed during pinion engagement.
3. Prolonged cranking of a diesel engine is possible without premature disengagement or the pinion repeatedly traveling in and out of mesh as resisting torque fluctuates.
4. A major source of excessive wearing of the starter pinion and ring gear teeth is eliminated.

OVERRUNNING CLUTCH

Starter motors are usually fitted with a one-way, over-running clutch to prevent the starter motor armature from spinning at higher engine speed and getting damaged, an overrunning clutch is used.

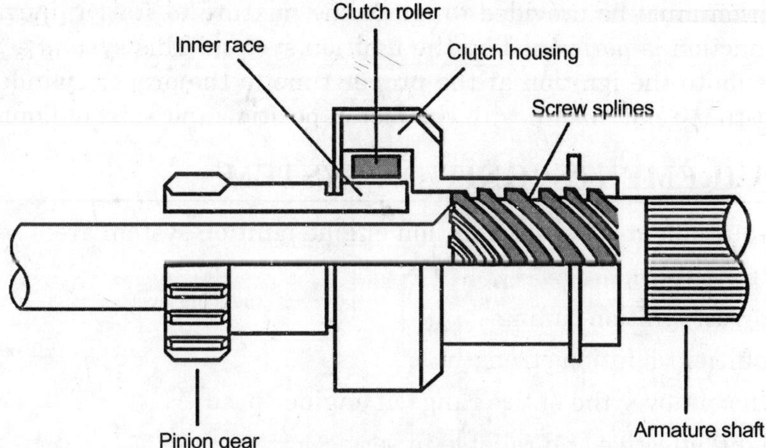

Fig.13.30: *Cut-view of overrunning clutch*

The overrunning clutch housing is internally splined to the armature shaft. The drive pinion turns freely on the armature shaft within the clutch housing. When the clutch housing is driven by the armature, the spring loaded rollers are forced into the small ends of their tapered slots and wedged tightly against the pinion barrel (Fig.13.31A). This locks the pinion and clutch housing solidly together, permitting the pinion to turn the flywheel and, thus, crank the engine.

When the engine starts, the flywheel spins the pinion faster than the armature. This releases the rollers, unlocking the pinion gear from the armature shaft. The pinion then freely spins on the armature shaft (Fig.13.31B). Once current flow is stopped, the pinion is then pulled away from the flywheel. The overrunning clutch is moved in and out of mesh by the starter drive linkage.

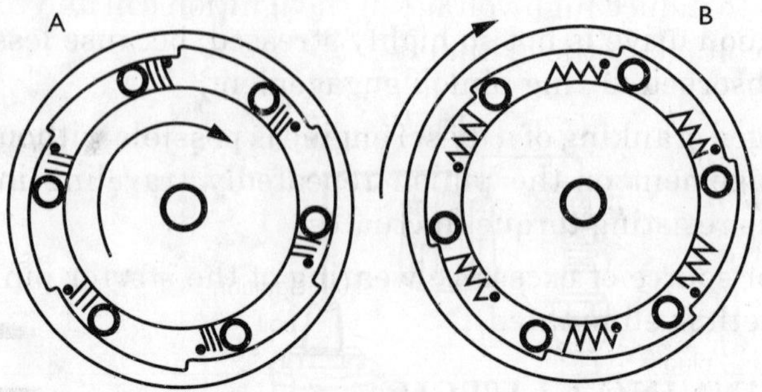

Fig.13.31: *Operation of overrunning clutch*

IGNITION SYSTEM

In petrol engine the mixture will not ignite by itself. Therefore a source of ignition must be provided to ignite the mixture to start it burning. This function is performed by the ignition system. This system serves to distribute the ignition at the proper time to the proper cylinder in correct time relationship with the piston position and start of ignition.

REQUIREMENTS IGNITION SYSTEM

The requirements of spark-ignition engine ignition system are:

1. Strong spark
2. Correct ignition timing
3. Sufficient ignition strength
4. Efficient over the entire range of engine speed
5. Light, effective and reliable in service

COMPONENTS OF (BATTERY) IGNITION SYSTEM

Battery

The battery is the power source for current to primary circuit when starting the engine or when the engine is idling. When the engine speed rises, the generator becomes the source of power.

Ignition coil

By passing the battery current through the primary winding of the ignition coil, a high voltage is induced in the secondary winding that is large enough to discharge amply large sparks across the spark plug electrode gaps.

Distributor

The functions of the distributor are to switch off the primary circuit current so as to induce high voltage in the ignition coil and to distribute the induced voltage to the cylinders.

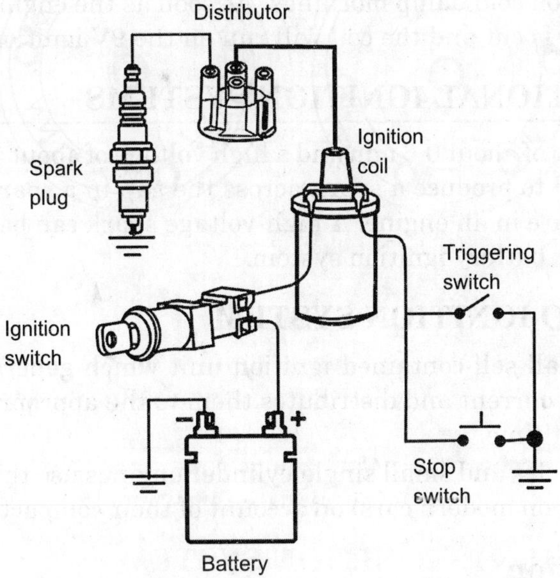

Fig.13.32: *Components of Ignition system*

Spark plugs

There are screwed into the combustion chambers, and form the gaps for the sparks. When the high voltage discharges across the gap, the resulting spark ignites the air-fuel mixture in the combustion chamber.

Ignition switch

The switch is located on the instrument panel in front of the driver's seat and serves to close the ignition circuit and to start and stop the engine.

Ballast resistor

Ballast is a resistor that has the purpose of reducing the voltage to the coil. Ballast resistor in not used with all the ignition systems.

The starter motor draws a huge current from the battery leaving less energy to spark plugs. The result is a weaker than normal spark which is not ideal for starting an engine. To overcome such a problem, ignition systems were changed to run a lower voltage coil (usually 9V), and these coils could still give the same output as the original 12V coils. In order to run such a coil, the 12V ignition feed runs through a ballast, reducing it to 9V at the coil. To assist starting, a 12V feed (usually from the starter) bypasses the 9V ignition feed, giving the 9V coil a 12V feed. The result is a better than normal spark which is ideal for starting, particularly on cold damp mornings. As soon as the engine has started, the 12V feed is cut and the coil will run on the 9V ignition feed.

CONVENTIONAL IGNITION SYSTEMS

A normal gap of about 0.6 mm and a high voltage of about 10,000-30,000 V is required to produce a spark across the gap in a spark plug that is under pressure in an engine. A high-voltage spark can be generated by a magneto or battery ignition system.

MAGNETO IGNITION SYSTEM

This is a small self-contained ignition unit which generates pulses of high-tension current and distributes them to the appropriate cylinders at the correct time.

Motor cycles and small single-cylinder engines use this system (but it is not used on modern cars) on account of their compact construction.

Construction

Magneto ignition system consists of a rotating pole wheel carrying a number of permanent magnets (Magneto rotor), the ignition armature with primary and secondary windings, the contact breaker points, the condenser and a cam rotating at crankshaft speed.

Operation

The alternating north and south poles of the rotating permanent magnets induce an alternating current in the primary winding. When the contact breaker points are closed, a current flows in the primary winding of the ignition armature and generates a magnetic field. The points are opened at the moment when the current flow is at its greatest, that is to say when the magnetic field in the primary winding has reached a maximum. The change of pole of magneto rotor causes the direction of the magnetic field in the ignition armature to change suddenly. This sudden reversal of the magnetic field induces a high voltage in the secondary winding. The ignition condenser, which is wired in parallel with the contact breaker, suppresses spark formation and encourages the magnetic field to decay rapidly. Since the cam revolves at crankshaft speed, a spark is produced during each revolution of the engine. This is ideal for the requirements of a single-cylinder two-stroke engine. In a single-cylinder four-stroke engine, a 'false spark' occurs every second revolution of the crankshaft, but takes place during the exhaust stroke and therefore has no effect on the engine's actual performance.

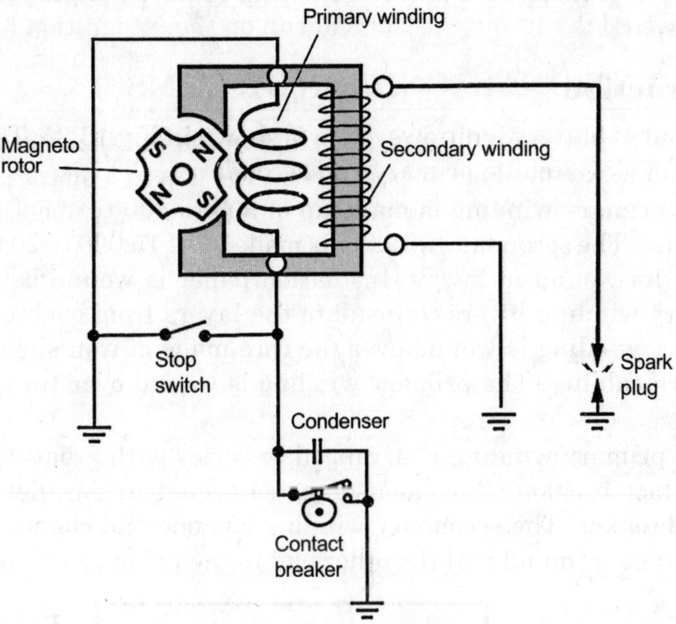

Fig.13.33: *Magneto ignition system*

Advantages

1. More reliable
2. Economical
3. More compact
4. Good spark energies at low cranking speed
5. Reduced cable requirements due to the absence of low tension cables

Disadvantages

1. Not suitable for multi-cylinder engines
2. Poor performance at starting

BATTERY IGNITION SYSTEM

Battery coil ignition system is not self-contained, like a magneto ignition system. A separate source of supply (6, 12 or 24V battery) is required to feed the primary winding. Compared with a magneto system, the battery ignition system makes engine starting much easier. Also it is simpler to control the maximum voltage to suit the conditions.

This system uses electrical energy produced by a battery or alternator to supply the low-tension ignition primary current. The ignition coil transforms the voltage to that required to produce a high-voltage spark.

Construction

The layout of battery ignition system is shown in Fig.13.35. The system consists of two circuits: primary and secondary.

The primary winding is made up of 300 to 400 turns of relatively heavy wire. The secondary winding is made up of 15,000 to 20,000 turns of fine wire wound in layers. Insulation paper is wound between the secondary winding layers to insulate the layers from each other. The secondary winding is wound over the core made of iron sheets having high permeability. The primary winding is wound over the secondary winding.

The primary winding is arranged in series with a battery, switch and contact breaker. A condenser is connected in parallel with the contact breaker. The secondary winding has one end connected to the high voltage terminal and the other end to the primary winding.

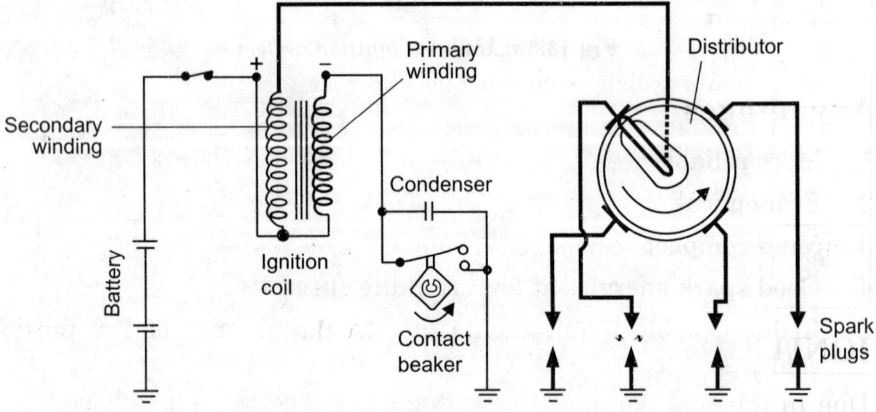

Fig.13.34: *Battery ignition system*

Operation

The current flowing through the primary coil magnetises the core and builds up a magnetic field around the core. When the primary current is suddenly cut off by the opening of the distributor contact points, the magnetic field collapses, resulting in producing voltage of 300 to 400

Volts in the primary winding through self-inductance and a voltage of 15,000 to 20,000 Volts in the secondary winding through mutual inductance.

Advantages

1. More reliable
2. Gives a much better spark for starting purposes
3. Easy maintenance
4. More suited for high speed engines

Disadvantages

1. Rely upon the battery
2. Contact points wear, need adjustment, prone to failure, bounce at high RPMs, and become wet and dirty.

Comparison of Battery Ignition and Magneto Ignition

S. No	Battery ignition	Magneto ignition
1.	Less moving parts, and require less maintenance.	Most of the parts moving and require more maintenance.
2.	Less cost	More cost
3.	Gives efficient spark at start	Gives weaker spark at start
4.	Gives weaker spark at high speeds and very high speed is not desired.	Gives efficient spark at high speed
5.	Efficient operation depends upon the condition of battery or generator.	Independent for its efficient operation
6.	Battery maintenance is required.	No battery maintenance problems.

IGNITION ADVANCE MECHANISMS

Due to the time lag needed for flame propagation after ignition, the mixture must actually be ignited before TDC. It is necessary to have some means to change (advance or retard) ignition timing so that it best matches engine speed, load, etc. For this purpose, ignition advancers are provided.

Two types of spark advancing mechanisms are provided in general:

1. Governor advancer – Advances in relation to engine speed
2. Vacuum advancer – Advances in relation to throttle valve opening

CENTRIFUGAL (OR) GOVERNOR ADVANCE MECHANISM

The spark advance during engine high speeds can be achieved by a centrifugal or governor advance mechanism. The governor advancer adjusts the ignition timing based on the engine speed. The governor therefore advances the ignition timing as the engine speed rises so that the maximum combustion pressure will always occur at a position 10° after TDC.

In other words, if the engine speed increase to higher rpm the time required for complete combustion of the air-fuel mixture is very less and hence it is necessary to increase the time to burn air-fuel mixture.

Construction and operation

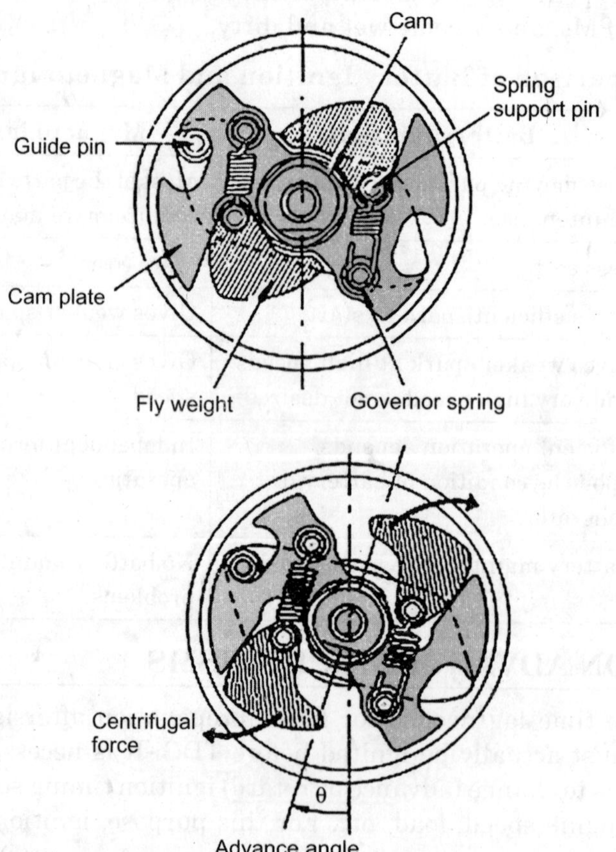

Fig.13.35: *Construction and operation of governor advancer*

The centrifugal advance mechanism consists of two pivoted weights in the lower housing of the distributor assembly. These are held towards the center of the distributor shaft by two independent springs. As the rotation speed increases, the weights are thrown outwards, due to centrifugal force, against the spring tension. This will turn the center shaft cam lobe to open the contact points prior to the normal timing and cause the advance sparking.

As the speed reduces to idle running, the centrifugal forces also reduces, thus the weights are pulled back to their original position by the springs tension. Thereby the distributor shaft lobe turns back to the original position to provide the ignition at initial timing (normal timing, usually 10°BTDC).

VACUUM SPARK ADVANCE MECHANISM

When the engine is running at partial throttle with average load, the air fuel ratio is very lean. As the mixture is lean which contains less fuel; it requires more time to complete the combustion process. In such case it is better to give ignition advance. The vacuum ignition advancer adjusts the ignition timing based on the varying vacuum in the intake manifold under varying engine loads, and advances the ignition timing accordingly.

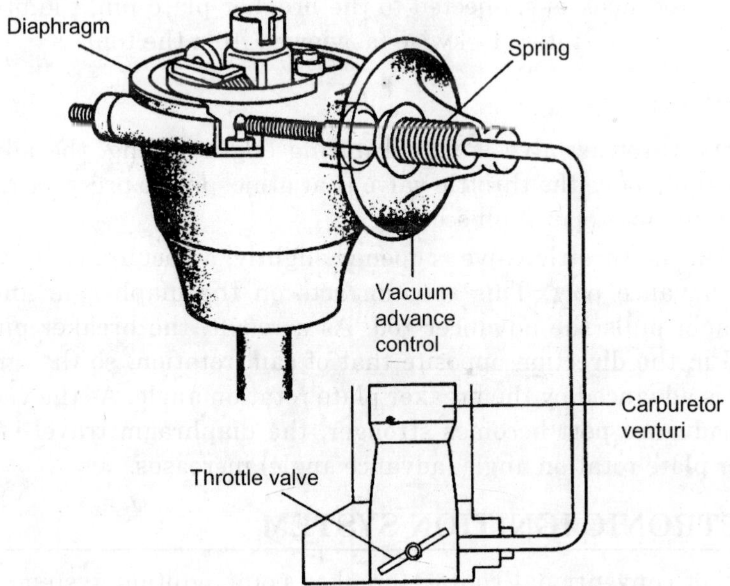

Fig.13.36: *Vacuum advance mechanism*

The vacuum ignition advancer therefore advances the ignition timing when the load is light to insure that the maximum combustion pressure will always occur at 10° after TDC.

Construction

The vacuum ignition advance mechanism consists of the vacuum advancer (diaphragm, diaphragm spring, advancer rod, etc.) and the distributor breaker (breaker plates, breaker points, stationary plate, etc).

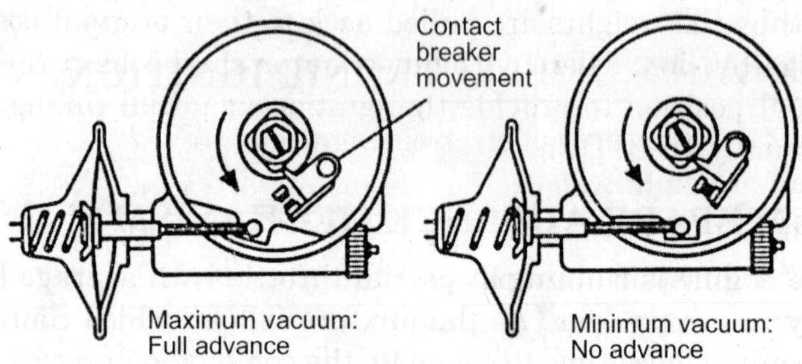

Fig.13.37: *Operation of vacuum advancer*

The vacuum advancer is divided into a vacuum chamber and an air chamber by the diaphragm. When the vacuum in the intake manifold is applied to the vacuum chamber, the advancer rod is pulled. Since the advancer rod hook is connected to the breaker plate pin, the breaker plate is rotated counter clockwise (as viewed from the top).

Operation

Since the throttle valve is closed during engine idling, the advance port located above the throttle valve is at atmospheric pressure and no vacuum advance operation occurs.

When the throttle valve is opened slightly, a vacuum is generated at the advance port. This vacuum acts on the diaphragm and the diaphragm pulls the advancer rod. As a result, the breaker plate is rotated in the direction opposite that of cam rotation, so the ignition timing is advanced by the breaker plate rotation angle. As the vacuum at the advance port becomes stronger, the diaphragm travel i.e. the breaker plate rotation angle (advance angle) increases.

ELECTRONIC IGNITION SYSTEM

In case of conventional contact breaker point ignition system, large amount of current is allowed to pass through the contacts. Thereby there is a chance of spark (arcing) at contacts causing damage to the contacts and also loss of high amount of current through the contacts. The spark produced also not powerful to ignite the lean mixture. The coil used in such system has to carry high amount of current to produce a powerful spark. This makes the size of the coil bulk and heavy. Later the electronic ignition has been introduced to avoid the problems in the conventional type of ignition.

Electronic ignition system uses electronic switching devices instead of mechanical contact breaker points to send a high voltage to the spark plug.

ADVANTAGES OF ELECTRONIC IGNITION

The general advantages of electronic ignition are:

1. Small and light weight
2. Higher ignition voltage
3. Longer duration of spark
4. Very accurate control of timing
5. Improves cold weather starting
6. Less maintenance
7. Longer service life
8. Less operating cost
9. Helps vehicles meet exhaust emission regulations and reduces fuel consumption.

DISADVANTAGES OF ELECTRONIC IGNITION

The common problems with electronic ignition systems are:

1. Hard to diagnose
2. Adjustment is not as accessible
3. More specific expertise is necessary
4. More susceptible to heat damage

TYPES OF ELECTRONIC IGNITION SYSTEM

Electronic ignition systems may be categorized as capacitor discharge or transistorized or distributor less.

CAPACITIVE DISCHARGE IGNITION SYSTEM

This system is not widely used and differs considerably from the standard and inductive systems. One of the problems with using a coil to provide the high voltage is the time required to build up and collapse the magnetic field before a spark can be produced. On eight or twelve cylinder engines this can be a major problem.

In the capacitor discharge system the supply to the coil is stored in a capacitor (C) inside an electronic control unit. The timing circuit (trigger) in Fig.13.38 could be either mechanical breaker points or a pick-up coil. A trigger suddenly releases the charge stored in the capacitor to the primary winding of the coil.

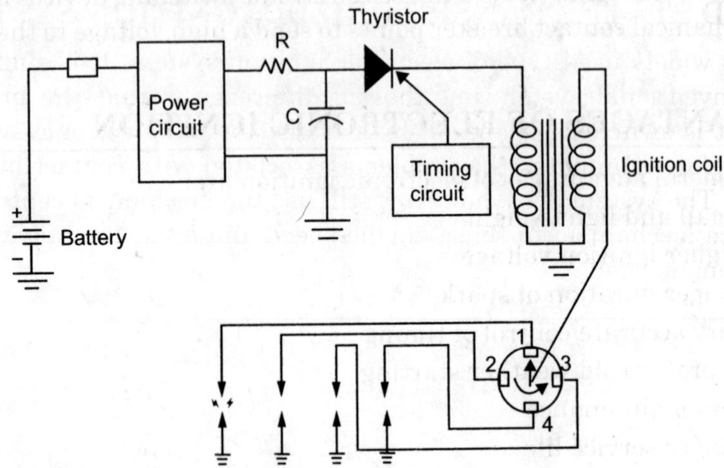

Fig.13.38: *Capacitive discharge ignition system*

Operation

In operation, an inverter converts the AC (Alternating Current) supply to DC (Direct Current) and feeds it to a step-up transformer to provide the 400 V needed by the capacitor. When a spark is required, the trigger releases the energy from the capacitor to the coil by activating a thyristor (an electronic switch). The sudden discharge of high voltage electrical energy causes a rapid build-up of the magnetic field around the coil and induces very high voltage in the secondary windings. This results in a spark of high intensity but short duration. Voltage transformation time is about one hundredth of the time taken by the standard system. This allows a much higher rate of sparks per second. However, the short duration of each individual spark renders the system unsuitable for most engines.

Advantages

1. Capacitor can be fully charged in a very short time (typically 1ms), only 20 microseconds (µs) are required to fully charge a capacitor.
2. Fast voltage rise (between 3 ~ 10 kV/µs) compared to typical inductive systems (300 ~ 500 V/µs).
3. Decreases spark plug fouling.
4. Increases sparkplug life.

5. Uses less current. This means less demand on the battery during starts and a potentially longer battery life.
6. Provides improved cold starting ability.

SOLID STATE TRANSISTORIZED IGNITION SYSTEM

This is widely used type of electronic ignition system. It is similar to the conventional system but the big difference is that the primary current is switched on and off by a transistor (an electronic switch). It therefore eliminates the problems associated with contact breaker points. The system does, however, still use the mechanical centrifugal advance mechanism to sense engine speed and a vacuum advance to sense engine load.

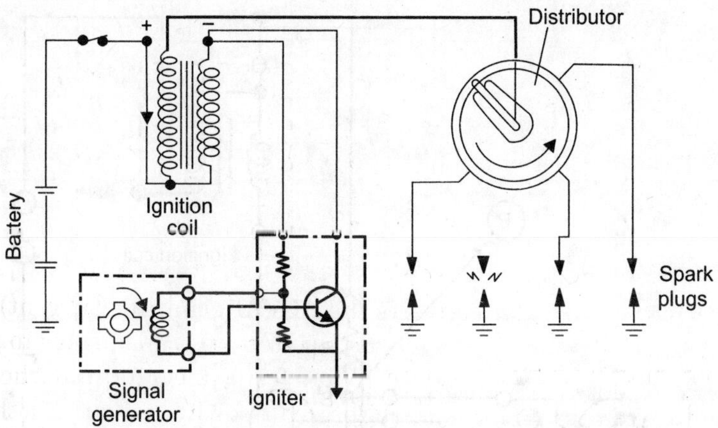

Fig.13.39: *Transistorized ignition system*

OPERATION

Engine stopped

A voltage is applied to point 'P' when the ignition switch is turned on. The voltage at point 'P' is kept below the base voltage required for transistor operation through voltage division by resistors R_1 and R_2. As a result, the transistor remains off while the engine is stopped, so no primary current flows in the ignition coil.

Engine running (Positive voltage generated in pick-up coil)

When the engine is cranked, the signal rotor of the distributor rotates, generating an AC voltage in the pick-up coil. If the generated AC voltage is positive, it is added to the voltage from the battery (applied to point

'P'), raising the voltage at point 'Q' (base voltage) above the transistor operating voltage, thus turning on the transistor.

As a result, the ignition coil primary current flows through the transistor from the collector (C) to the emitter (E).

Engine running (Negative voltage generated in pick-up coil)

When the AC voltage in the pick-up coil is negative, this voltage is added to the voltage at point 'P', so the voltage at point 'Q' decreases below the transistor operating voltage, turning off the transistor. As a result, the ignition primary current is turned off and a high voltage is induced in the secondary coil.

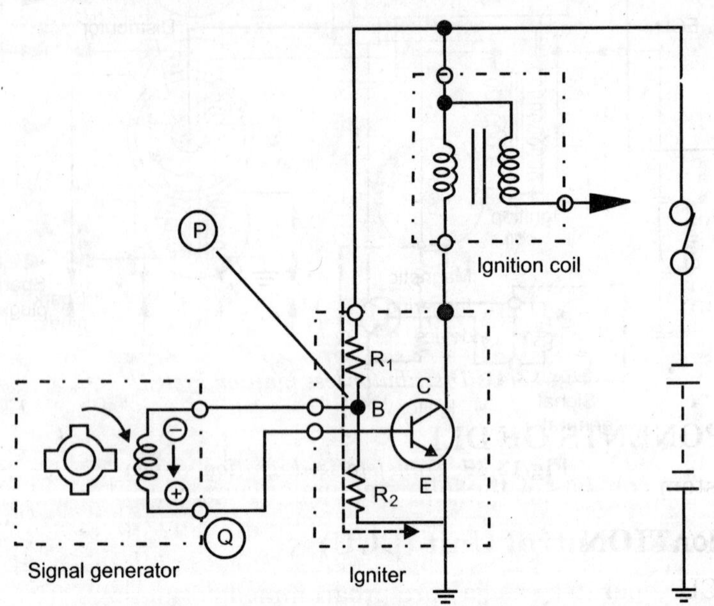

Fig.13.40: *Circuit of transistorized ignition system*

Advantages

1. Delivers much higher voltage
2. Plug life at least doubles
3. Because the points are scrapped, the ignition timing is stable over the full operating range.
4. No moving parts to fail or bounce.
5. Longer service between tune-ups
6. Better fuel economy
7. Safer and more reliable.

DISTRIBUTOR LESS IGNITION (DLI) SYSTEM - FULLY ELECTRONIC

In past transistorized ignition systems, the high voltages generated by a single ignition coil were distributed to each spark plug through the distributor. DLI is a spark-ignition engine ignition system which uses no distributor. This system uses one ignition coil for every two spark plugs. The electronic control unit (ECU) distributes the primary current to each ignition coil directly, causing each spark plug to spark.

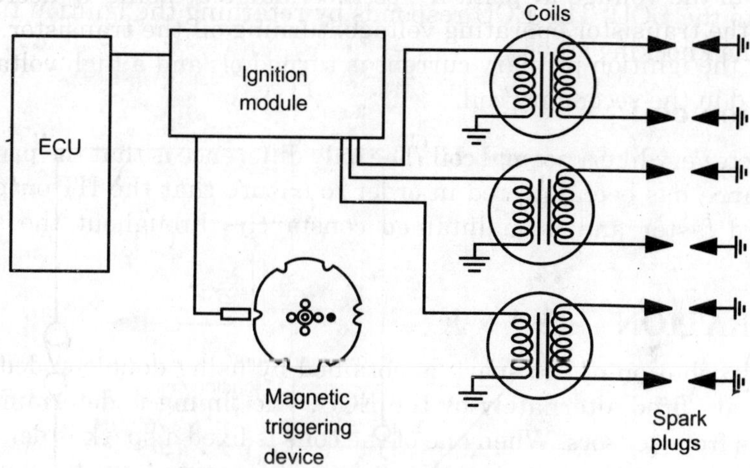

Fig.13.41: *Distributor less ignition system*

COMPONENTS OF DLI

This system consists of the following major components.

Electronic Control Unit (ECU)

The ECU operates as a pre-programmed computer. It gathers information from a variety of sensors interprets that information and adjusts the ignition timing accordingly. Its two main functions are to control the ignition timing electronically using information from sensors and to control the ignition coil output. The main advantage is that the ignition timing is controlled by the ECU instead of the centrifugal and vacuum advance mechanisms.

Crankshaft sensor

This determines the crankshaft speed and position. As the crankshaft rotates, the sensor produces pulses, which are fed to the ECU. From this the ECU can determine the TDC position and the engine speed. These inputs are used by the ECU to calculate basic ignition timing.

Coolant temperature sensor

The ECU uses the information from coolant temperature sensor to adjust ignition timing according to engine operating temperature.

Knock sensor

This provides a small input voltage to the ECU, which is proportional to the vibration sensed in the engine block. The vibrations set up by combustion detonation are detected by the sensor and the information is fed to the ECU. The ECU responds by retarding the ignition timing until the knocking stops.

Ignition coil

This acts very like a normal coil. The only difference is that the primary resistance has been lowered in order to ensure that the HT output is reached faster and is maintained constantly throughout the speed range.

OPERATION

The distribution of the spark is obtained by using double-ended coils which are fired alternately by the ECU. The timing is determined by signals from sensors. When one of the coils is fixed a spark is delivered to two engine cylinders. For the majority of engine layouts where an even number of cylinders is used, two of the pistons will rise and fall within the cylinders at the same time. The spark delivered to the cylinders on the compression stroke will ignite the mixture as normal. The spark produced in the other cylinder will have no effect, as this cylinder will be just completing its exhaust stroke.

Because of the low compression and the exhaust gas in the waste spark cylinder, the voltage used for the spark to jump the gap is only about 3000 Volts. The spark produced in the compression cylinder is therefore not affected.

ADVANTAGES

This system has the following advantages:

1. High tension cords can be shortened, thus reducing noise.
2. Reduction in the number of mechanical components improves reliability.
3. Ignition timing can be controlled over a wider range.
4. More precise timing
5. Timing curve is more adjustable

6. More powerful spark
7. Fewer moving parts
8. No distributor to accumulate moisture and cause starting problems
9. No distributor to drive thus providing less engine drag

SPARK PLUG

The high-voltage current from the distributor generates a high-temperature spark between the center and ground (earth) electrodes of the spark plug to ignite the compressed air-fuel mixture.

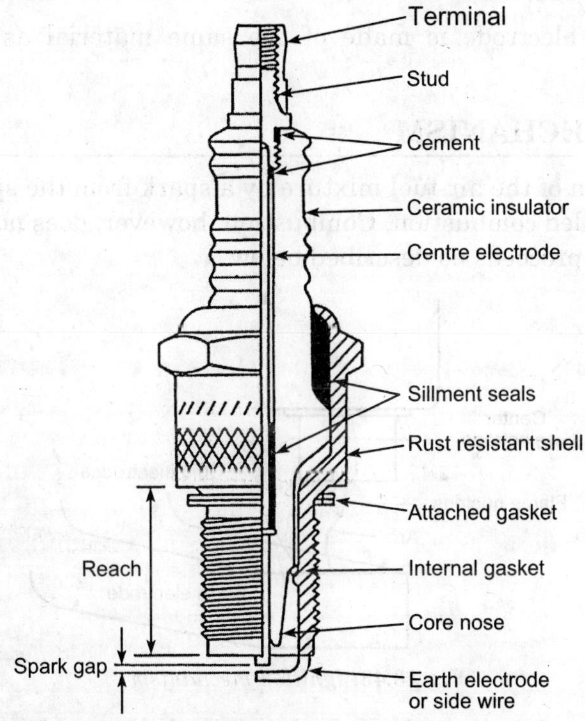

Fig.13.42: *Components of a typical spark plug*

PARTS OF A SPARK PLUG

The spark plug consists mainly of the insulator, casing, center electrode, and ground electrode.

Insulator

The insulator holds the center electrode and serves as the insulation between the center electrode and casing. The insulator is made of high-purity alumina porcelain having excellent heat resistance, mechanical strength, dielectric strength at high temperatures, and thermal conductivity.

Casing

The casing supports the insulator while at the same time serving as the means for mounting the spark plug to the engine.

Center electrode

The center electrode generates the spark with the ground electrode.

Ground electrode

The ground electrode is made of the same material as the center electrode.

SPARK MECHANISM

The explosion of the air-fuel mixture by a spark from the spark plug is generally called combustion. Combustion, however, does not occur in a moment but proceeds as described below.

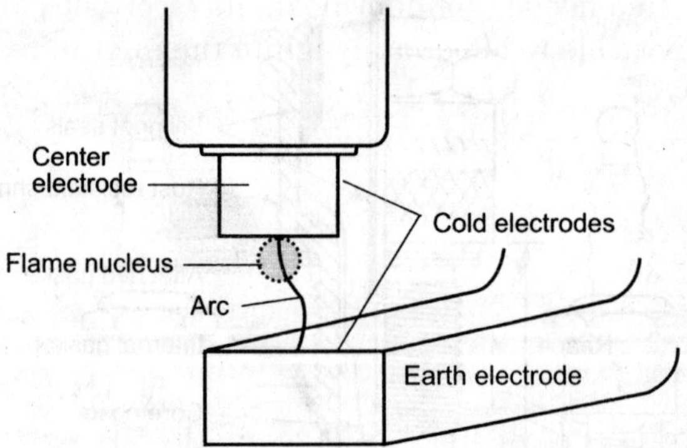

Fig.13.43: *Ignition mechanism*

As the spark travels through the air-fuel mixture from the center electrode to the ground electrode as shown in the Fig.13.43, the air fuel mixture along the path of the spark is activated and a so-called "flame nucleus" is formed. The molecules of the air-fuel mixture in the immediate vicinity of this flame nucleus are activated by the propagation of the heat of the spark and are pushed outward away from the nucleus by the shock wave created by the spark, activating the molecules further out from the center of combustion. These molecules then become part of the flame nucleus, which now has enough energy to continue to propagate outwards in this way by itself.

SPARK PLUG REACH

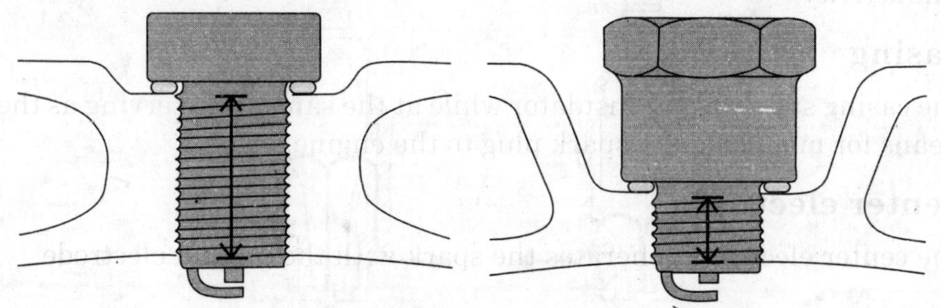

Fig.13.44: *Spark plug reach*

Spark plug reach refers to the length of the shell from the contact surface at the seat to the bottom of the shell, including both threaded and non-threaded sections.

Long-reach plug is used where the cylinder head section is very deep. In a thinner head such a plug would project too far into the combustion chamber. The exposed plug threads act as hot spot and cause preignition. Short-reach plug is used where the cylinder head section is thin. In a deeper cylinder head, its electrodes are in a pocket and the arc is not able to adequately ignite the mixture.

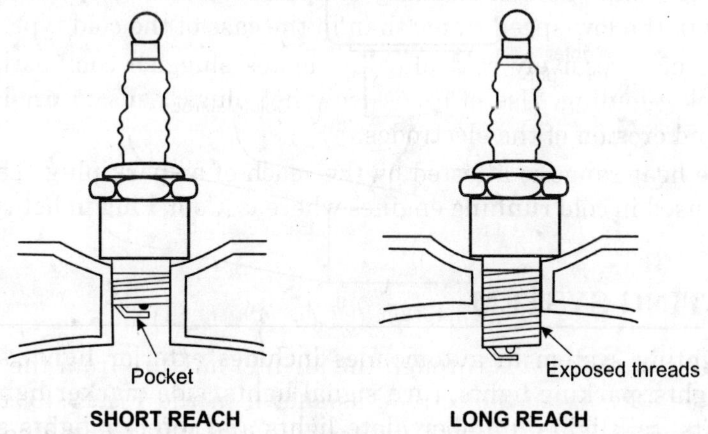

Fig.13.45:

HOT AND COLD PLUGS

The heat range of a spark plug refers to amount of heat radiated by the spark plug. It is determined by the heat radiated distance before escaping to the spark plug shell. Spark plug that radiates more heat is called "cold plug", because the plug itself stays cooler while one that radiates less heat is called a "hot plug", because it keeps its heat in.

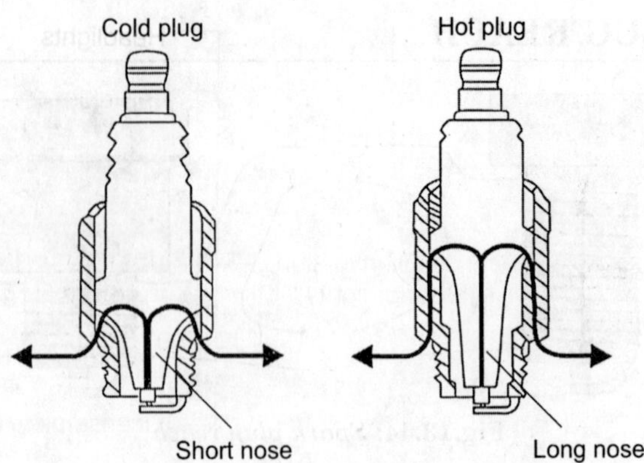

Fig.13.46: *Cold and hot plugs*

The lower operational limit of a spark plug is the self-cleaning temperature (450°C), while the pre-ignition temperature (950°C) is the upper limit. Spark plugs perform best when the center electrode temperature is between about 450 and 950°C.

Since the hot type plug has a longer insulator nose, the surface area exposed to the flame is large, the radiation route is long and heat radiation is small. As a result, the center electrode temperature rises fairly high and the self-cleaning temperature can be attained more quickly in the low-speed range than in the case of the cold type.

Use of excessively cold plugs promotes sluggish combustion and spark plug fouling. Use of excessively hot plugs leads to pre-ignition and rapid erosion of the electrodes.

The heat range is affected by the reach of a spark plug. The 'Hot' Plug is used in cold running engines whereas 'Cold' Plug in hot running engines.

LIGHTING SYSTEM

The lighting system in automobiles includes exterior lights such as head lights, parking lights, turn signal lights, side marker light, back up lights, tail lights, number plate lights and interior lights such as instrumental-panel lights, map lights, various warning and indicator lights and courtesy lights which turn on when the car door is opened.

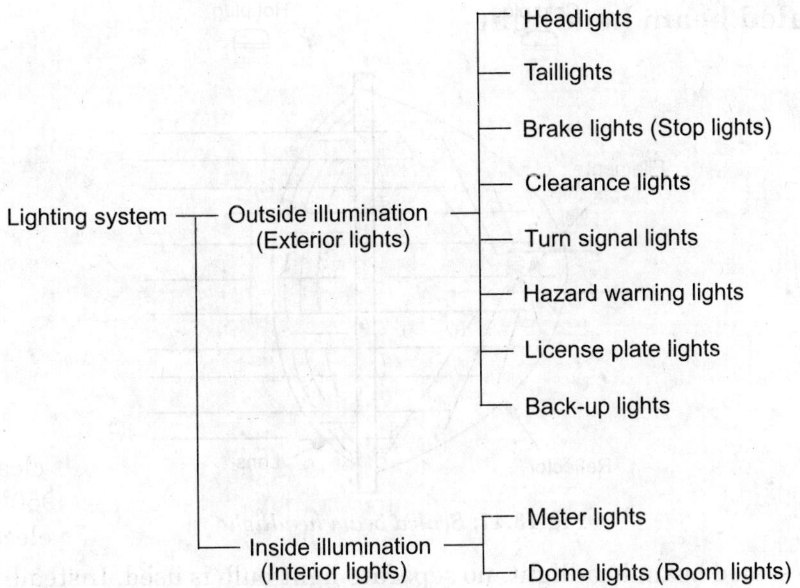

FUNCTIONS

The lighting system of a commercial vehicle has the following functions:

- Illumination of the road
- Defining the vehicle outline
- Indicating intended movements
- Warning other road users
- Indicating operating states.

EXTERIOR LIGHTS

HEADLIGHT

The headlight system is a lighting system to illuminate the road in front of the vehicle. Generally, high-beam and low-beam lights are provided and can be switched from one to the other by the dimmer switch. The high beam (upper or far) filament is focused to produce a high intensity light pattern to illuminate the road a greater distance (about 100 m) from the vehicle for night driving. The dipped or low beam produces a light pattern that illuminates the road a certain distance (about 40 m) from the vehicle. The low beam is low enough to prevent glare to oncoming vehicles.

Two types of headlights are used in vehicles: sealed-beam type and semi-sealed beam type.

Sealed beam headlight

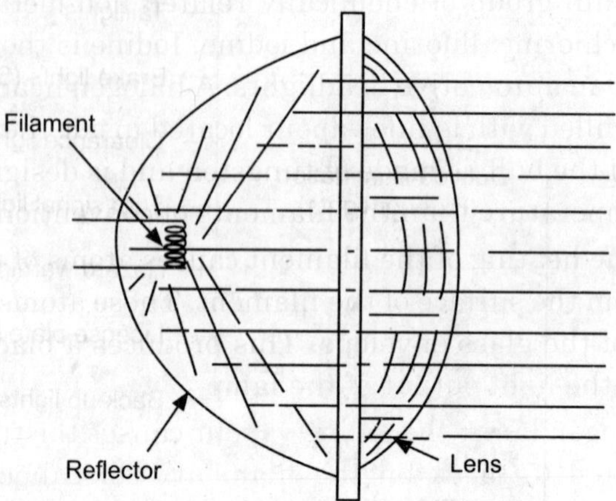

Fig.13.47: *Sealed beam headlight*

In a sealed beam headlight, no separate light bulb is used. Instead, the whole assembly itself is a light bulb. The filament is placed in front of a reflector, to which a glass lens is welded. The reflector directs the lights from the filament through the lens. The lens focuses the light from the reflector into the desired pattern.

Semi-sealed beam headlight

The difference between the sealed beam and semi-sealed beam headlight is its design, which allows bulb replacement. Since the bulb can easily be replaced, it is not necessary to replace the whole headlight assembly if a filament burns out. Furthermore, when a bulb is replaced, the orientation (direction and angle) is not likely to change.

Halogen headlights

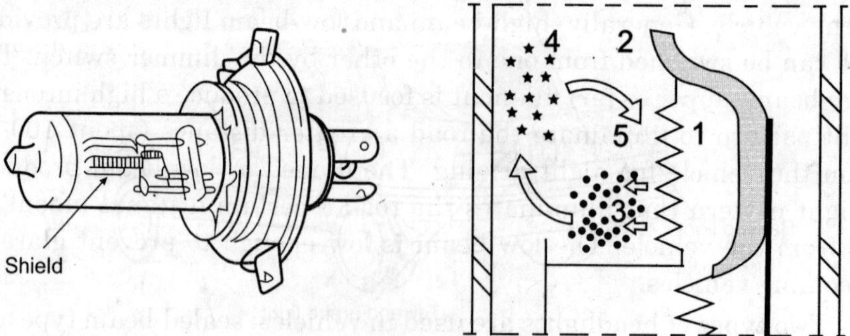

Fig.13.48: *Halogen headlight*

Halogen headlights are used on some vehicles. Halogen is the name given to a certain group of chemically related non-metallic elements, which include chlorine, fluorine and iodine. Iodine is the halogen most commonly used in automotive headlights. A halogen headlamp consists of a small bulb filled with iodine vapour located inside the light housing. The filament of the bulb is made of tungsten and is designed to operate at a higher temperature than the filaments of conventional headlights. In any lamp, the heating of the filament causes atoms of tungsten to be released (3) from the surface of the filament. These atoms are deposited on the inside of the glass envelope. This produces a blackening effect, which reduces the light output of the lamp.

With a halogen lamp, the iodine vapour causes the tungsten atoms to be redeposited (5) back on the filament. This almost eliminated blackening of the lamp. Therefore, the lamp produces a greater light output. They emit much whiter and brighter light than other sealed beam headlights.

SIGNAL LIGHTS

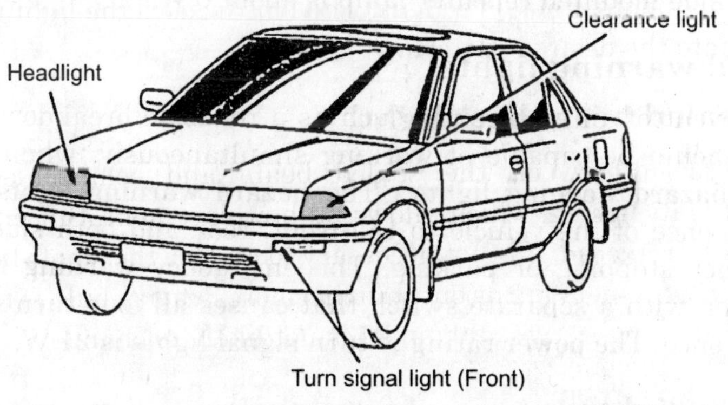

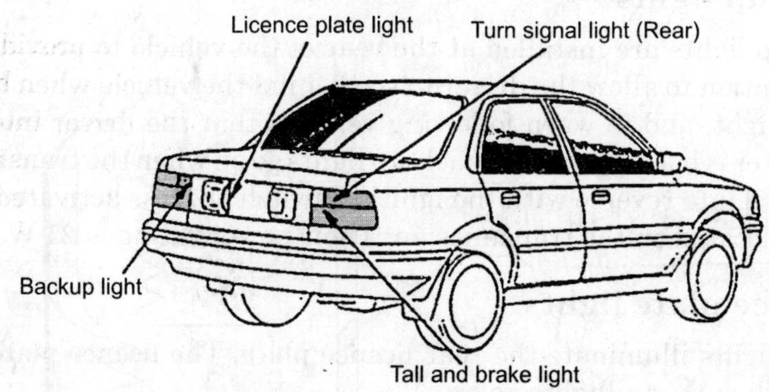

Fig.13.49: *Signal lights*

Clearance lights and taillights

These are low-intensity lights to indicate the presence and width of a vehicle at night to other vehicles in the front or behind. Those in the front are called clearance-lights and those in the back are called taillights. The power rating of these lights is 5 W.

Brake lights (Stop lights)

Brake lights are installed at the rear of the vehicle to prevent rear-end collisions by indicating to following vehicles that the vehicle is braking. Brake lights come on when the driver presses the brake pedal. The stoplights are about three times as bright as taillight. The stoplights have power rating of 21 W.

Turn signal lights

Turn signal lights are used to let other drivers know that you are about to make a turn. They flash 60 to 120 times per minute. This blinking (flashing) controlled by a flasher unit in the circuit. Bulb rating should be within the range of 15 to 36 W for each of the front and rear lamps, and the side mounted repeater lamp is about 6 W.

Hazard warning lights

In the event of an emergency, such as a roadside breakdown, all the lamps should be capable of working simultaneously, when they are called "hazard warning lights". The hazard warning lights indicate the existence of the vehicle to the front, rear and both sides during emergency stopping or parking. The emergency warning lights are turned on with a separate switch that causes all four turn signals to flash at once. The power rating of turn signal lights is 21 W.

Back-up lights

Back-up lights are installed at the rear of the vehicle to provide extra illumination to allow the driver to see behind the vehicle when backing up at night, and to warn following vehicles that the driver intends to back up or is backing up. The back-up lights go on when the transmission is shifted into reverse with the ignition switch on. It is activated by the switch located on the transmission. Their power rating is 21 W.

Licence plate lights

These lights illuminate the rear licence plate. The licence plate lights go on when the taillights go on.

INTERIOR LIGHTS

The interior light illuminates the interior of the passenger compartment, but is designed so as not to dazzle the driver at night. A car's interior can have up to 20 lamps.

The bulbs that light when a door is opened are called courtesy lights. Instrument panel lights are used to illuminate the meters on the instrument panel at night, to enable the driver to read the meters and gauges quickly and easily while driving. The instrument panel lights go on when the taillights go on.

The Instrumental panel warning lights on the dashboard are used to warn the driver of a car with problem. These indicates unclosed door, low engine oil pressure, an under charging alternator, high coolant temperature, problems in the brake system and other potentially harmful situations.

HEADLIGHT AIMING

Headlight aiming is a critical service item on cars. The headlights must be aimed correctly. If they are aimed too high or to the left, they might blind an oncoming driver and cause a serious accident. Incorrect aiming can also reduce the driver's ability to see the road properly. This could also lead to an accident.

Headlight aim should be checked and adjusted:

- Every 12 months
- When a headlamp bulb is replaced
- When there are special situations of vehicle rear loading, or
- When front end sheet metal is repaired or adjusted.

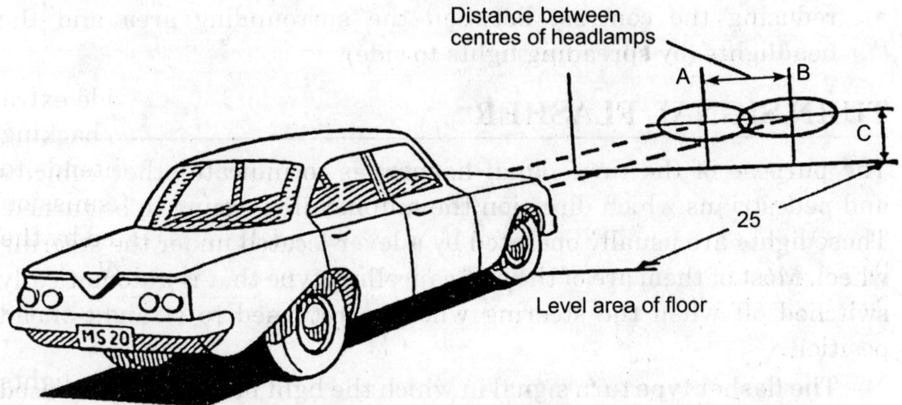

Fig.13.50: *Headlamp adjustment*

There are several methods of checking the aiming of headlights. These include the use of an aiming screen, a mechanical aimer and an optical aimer. The simplest method uses a screen set 25 feet (7.6 m) in front of the vehicle and a perfectly level floor. With the car aligned perpendicular to the screen, the low beam and high beam can be checked. Note that the centrelines of the lights and the centreline of the car are indicated in Fig.13.50. As shown are the high-intensity area (hot-spot) reference lines. The low beams are adjusted so that the hot spot is below and to the right of the reference lines. The high beams are centered on the vertical and horizontal centrelines of the lamps.

HEADLIGHT DAZZLING

When two vehicles approach each other, the light intensity of the headlights falling on the eyes of the drivers will produce a sort of blindness temporarily. Such phenomenon is known as headlight dazzling. In other words, headlight dazzling is defined as the brightness which causes interference with vision. They blind oncoming traffic and are a major cause for accidents at night.

The three factors which govern dazzling are brightness, contrast, and the angle subtended by the light area on the eye.

An intensity bright area with very dark surroundings, does not give dazzling (For example, an electric bulb filament when viewed at night). The dazzling is increased when the same bulb is viewed in a reflector.

Dazzling can be prevented by

- reducing the brightness of headlights (by dimming)
- stopping high intensity light from entering the eyes of on-coming driver
- reducing the contrast between the surrounding area and the headlights (by spreading lights to side)

TURN SIGNAL FLASHER

The purpose of the turn signal flashers is to indicate other vehicles and pedestrians which direction the automobile is going to be turned. These lights are usually operated by a lever located under the steering wheel. Most of them are of the self-cancelling type that is automatically switched off when the steering wheel is returned to straight-ahead position.

The flasher type turn signal in which the light blinks is mostly used at present. This type is usually made to serve also as parking light. The flasher light when in operation, flashes on-and-off from 50 to 120 times a minute. An indicator light on the instrument panel or clicking sound is normally provided to indicate (during turning) or warn (after turning) the driver that the flasher lights are operating.

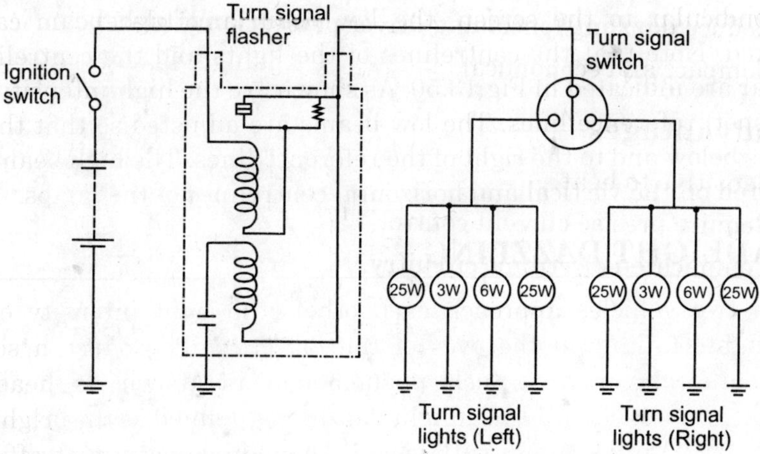

Fig.13.51: *Turn signal lasher circuit*

Working

The flashing is accomplished by a switch that opens and closes periodically. The switch is usually a condenser relay, but a bimetal type switch is sometimes used. The condenser relay is comprised of condenser and a relay having two coils wound in opposite directions. The charging and discharging actions of these components open and close the relay contacts to cause on-and-off current flow to the flasher lights.

LED HEADLIGHTS

Currently LED (Light-emitting diodes) headlamps are available on some vehicles. LEDs are semiconductors that change electrical energy into light. An individual LED does not produce enough light to serve as a headlamp. Ten to twenty LEDs are needed for ample forward lighting.

Advantages of LED headlights

There are many reasons for using LEDs in headlights:

1. Require up to 40% less power than traditional lighting sources. This improves a vehicle's fuel economy.
2. Provide a whiter light than xenon.
3. Mercury-free, making them environmental friendly.

4. Reliable. Never need to be replaced.
5. Resistant to shock and vibration.
6. Turn on quicker than ordinary bulbs.
7. Reduce dazzling (oncoming driver perception of glare)
8. Compact and economical.

Disadvantages

1. Sensitive to heat.
2. Require precise current control
3. Complicated electronic circuitry.

Instruments and Accessories

DASHBOARD INSTRUMENTS

Instruments are arranged on the instrument panel in front of the driver's seat to enable the driver to easily check the vehicle condition. The instrument panel tells the driver of detailed and present conditions of the vehicle by means of meters (gauges) and lights. It generally includes the following meters, gauges and indicator lights.

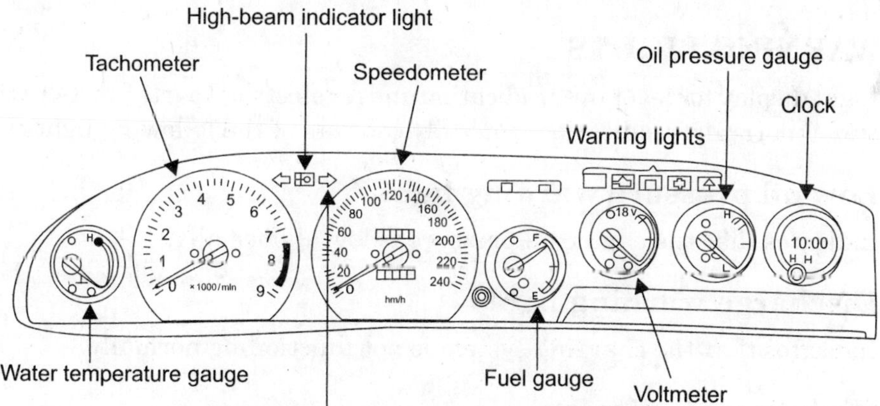

Fig.14.1: *Dashboard instruments*

METERS

Meters may be of the analog type (with an indicator needle) or the digital type (with a numerical display). Here we will focus on explaining the analog meters.

Speedometer

Speedometer indicates the vehicle speed. It also has odometer indicating the distance travelled, and the trip meter, which can be reset to 'zero' as desired.

Tachometer

A tachometer measures the engine speed and indicates the revolutions per minute. Electrical tachometers are most commonly used.

Voltmeter

Voltmeter indicates the battery voltage or alternator output voltage.

Oil pressure gauge

Indicates the engine oil pressure.

Fuel gauge

Fuel gauge indicates the fuel level in the fuel tank.

Water (engine coolant) temperature gauge

It indicates the engine coolant temperature.

WARNING LIGHTS

Light display for level-over indication and for electrical parts functioning and abnormality indication generally consists of the following lights:

Low oil pressure warning light

Indicates that the engine oil pressure is low abnormally.

Discharge warning light

Indicates that the charging system is not functioning normally.

High beam indicator

Indicates that the headlights are on high beam.

Turn signal indicators

Indicate that the right or left turn signal lights are flashing.

Hazard warning indicator

Indicate that both right and left turn signal lights are flashing.

Low fuel level warning light

Indicates that the remaining fuel is insufficient.

Brake system warning light

Indicates that the parking brake is applied or brake fluid is insufficient.

Open door warning light (courtesy light)

Indicates that a door is not fully closed.

SPEEDOMETER

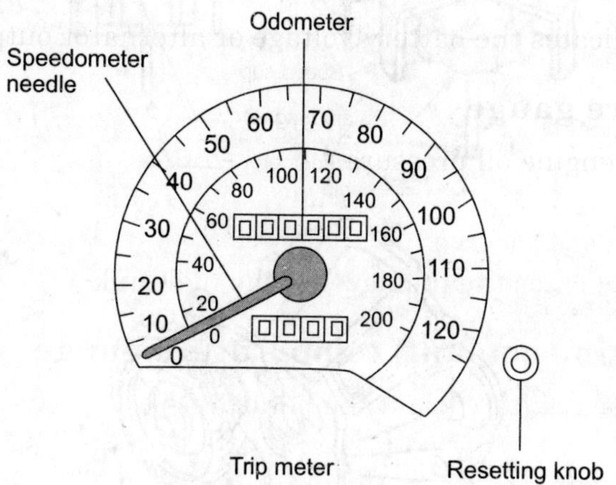

Fig.14.2: *Speedometer*

The speedometer indicates the speed at which the automobile is running. It is usually made in combination with the odometer that registers the total distance travelled. Many speedometers are further provided with a trip meter that can be reset to zero by the driver.

The speedometer is driven by a gear mounted on the rear end of transmission output shaft, the motion being transmitted to the speedometer by means of a flexible cable. The speed is indicated in kilometres (or miles) per hour.

There are many types of speedometers, such as the centrifugal type, magnetic type, and digital type. The magnetic type and digital type speedometers are most commonly used today.

Magnetic speedometer

A magnetic speedometer uses a magnet to indicate the vehicle speed based on wheel revolutions. The revolutions are transmitted to the magnetic speedometer by a worm gear from the differential gear to the transaxle (on front-engine front-drive vehicles) or from the output shaft of the transmission (on front-engine, rear-drive vehicles). The rotary magnet is driven by a flexible cable. The speed cup fixed on the pointer shaft rotates with the pointer. As the rotary magnet rotates, a vortex magnetic field is generated to act upon the speed cup to drive the pointer.

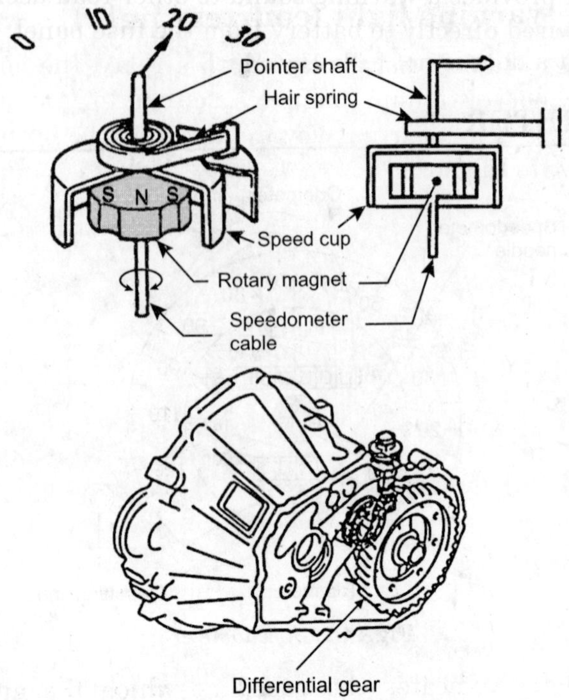

Fig.14.3: *Magnetic speedometer*

As the vehicle speed increases, the speed cup torque increases and indicates the speed by the balance with the hair spring force. When the vehicle stopped, the hair spring keeps the pointer at the zero position.

Digital speedometer

A small slotted wheel is rotated by the speedometer cable. A light-emitting diode (LED) and phototransistor are positioned so that they face each other, separated by the slotted wheel. As the wheel rotates, light from the LED reaches the phototransistor each time a slot in the wheel lines up with the LED. The phototransistor produces a voltage pulse each time a slot passes the LED. The microcomputer receives these voltage pulses from the sensor. At the same time, the internal clock provides a time reference. The microcomputer counts the number of voltage pulses received in a certain period of time and calculates the vehicle speed in miles per hour or kilometres per hour. The number calculated is converted to the voltages necessary to operate the digital display.

HORN

The horn is an electromagnetic vibrating device, which is operated by the driver. It provides a warning sound to other road users. The horns are usually wired directly to battery from the fuse panel. The majority of automobiles utilize a horn relay. With a relay, the horn button on the steering wheel or column completes a circuit to ground which closes a relay, and the heavy current flow required by the horn then travels from the relay to the horn.

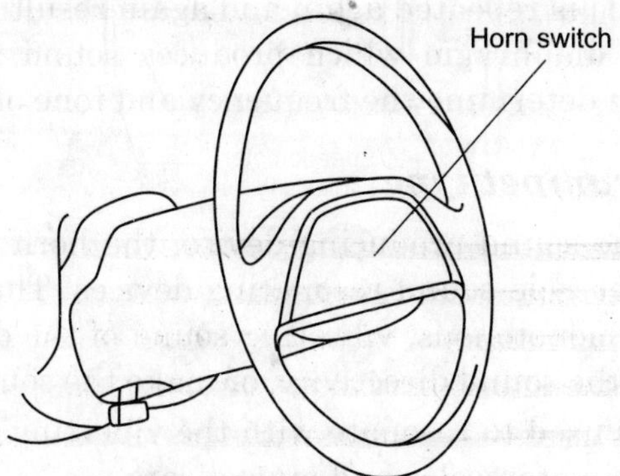

Fig.14.4: *Horn switch*

Horns are manufactured in several different tones or frequencies ranging between 1800 and 3550 Hz. Car manufacturers can select various horn tones for a particular car sound. When two horns are used, each has a different tone when operated separately, yet the sound combines when both are used. The horns can be roughly classified into the electric type, air type, and vacuum type.

Electric horn

Electric horn operates by vibrating a metal diaphragm with an electromagnet, and can be classified into vortex type, trumpet type, and Bosch type.

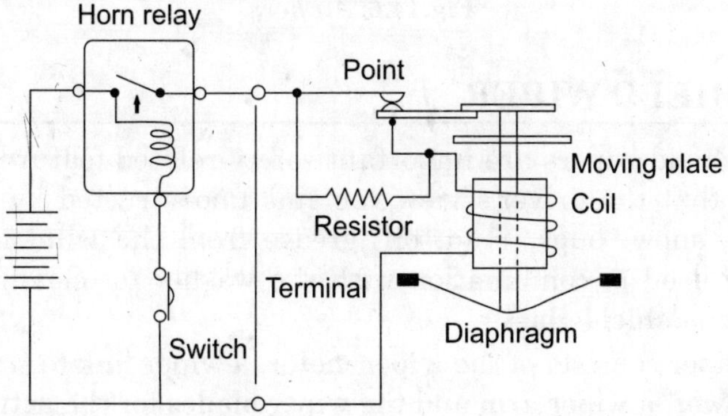

Fig.14.5: *Electric horn*

It consists of coil winding, armature, diaphragm, and contact points. When the horn button is depressed, current starts flowing through the field coil produces a magnetic field that pulls the armature with the diaphragm and opens the contact points due to which circuit is broken. So that current cannot flow. The magnetic field in the coil collapses and the diaphragm is released. The contact points are closed then. The cycle again starts which is repeated again and again resulting in continuous vibration of the diaphragm which produces sound. The design and shape of the horn determine the frequency and tone of the sound.

Vortex and trumpet type

In addition to the sound producing device, the horn is provided with vortex or trumpet type sound resonating devices. The purpose of this is to change the monotonous, vibrating sound of the diaphragm into a richer tone, give the sound directivity, or make the sound carry further.

A vibrator is used to resonate with the vibrating sound. This type is mainly used in motorcycles and midget cars.

Air horn

Air horns are used by large trucks and other big motor vehicles. These horns operate on compressed air from the air reservoir that vibrates the diaphragm to produce the warning sound.

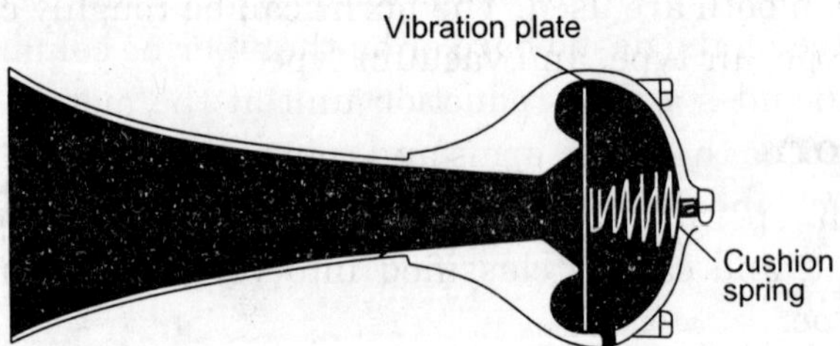

Fig.14.6: *Air horn*

WINDSHIELD WIPER

The windshield wipers are important safety-related features installed to ensure that the driver's view remains unobstructed by wiping off rainwater, snow, bugs, mud, oil, grease from the windshield. It is frequently used in combination with the washer for cleaning off dirt from the windshield glass.

The wiper consists of the wiper motor, a wiper link to transmit the driving power, a wiper arm and the wiper blades for the actual wiping.

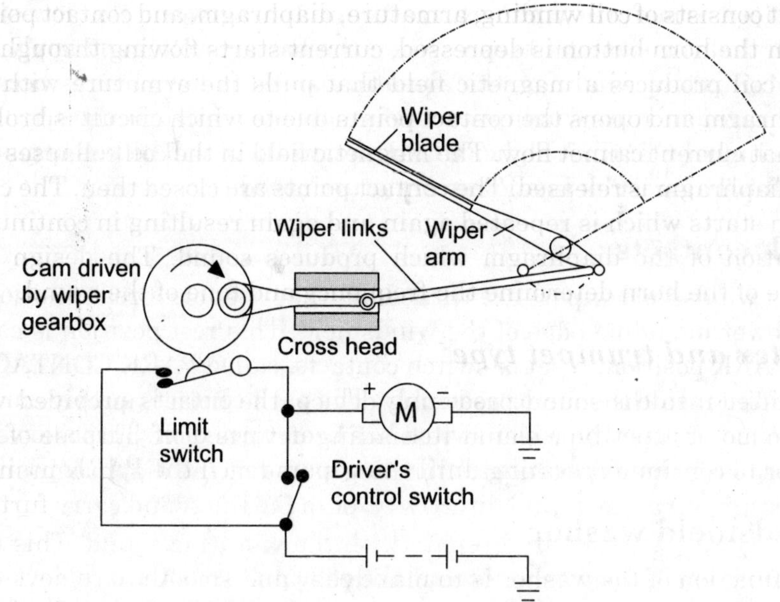

Fig.14.7: *Windshield wiper mechanism*

Most vehicles have two wiper arms for the windshield. Some have an additional wiper for the rear window. Windshield wipers are operated by a wiper switch (controlled by the driver), a wiper motor and a mechanism which links the motor to the wiper arms. Wiper switches can be mounted on the dash board or on the steering column. The wiper mechanism includes a gear reduction unit at the motor. A crank and linkage connect to the wiper arms and convert the rotary motion of the motor into back and forth movement of the wiper arms.

Wiper motor

Most windshield wiper motors are two-speed electric motors with permanent field magnets. Three brushes contact the commutator: a high-speed brush, a low-speed brush and a common brush. The speed of the motor is controlled by the number of armature windings that are connected in the circuit. Fewer armature windings are connected between the high-speed brush and the common brush. When battery voltage is applied to fewer armature windings, the armature speed is high. This is because with fewer armature windings, there is less magnetism in the armature and a lower counter electromotive force (CEMF). Because the CEMF opposes battery voltage, a lower CEMF in the armature means there will be greater armature current. Greater armature current means a higher armature speed.

More armature windings are connected between the low-speed brush and the common brush. When the battery voltage is applied to more armature windings, the armature speed is low. This is because with more armature windings, there is more magnetism in the armature and a higher CEMF. A higher CEMF means there will be less armature current. This results in a lower armature speed.

Park contacts

When windshield wipers are turned off, they assume a rest position at the lower moulding edge of the windshield. This rest position is called the PARK position. A set of switch contacts, called PARK CONTACTS, is located inside the motor assembly. The park contacts supply current to the motor after the wiper switch has been turned off. This allows the motor to continue operating until the wipers reach the PARK position.

Windshield washer

The function of the washer is to make easy and smooth to remove dust and insects from the windshield and rear glass with washer fluid. The electric type washer is most commonly used. The electric type washer consists of the washer tank, washer motor and nozzles.

POWER WINDOW

Power windows are electric windows which can be raised and lowered at the touch of a button rather than the traditional handle which has to be turned manually for raising and lowering car windows. In hand-crank window, turning the winding handle rotates the two parallel links which move the window carrier up and down squarely and smoothly.

Power windows are now common on most vehicles. A car with power window is slightly higher in price as compared to the car with normal windows.

Mechanism

Force is transmitted from the motor to the window glass will vary by manufacturer and vehicle model. They will usually comprise a gearbox and scissor-acting lever mechanism. Alternatively a flexible rack or cable assembly can be used as shown in Fig.14.8.

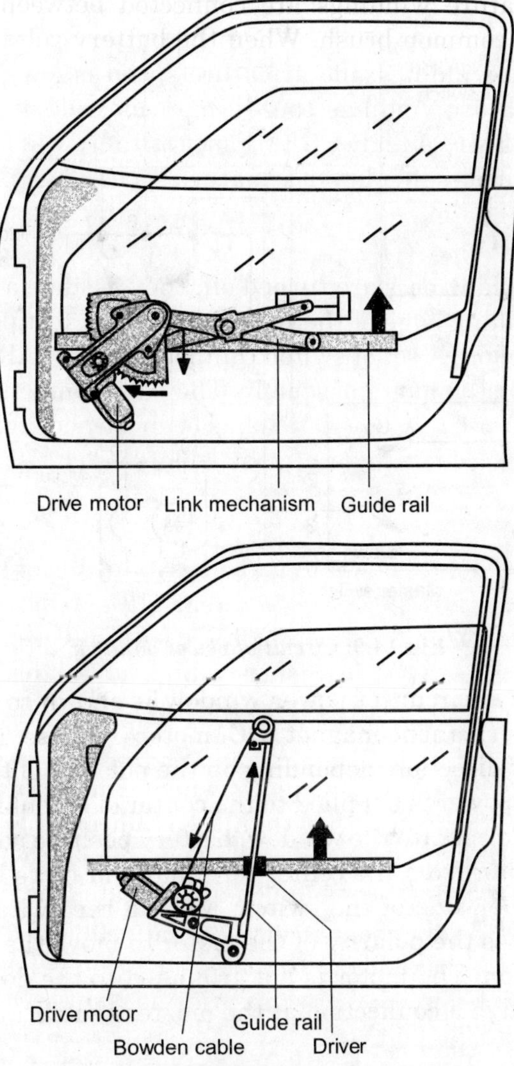

Fig.14.8: *Power window mechanism*

Working of power windows

Power windows work only when the car is on start mode. When the switch is activated it completes a circuit, sending electricity to the motor which then moves the window up or down.

The motors are controlled by individual switches at each window or by a master switch, usually located on the driver's door. Power for the individual switches is distributed by the master switch. The two switches are wired in series, so a problem with either of them will prevent the window from operating.

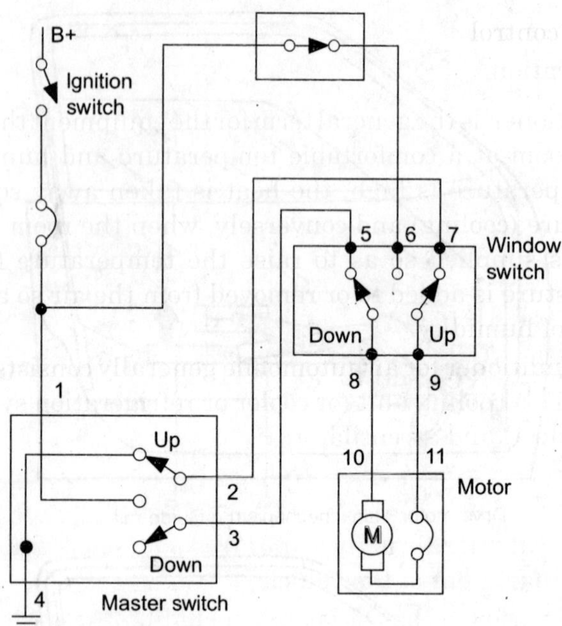

Fig.14.9: *Circuit of power window*

The electrical circuit of power window is shown in Fig.14.9. Most systems use permanent-magnet DC motors. These motors can be driven in either direction, depending on the polarity of the wires to the motor. Battery power is supplied to the center of a double-pole, double-throw switch. To operate the window, battery positive power is directed through the motor from the center terminal and flows back to ground through the other side of the switch. Moving the switch in the other direction reverses the polarity of the motor to move the window in the opposite direction. The motor is not grounded to the door but receives its ground through a connection at the master switch.

Advantages

The advantages provided by these automatic car windows are:

1. Ensure passengers (especially children) safety.
2. Driver attention is not distracted.
3. Much easier to operate than the hand-crank windows.

AIR-CONDITIONER

An air-conditioner is an apparatus for:

- temperature control
- air circulation control

- humidity control
- air purification

'Air conditioner' is the general term for the equipment that maintains the air in a room at a comfortable temperature and humidity. When the room temperature is high, the heat is taken away so as to lower the temperature (cooling) and conversely, when the room temperature is low, heat is supplied so as to raise the temperature (heating). In addition, moisture is added to or removed from the air so as to attain a comfort level of humidity.

The air conditioner for an automobile generally consists of a heating unit (or heater), a cooling unit (or cooler or refrigeration system) with a moisture remover, and a ventilator.

HEATER

A device that heats the air in the car or fresh outside air drawn into the passenger compartment of the vehicle.

There are various types of heater, including the hot water heater, combustion heater and exhaust heater, but ordinarily the hot water heater is used.

Principle

In the hot water heater system, engine coolant is circulated through the heater core to make the heater core hot. Then a blower is used to blow cold air through the hot heater core to heat the air.

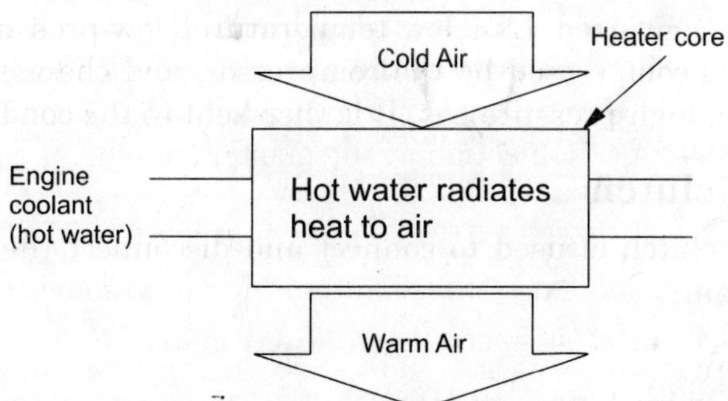

Fig.14.10: *Principle of heater*

Naturally, since the engine coolant serves as the source of heat, the heater core will not get hot as long as the engine is cold, and the temperature of the air blown through the heater core will therefore not rise.

COOLER (REFRIGERATION SYSTEM)

Cooling (Refrigeration) is the process by which air is cooled, cleaned and dehumidified before entering, or re-entering, the passenger compartment. Basically, an air conditioning system removes heat from the passenger compartment by absorbing it and carrying it outside, where it is released into the atmosphere.

The basic components of an automobile refrigeration system consist of the compressor, condenser, receiver/dryer, expansion valve, and evaporator. In addition, there are other devices and functions which help bring out the system's full potential, such as anti-frosting, engine stall prevention, engine idle-up, etc.

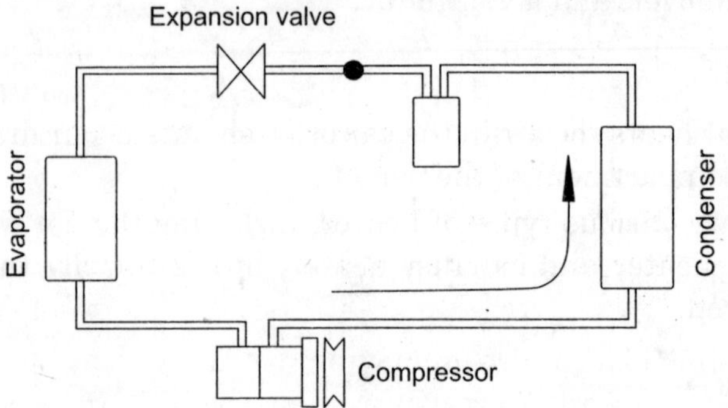

Fig.14.11: *Components of refrigeration system*

Compressor

After being converted into low-temperature, low-pressure gas, the refrigerant is compressed by the compressor, and changed into high-temperature, high-pressure gas. It is then sent to the condenser.

Magnetic clutch

A magnetic clutch is used to connect and disconnect the compressor from the engine.

Condenser

The condenser is used for the purpose of cooling and robbing the heat from the refrigerant gas, which has been compressed by the compressor into high-temperature, high-pressure gas, so as to change this gas into liquid refrigerant.

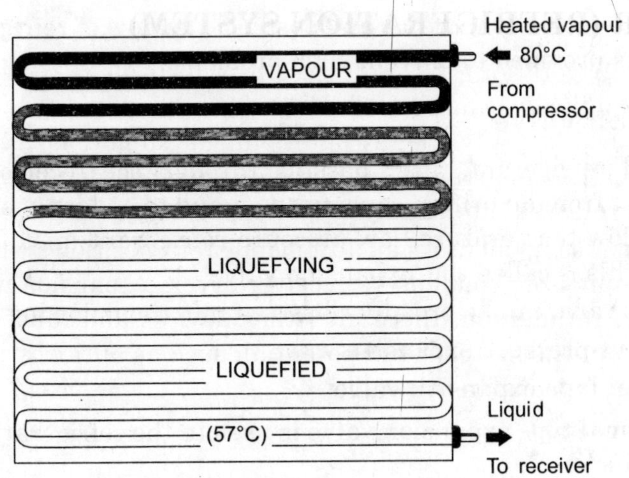

Fig.14.12: *Condenser*

Receiver/Dryer

A filter and desiccant sealed in a container, called the receiver and dryer, has been provided for the purpose of (1) temporarily storing the refrigerant liquefied by the condenser so as to allow a supply conforming with the cooling load; and (2) removing the dirt and moisture that could be injurious to the refrigeration system if allowed to get in. A sight glass is provided on the top of the receiver/dryer to allow the flow of the refrigerant to be viewed.

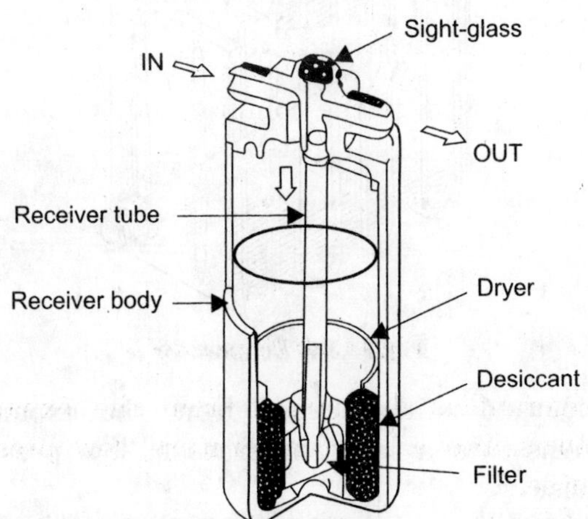

Fig.14.13: *Receiver*

If the refrigerant should contain moisture, this moisture will tend to corrode the functional parts. It could also freeze in the expansion valve orifice and plug the orifice, blocking the refrigerant passage, or it could freeze in the evaporator and obstruct the flow of refrigerant. The desiccant is provided to prevent such problems.

Expansion valve

The liquid refrigerant, after passing through the receiver/dryer, is injected out from an orifice, causing the liquid to suddenly expand and become a low-temperature, low-pressure refrigerant mist. The device that does this is called the 'expansion valve'.

Expansion valve can be broadly classified into the following types:

- constant-pressure expansion valve
- thermal type expansion valve

A thermal type expansion valve is used in the refrigeration system of most vehicles.

Evaporator

The purpose of the evaporator is the opposite of that of the condenser. The state of the refrigerant just behind the expansion valve is 100% liquid.

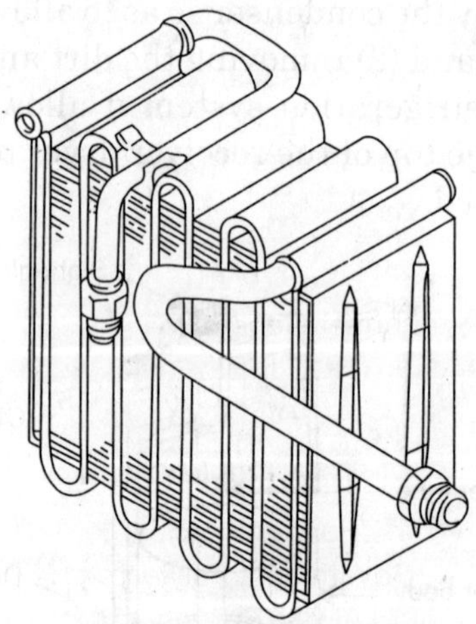

Fig.14.14: *Evaporator*

Once the refrigerant is discharged from the expansion valve, it immediately turns into a low-temperature, low pressure mist and starts to vapourise.

The evaporator is made of aluminium. Like the condenser, the evaporator is of very simple construction, but it is the most important component of the refrigerating system. The construction and the operating condition of the evaporator has a great effect on the efficiency of the refrigeration system.